D1660959

Wind Energy and Wildlife Interactions

Johann Köppel
Editor

Wind Energy and Wildlife Interactions

Presentations from the CWW2015 Conference

Editor
Johann Köppel
School VI—Planning Building
Environment, Environmental Assessment
and Planning Research Group
Berlin Institute of Technology
Berlin
Germany

ISBN 978-3-319-51270-9 ISBN 978-3-319-51272-3 (eBook)
DOI 10.1007/978-3-319-51272-3

Library of Congress Control Number: 2016960717

Printed on acid-free paper

This Springer imprint is published by Springer Nature
The registered company is Springer International Publishing AG
The registered company address is: Gewerbestrasse 11, 6330 Cham, Switzerland

From the Editor—Perspectives in 'Wind Energy and Wildlife Interactions'

In March 2015, the Berlin Institute of Technology hosted the Conference on Wind energy and Wildlife impacts (CWW15), offering a platform to national and international participants to showcase the current state of knowledge in wind energy's wildlife implications. The CWW15 was dedicated to synoptical and specific contributions in the last decade, including offshore research with multi-year studies. Previous European conferences as the first CWW 2011 (Trondheim, Norway) and the Conference on Wind power and Environmental impacts (CWE 2013 Stockholm, Sweden) provided valuable opportunities for international exchange and future collaborative research. In addition, improved research methodologies can be presented up front, with the goal of identifying what we have learnt thus far as well as which predominate uncertainties and gaps remain for future research. The CWW15 not only provided new insights to already substantial research in understanding and mitigating impacts on wildlife and their habitats, but also revealed new interesting topics and current challenges both in Germany and internationally. While we will never become absolutely aware of potential impacts and ecosystem reactions nor can we mitigate these impacts 100%, these platforms aim to understand what is known and unknown internationally, and begin to look at alternatives —more specifically accepting uncertainties and taking more adaptive approaches.

Based on 162 submitted abstracts, the conference was organized in two parallel streams over three days for all researchers involved in wind energy development and wildlife protection. The CWW15 hosted 65 oral presentations in plenary and parallel sessions, 54 poster presentations, a panel discussion and workshop, as well as an innovation exposition showcasing current field technologies. As more transdisciplinary approaches are needed more than ever before, the CWW brought together 32 attendees from non-governmental organizations, 46 from academia, 58 from developers and industry, 76 from agencies, and 181 attendees from various consultancies. We were overwhelmed by those interested in the conference with nearly 400 delegates attending from 33 different countries, proving wind energy and wildlife effects have become a topic of growing international interest also to those who will be substantially increasing their generation of wind energy based electricity during the next decade (Figs. 1 and 2).

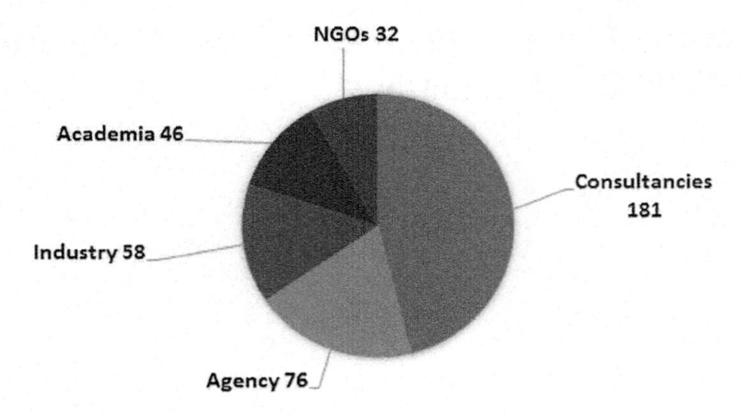

Fig. 1 Where does the knowledge come from?—affiliations of CWW15 attendees

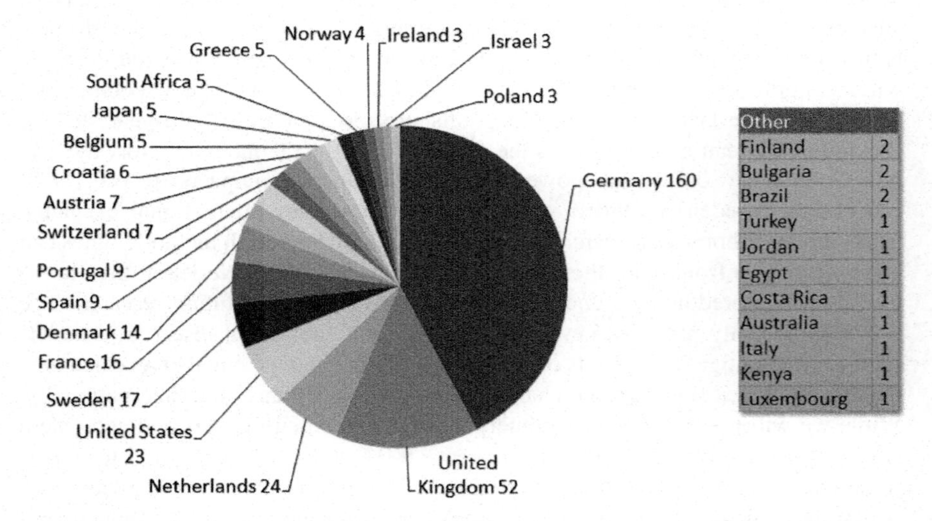

Other	
Finland	2
Bulgaria	2
Brazil	2
Turkey	1
Jordan	1
Egypt	1
Costa Rica	1
Australia	1
Italy	1
Kenya	1
Luxembourg	1

Fig. 2 International knowledge exchange—gathering the state of knowledge and perspectives on a global scale helps reduce redundancies and can help speed up learning processes

The parallel sessions included seven topics: Planning and Siting; Monitoring and Long-term Effects; Mitigation, Compensation, and Effectiveness of Measures; Landscape Features and Gradients; Collision Risk and Fatality Estimations; Species Behavior and Responses; and Tools and Technology. This book covers most of these topics, with adding 'Future Research' and 'Knowledge Platforms.' The following chapters are arranged based on their topic, varying in research from birds, bats, and offshore research, to possible tools or models which can be used in improving research efficiencies. Authors have conducted research in various parts of Europe (Germany, England, Portugal, etc.), in the USA, and as well in the Middle East/northern Africa.

Our Chapter "Red Kites and Wind Farms—Telemetry Data from the Core Breeding Range" and thus first topic, 'Species Behavior and Behavioral Responses' focuses on Germany's strongly relevant and protected red kite (*Milvus milvus*). More than half of the world's population of red kites are found in Germany and are the second most often reported victims of collisions with wind turbines. Between 2007 and 2010, a telemetry study of red kites in a core breeding ground in Sachsen-Anhalt took place, collecting data on flight heights, habitat preferences, and wind turbine proximities to nests. While results showed red kites not being affected via displacement by turbines, authors Hermann Hötker, Kerstin Mammen, Ubbo Mammen, and Leonid Rasran advocated implementing buffer zones around nest sites in order to reduce substantially collision risk as well as minimizing attractive features (e.g., harvests, clearing) around wind turbines.

Chapters "Unforeseen Responses of a Breeding Seabird to the Construction of an Offshore Wind Farm"–"A Method to Assess the Population-Level Consequences of Wind Energy Facilities on Bird and Bat Species" cover 'Collision Risk and Fatality Estimations.' Authors Thomas Grünkorn, Jan Blew, Oliver Krüger, Astrid Potiek, Marc Reichenbach, Jan von Rönn, Hanna Timmermann, Sabrina Weitekamp, and Georg Nehls discuss a large-scale, multi-species assessment of avian mortality rates at wind turbines in agricultural landscapes of northern Germany. Systematically searching via line transects over 55 randomly selected wind farm seasons, 285 fatalities were found to be large- to medium-bodied, mostly local bird species (Chapter "A Large-Scale, Multispecies Assessment of Avian Mortality Rates at Land-Based Wind Turbines in Northern Germany"). In the USA, a method assessing population-level consequences of wind energy on bird and bat species was created and is discussed in Chapter "A Method to Assess the Population-Level Consequences of Wind Energy Facilities on Bird and Bat Species" (Jay Diffendorfer, Julie Beston, Matthew Merrill, Jessica Stanton, Margo Corum, Scott Loss, Wayne Thogmartin, Douglas Johnson, Richard Erickson, and Kevin Heist). Through forecasting annual mortality rates based on scenarios of future increased market penetration of wind energy, these forecasts incorporate uncertainty from several sources helping decision makers understand potential impacts and develop avoidance, minimization, and mitigation measures. Chapter "Unforeseen Responses of a Breeding Seabird to the Construction of an Offshore Wind Farm" (Andrew Harwood, Martin Perrow, Richard Berridge, Mark Tomlinson, and Eleanor Skeate) extends offshore, using visual tracking over a five-year period detailing flight paths and behavior of terns looking at an 88 wind turbine facility site before, during, and after construction off the UK's western coast.

The following two chapters cover 'Landscape Features and Gradients,' with authors Hendrik Reers, Stefanie Hartmann, Johanna Hurst, and Robert Brinkmann, presenting comprehensive findings of bat activity at nacelle heights over forests in Germany (Chapter "Bat Activity at Nacelle Height Over Forest") and Allix Brenninkmeijer and Erik Klop discussing the location and spatial design of wind facilities as well as power line interactions of two Dutch wind facilities affecting bird mortality (Chapter "Bird Mortality in Two Dutch Wind Farms: Effects of Location, Spatial Design and Interactions with Powerlines"). Using 193 nacelle surveys generating acoustic datasets, research showed no significant difference in bat activity between forests and open landscapes, but strong regional differences of activity. Research at the Dutch wind farms showed turbine location (between wind facilities and within the facility) is a major impact on gulls, while the low visibility of power lines affects passerines and waterfowl.

In terms of 'Mitigation, Compensation, Effectiveness of Measures,' we provide two land-based chapters regarding shutdown on demand for birds and curtailment algorithms for bats, with a third chapter discussing offshore pile-driving mitigation. Research in Portugal by Ricardo Tomé, Filipe Canário, Alexandre H. Leitão, Nadine Pires, and Miguel Repas analyzed radar-assisted shutdown on demand of wind turbines when soaring birds are in the area, showing that the use of radar, SCADA access ('Supervisory Control & Data Acquisition'), and experience can lead to zero soaring bird mortality while decreasing losses in electricity generation (Chapter "Radar Assisted Shutdown on Demand Ensures Zero Soaring Bird Mortality at a Wind Farm Located in a Migratory Flyway"). In Germany, Oliver Behr, Robert Brinkmann, Klaus Hochradel, Jürgen Mages, Fränzi Korner-Nievergelt, Ivo Niermann, Michael Reich, Ralph Simon, Natalie Weber, and Martina Nagy discuss a model-based approach based on Korner-Nievergelt et al. (2011) curtailing bat mortality site-specific using acoustic monitoring at the nacelle (Chapter "Mitigating Bat Mortality with Turbine-Specific Curtailment Algorithms: A Model Based Approach"). Chapter "Is There a State-of-the-Art to Reduce Pile-Driving Noise?" examines the question looking at Noise Mitigation Systems (NMS) such as single and double bubble curtains (BCs), Hydro Sound Dampers (HSDs), and screens (IHC-NMS) in German territorial waters by Michael Bellmann, Jan Schuckenbrock, Siegfried Gündert, Michael Müller, Hauke Holst, and Patrick Remmers.

Chapter "The Challenges of Addressing Wildlife Impacts When Repowering Wind Energy Projects" looks at the challenges of repowering in the context of wildlife impacts under the topic of 'Monitoring and Long-Term Effects.' Shawn Smallwood (see 'dialogue' box) states a shift in research toward repowering is necessary as wind turbines have begun to exceed their operational life spans. While repowering could reduce animal fatalities by 60–90%, reduce the number of turbines on the landscape through improved siting, and provide an opportunity to move power lines underground, new challenges are presented such as wider roads, larger pads, increase in bat mortality due to slower cut-in speeds, and new species mortality as turbine heights have extended higher.

A Competition with Wind as a Resource—Dialogue from Shawn Smallwood

Some wildlife species have adapted to using wind resource areas for foraging, socializing, and migration. Now we have come along converting this resource for our use; wind turbines are simply our means of grabbing energy in which wildlife has already been exploiting for millions of years.

When we discuss 'wind–wildlife interactions', *we are describing the conflict between wind energy and wildlife, and understanding the interaction between wildlife and wind is critical for understanding wildlife impacts caused by wind turbines. In addition, as wind is strongly affected by landscape, so technically we are discussing wind–landscape–wildlife relations.*

Today, we take into account wind turbines' attraction factors and barrier effects—some birds appear to be directing their activities toward wind turbines, and bats are certainly doing this. Some birds shift their flights to avoid wind turbines and others abandon habitat after wind turbines are installed—these are all interactions with wind turbines.

Another interaction, rarely discussed, is the scavenger communities' responses to a new food supply around wind turbines. When watching with my thermal camera, the onset of darkness is like a dinner bell for skunks, coyotes, foxes, badgers, and others, all of whom come running to the wind turbine pads for the insects, birds, and bats dropped by wind turbines.

Our insertion of wind turbines into the heart of aero-ecosystems shifts age-old balances, costing some species to the benefit of others. We are competing for wind as a resource. Sharing wind with wildlife requires greater understanding of wind, the moderating influences of landscape, and how wildlife use wind and landscape before and after wind turbines are installed.

The following four chapters deal in 'Planning and Siting' both land-based and offshore. Alkis Kafetzis, Elzbieta Kret, Dora Skartsi, Dimitris Vasilakis, and Ioli Christopoulou present a case study conducted in an area of high ecological value in Thrace, Greece, where a large population of vultures and large birds of prey migrate and breed. This area also has high wind potential, where 181 wind turbines are planned. The case study used data generated from long-term research in the area on specific bird species and sensitivity mapping in identifying areas appropriate for sustainable wind development, as well as establishing exclusion zones set aside for biodiversity (Chapter "Wind Farms in Areas of High Ornithological Value—Conflicts, Solutions, Challenges: The Case of Thrace, Greece"). In Germany, a review of current knowledge, insights, and remaining gaps on the impacts of wind turbines on birds is presented in Chapter "Wind Turbines and Birds in Germany—Examples of Current Knowledge, New Insights and Remaining Gaps"—stating current planning processes in the country use a mixture of scientifically based knowledge and precautionary presumptions. Marc Reichenbach emphasizes the importance of best available science in planning recommendations and provides additional suggestions for the German planning process. Chapter "Introducing a New Avian Sensitivity Mapping Tool to Support the Siting of Wind Farms and Power Lines in the Middle East and Northeast Africa" introduces an avian sensitivity mapping tool by Tristram Allinson to support the rapid expansion of renewable energy technologies in the Middle East through Northeast Africa. Developed by BirdLife International as part of the 'Migratory Soaring Bird Project,' this tool helps deal with the risk of cumulative effects in wind facility siting along one of the world's most important migratory flyways. A fourth chapter within 'Planning and Siting' goes offshore, with Maarten Platteeuw, Joop Bakker, Inger van den Bosch, Aylin Erkman, Martine Graafland, Suzanne Lubbe, and Marijke Warnas developing "A Framework for Assessing Ecological and Cumulative Effects (FAECE) of Offshore Wind Farms on Birds, Bats and Marine Mammals in the Southern North Sea". Based on the DPSIR (Driving forces, Pressures, States, Impacts, Responses) approach, FAECE determines parameters of offshore wind facilities (OWFs) and other platforms to establish sustainable development. The case study shows that effective mitigation measures are required in order to line up with EU legislation via permits (e.g., maximum underwater noise levels, and mitigation measures during migration seasons).

This book's conclusion provides two chapters within 'Future Research and Online Platforms'. Chapter "Future Research Directions to Reconcile Wind Turbine—Wildlife Interactions" brought researchers from the CWW15 Scientific Committee together (Roel May, Andrew Gill, Johann Köppel, Rowena Langston, Marc Reichenbach, Meike Scheidat, Shawn Smallwood, Christian Voigt, Ommo Hüppop, and Michelle Portman) to provide an overview of remaining key knowledge gaps learned since the conference, possible future research directions, and their significance for management and planning in wind energy development. Finally, Chapter "Sharing Information on Environmental Effects of Wind Energy Development: WREN Hub" details an online knowledge management system which brings together the community seeking to collect and disseminate relevant information regarding wind energy development. Authors Andrea Copping, Luke Hanna, and Jonathan Whiting describe how WREN Hub helps identify key knowledge as well as open questions for researchers, regulators, developers, and key stakeholders to use for future projects.

All these findings show wind energy development effects on wildlife strongly rest on site-species-season specificity and even turbine specificity, whether land-based or offshore. This emphasizes the need for more site-specific Before-After Control-Impact (BACI) research to understand local and regional species behavior and habitats when developing a wind area. The CWW2015 showed a number of factors, such as knowledge on 'VIP species' particularly raptor species such as the golden eagle (*Aquila chrysaetos*) and red kite (*Milvus milvus*). Extensive research effort has been made in order to bridge some of these remaining gaps, such as the development of complex search protocols (Grünkorn et al. 2009) as well as fatality estimation approaches (Bispo et al. 2013). There is also a shift in focus to better understand local population sizes versus migratory species (Voigt et al. 2012), investigate the correlation between pre-construction and post-construction observations (Arnett and Baerwald 2013), as well as recognizing indirect mortality and off-site deaths (Grodsky et al. 2011). Many presentations at the CWW15 covered offshore research as well, with particular focus into wind facility construction impacts on seabirds (Perrow et al. 2015), fish (Debusschere et al. 2015), benthos (Janßen et al. 2015; Coolen et al. 2015), and mammals (Russell et al. 2015), showing promise of further continued research in these fields.

Mitigation strategies are increasingly being used within the last decades in reducing impacts to wildlife both onshore and offshore. Thus, the discussion of the mitigation measures used and their efficiency have grown exponentially. As seen in Gartman et al. (2016a, b), we have collected the current state of knowledge of mitigation measures with similar methodology to Schuster et al. (2015), categorizing and providing an analytical outlook for mitigation research in the future.

While significant research in the last few years has begun focusing on the efficacy of mitigation efforts, we still have considerable headway for the future. One of the greater challenges is the applicability and practicability, for example, for small-scale and detailed mitigation efforts, as coordination between numerous stakeholders involved can become a timely challenge, as well as any legal compliance measures so as to avoid potential environmental or wildlife violations. There is a small incentive lost as we lack sufficient empirical studies to prove the importance, and thus acceptance, of mitigation measures much less the combination of measures to be allocated jointly. This is also said in evaluating or monitoring reports as we lack proper transparency between researchers and facilitators to further research in understanding the implementation and thus effectiveness of mitigation efforts.

Another point of inquiry pertains to possible habituation or species familiarity to mitigation measures, causing the lowered risk for collision or mortality to become uncertain. May (2015) states, 'continuous exposure over times to the presence of wind turbines in an area may lead to learning by birds, either resulting in functional habitat loss or reduced risk perception (increased risk) and increased habituation.' This concern can affect wind facilities at a spatiotemporal level (e.g., undermining the need for buffer zones) or at a technical level (e.g., resources on acoustic or visual devices). Offshore for birds who are reluctant to enter wind facilities, good or improved foraging conditions might enhance the habituation of these birds, as seabirds are known to exploit areas with high and predicable food supply (Vanermen et al. 2015).

The most plausible action in mitigating impacts is to consequently use a combination of measures along the whole range of the mitigation sequence from macro- and micro-siting to deterrents to shutdowns, including compensatory mitigation as a last feasible option when spatial resources become limited. This 'last resort' deserves further assessment as compensation should be necessary since some effects will always remain. To conclude, good practices of avoidance and reduction have been developed and are in wide precautionary use, yet we must continue to assess the effectiveness of mitigation strategies.

Wind energy planning can vary in site characteristics, various species and species' behavior, and differing facility and turbine characteristics—adaptive management can be an applicable strategy dealing with these variances while ensuring mitigation efforts are continually monitored and updated to minimize uncertainties in the future. Adaptive management is a, 'natural, intuitive, and potentially effective way to make decisions in the face of uncertainties' (Williams and Brown 2014). This flexible, iterative decision process allows outcomes in wind energy development and operations to become better understood through monitoring, allowing for adaptations of future management actions —a 'learning while doing' approach.

However, certain parameters must be agreed upon for adaptive management to be successful, regarding responsibilities ('Who does what?'), monitoring ('What methods and time frame?'), tiers ('What events trigger adjustment?'), and measures themselves ('What adjustments are needed?'). All agreements need to be settled by all involved stakeholders; thus, developing an administrative framework or advisory council is greatly recommended. Adaptive management was developed in the USA in the 1970s primarily for forestry and fishery management when it was unknown how much of a resource they could extract without unbalancing the ecosystem. Today, it has developed further, with the USA providing adaptive management guidelines and its possible implementation within its legal framework (e.g., Biological Opinion, Avian and Bat Protection Plan, Habitat Conservation Plan).

While adaptive management seems to be promising from academics and nature conservationists, there is a concern from agencies and industry regarding additional costs and planning insecurity. Conclusively, promoting adaptive management has to come with incentives for investors, such as to establish boundaries of adaptive management; for example, instituting limits on shutdown periods which could control financial uncertainties for developers. Even permitting authorities could benefit in that the responsibility of development and implementation would be dispersed to all involved stakeholders, and thus, improved transparency can lower the risks of negative impacts.

It is widely recognized that renewable energy, and in particular wind energy, has a number of beneficial impacts on society and the environment, diversified energy supply, increased regional and rural development possibilities, domestic industry, and employment opportunities as well as having minimal construction and no emission contributing to climate change (Dai et al. 2015).

Today, research has provided insight to more efficient methods and instruments, but the application of these into policy and into practice needs to be enhanced. A major paradigm shift is required, through transdisciplinary and interdisciplinary collaboration providing solutions to close the science-policy-practice gaps. Improved transparency and cooperation from proponents in providing monitoring results, data collection, and analyses is needed, as mitigation measures are only as effective as the willingness for proponents and conservationists to work jointly, to understand and use measures most applicable on a case-by-case basis. On a whole, planners and developers, researchers, government officials, and the public need to proficiently come together to ensure that projects are continually (and in the future) avoiding, reducing, and compensating for environmental and social impacts.

Research at the Berlin Institute of Technology has identified predominate hypotheses on potential effects wind energy has on wildlife species groups, and what are the current mitigation practices. These next chapters provide crucial research improving these methods and instruments in varying areas and for various species in this context, and future directions.

The CWW2015 conference would not have been possible without the comprehensive support of all our working group's researchers and students. Marie Dahmen compiled the figures (participants' affiliations and nationalities). Since then, Jessica Weber and Rebecca Wardle helped tremendously to make this book possible, reaching out to the authors, formatting, proofreading, polishing non-native speakers' writing, and keeping amendments records. Victoria Gartman kindly helped with the editorial, and Nora Sprondel with the final proofreading. Last but not least, I'm very grateful to Margaret Deignan for having motivated and supported the making of this book.

Victoria Gartman, Johann Köppel
Environmental Assessment and Planning Research Group
Berlin Institute of Technology (TU Berlin)
Berlin, Germany

References

Arnett E, Baerwald E (2013) Impacts of wind energy development on bats: implications for conservation. In: Adams RA, Pedersen SC (eds) Bat evolution, ecology, and conservation. Springer, New York, pp 435–456

Bispo R, Bernardino J, Marques T, Pestana D (2013) Discrimination between parametric survival models for removal times of bird carcasses in scavenger removal trials at wind turbines sites. In: Lita da Silva J, Caeiro F, Natário I, Braumann CA (eds) Advances in regression, survival analysis, extreme values, markov processes and other statistical applications. Springer, Berlin, Heidelberg, pp 65–72

Coolen J, Lindeboom H, Cuperus J, van der Weide B, van der Stap T (2015) Benthic communities on old gas platforms as predictors for new offshore wind farms. In: Köppel J, Schuster E (eds) Conference on wind energy and wildlife impacts. Book of abstracts

Dai K, Bergot A, Liang C, Xiang WN, Huang Z (2015) Environmental issues associated with wind energy—a review. Renew Energy 75:911–921

Debusschere E, Bolle J, Blom E, Botteldooren D, de Coensel B, Glaropoulos A, Hostens K, Papadakis V, Vercauteren M, Vandendriesche S, Vincx M, Wessels, Degraer S (2015) Offshore pile-driving and young fish, a destructive marriage? In: Köppel J, Schuster E (eds) Conference on wind energy and wildlife impacts. Book of abstracts, p 27

Gartman V, Bulling L, Dahmen M, Geißler G, Köppel J (2016a) Mitigation measures for wildlife in wind energy development, consolidating the state of knowledge—part 1: planning & siting, construction. J Environ Assess Policy Manage 18(3):1650013-1-45. doi:10.1142/S1464333216500137

Gartman V, Bulling L, Dahmen M, Geißler G, Köppel J (2016b) Mitigation measures for wildlife in wind energy development, consolidating the state of knowledge—part 2: operation, decommissioning. J Environ Assess Policy Manage 18(2):1650014-1-31. doi:10.1142/S1464333216500149

Grodsky S, Behr M, Gendler A, Drake D, Dieterle B, Rudd R, Walrath N (2011) Investigating the causes of death for wind turbine-associated bat fatalities. J Mammal 92(5):917–925

Grünkorn T, Diederichs A, Poszig D, Diederichs B, Nehls G (2009) Wie viele Vogel kollidieren mit Windenergieanlagen?: How many birds collide with wind turbines? Natur und Landschaft 84(7):309–314

Janßen H, Hinrichsen HH, Augustin C, Kube S, Schröder T, Zettler M, Pollehne F (2015) Offshore wind farms in the southwestern Baltic Sea: a model study of regional impacts on oxygen conditions and on the distribution and abundance of the jellyfish *Aurelia aurita*. In: Köppel J, Schuster E (eds) Conference on wind energy and wildlife impacts. Book of abstracts

Korner-Nievergelt F, Korner-Nievergelt P, Behr O, Niermann I, Brinkmann R, Hellriegel B (2011) A new method to determine bird and bat fatality at wind energy turbines from carcass searches. Wildl Biol 17(4):350–363

May R (2015) A unifying framework for the underlying mechanisms of avian avoidance of wind turbines. Biol Conserv 190:179–187

Perrow M, Harwood A, Berridge R, Skeate E (2015) Avoidance of an offshore wind farm by a breeding seabird and its implications for the off-shore renewables industry. In: Köppel J, Schuster E (eds) Conference on wind energy and wildlife impacts. Book of abstracts, p 52

Russell D, Hastie G, Janik V, Thompson D, Hammond P, Matthiopoulos J, Jones E, Moss S, McConnell B (2015) The effects of construction and operation of wind farms on harbour seals (*Phoca vitulina*). In: Köppel J, Schuster E (eds) Conference on wind energy and wildlife impacts. Book of abstracts, p 59

Schuster E, Bulling L, Köppel J (2015) Consolidating the state of knowledge: a synoptical review of wind energy's wildlife effects. Environ Manage 56(2):300–331

Vanermen N, Onkelinx T, Courtens W, van de Walle M, Verstraete H, Stienen E (2015) Seabird avoidance and attraction at an offshore wind farm in the Belgian part of the North Sea. Hydrobiologia 756(1):51–61

Voigt C, Popa-Lisseanu A, Niermann I, Kramer-Schadt S (2012) The catchment area of wind farms for European bats: a plea for international regulations. Biol Conserv 153:80–86

Williams B, Brown E (2014) Adaptive management: from more talk to real action. Environ Manage 53(2):465–479

Contents

Part I
Species Behavior and Responses

Red Kites and Wind Farms—Telemetry Data from the Core Breeding Range

Hermann Hötker, Kerstin Mammen, Ubbo Mammen and Leonid Rasran

Abstract Red Kites (*Milvus milvus*) are the second most often reported species in relation to collisions with wind turbines in Germany. Germany houses more than half of the world's population of Red Kites and, therefore, has a high international responsibility for the protection of this species. The German Federal Ministry of the Environment, Nature Conservation and Nuclear Safety funded a field study to investigate why Red Kites and other birds of prey frequently collide with wind turbines, and which risk mitigation measures are most appropriate. The study took place in the core of the Red Kite global breeding range in Sachsen-Anhalt between 2007 and 2010. Ten breeding adult Red Kites were equipped with radio tags (seven birds) or GPS satellite transmitters (three birds). Each bird was tracked for one or two breeding and non-breeding seasons. Data on flight height and habitat preference were collected by visual observations. The collision risk was modeled in relation to the nest's proximity to wind turbines. It was found that Red Kites spent most of their time close to their nests. Most (54%) of the fixes were located within a radius of 1000 m around nests. It is important to note that the data did not indicate displacement of Red Kites by wind farms. Red Kites frequently visited wind farms for foraging and spent about 25% of their flight time within the swept heights of rotors of the most common wind turbines present in the study sites. The probability of closely approaching a wind farm significantly decreased with the distance between wind turbines and nests. Furthermore, the collision probability model predicted a sharp decrease of collision risk with increasing distance from the nest. The results clearly indicate that implementing buffer zones around nest sites reduces collision risk.

Keywords Red Kites · Wind farms · Telemetry data · Collision · Buffer zones

H. Hötker (✉) · L. Rasran
Michael-Otto-Institut im NABU, Bergenhusen, Germany
e-mail: Hermann.Hoetker@NABU.de

K. Mammen · U. Mammen
ÖKOTOP GbR, Halle (S), Germany

J. Köppel (ed.), *Wind Energy and Wildlife Interactions*,
DOI 10.1007/978-3-319-51272-3_1

Introduction

Onshore wind farms pose two major threats to birds and bats: (1) displacement due to habitat loss; and (2) collisions with individual wind turbines. While not all bird species are equally affected, birds of prey seem particularly vulnerable (Langston 2002; Smallwood 2007; Drewitt and Langston 2008; Smallwood and Thelander 2008). It has been well documented that many raptors are killed due to collisions with wind turbines, and several studies have indicated repercussions on local populations (Carrete et al. 2010; Garvin et al. 2011; Dahl et al. 2012). However, little is known whether these losses of individuals are relevant on a population level. Life-history theory though predicts that long-lived species such as raptors are highly sensitive to an increasing adult mortality (Sæther and Bakke 2000).

Few systematic studies on bird collisions with wind turbines in Germany have been published, yet it has become evident that raptors are among the birds with the highest mortalities due to collisions with wind turbines. Since 2002, the State Bird Observatory of the Brandenburg Environmental Agency has collected data on bird and bat collisions with wind turbines all over Germany (Dürr 2004). The species most often reported are: Common Buzzard *Buteo buteo* (290 victims, as at 11th December 2014), followed by Red Kite *Milvus milvus* (252 victims) and White-tailed Eagle *Haliaeetus albicilla* (102 victims) (Landesamt für Umwelt 2014). The fact that Red Kites are found high on the list deserves special attention, because regional populations are declining in most of its breeding range and the species is classified globally as "Near threatened" (IUCN 2014). The high number of reported fatalities at wind turbines has recently been identified as a major source of human-induced mortality (Knott et al. 2009).

Germany hosts more than 50% of the global breeding population of Red Kites (BirdLife International 2015), and hence should be responsible for protecting this species. A preliminary analysis (Mammen et al. 2013) showed that many adult Red Kites (older than two years) in Germany were killed by colliding with wind turbines (57 out of 63 cases). However, one and two-year-old Red Kites seemed less affected. Many of these collisions took place during the breeding season and caused both the loss of a partner and the loss of the brood. Bellebaum et al. (2013) modelled the numbers of wind turbine collisions of Red Kites in Brandenburg, Germany, and found that collisions were responsible for a 3.1% decline in the local post breeding population. They state that the mortality of Red Kites due to wind farms is approaching critical thresholds with respect to population growth. Schaub (2012) modelled population growth of Red Kites based on the Swiss population, and found that the population growth rate declined with an increasing number of wind turbines. The model also showed a clear effect of distance between wind turbines and nest location.

In 2007, the federal states' Bird Observatories Working Group recommended a minimum distance of 1000 m between wind turbines and Red Kites' nests (Länder-Arbeitsgemeinschaft der Vogelschutzwarten 2007). Although this recommendation has been widely accepted, local planning authorities approved numerous wind farm projects that were developed within this threshold distance. Furthermore, several wind farm developers whose project proposals had been refused because the turbines were proposed within the 1000 m recommendation took legal action. However, it should also be mentioned that at the time the Working Group published their recommendation, there was little scientific evidence that 1000 m was an effective safety radius around a Red Kite's nest. As a consequence, the Michael-Otto-Institute at NABU, the consultancy firm BioConsult SH and the Leibniz Institute for Zoo and Wildlife Research conducted a study on the interaction of birds of prey and wind farms (Hötker et al. 2013a). This was funded by the German Federal Ministry of the Environment, Nature Conservation and Nuclear Safety (BMUB): "Raptors and Wind Turbines: Analysis of Problems and Solutions Proposed (FKZ: 0327684/0327684A/0327684B)". This paper presents the results of the telemetry studies on Red Kites, which were part of this study. It focuses on activity patterns of Red Kites in relation to nest and wind farm locations, and it also estimates collision risk in relation to distance between nest and wind farm.

Method

The study took place at four wind farms in Sachsen-Anhalt, the federal state holding the highest Red Kite breeding density in Germany (Gedeon et al. 2015). The study sites were in the core breeding range of Red Kites worldwide. All study sites predominantly contained intensively-managed arable land (for details see Mammen et al. 2013).

From 2007 to 2009 VHF transmitters (VHF TW3-single celled tag, Biotrack Ltd, Wareham, Dorset, UK, weight 22 g) were attached to seven adult Red Kites (three males, four females). The tagged birds were hand-tracked using a YAESU VR 500 receiver with a HB9CV antenna, and a Sika-Receiver (Biotrack Ltd.) with a 3-Element-YAGI-Antenna (Lintec). The tagged Red Kites were usually tracked until they could be visually observed. The position of the bird was recorded on a map with a precision of ±15 m. On a few days Red Kite locations were determined by cross bearing without having visual contact with the birds. Fixes were evenly distributed over the daily activity periods of Red Kites. The minimum interval between two fixes of the same bird was 15 min. In 2010 GPS satellite transmitters (ARGOS PTT) were attached to three additional adult Red Kites (two males, one female). The precision of the fixes was ±15 m. The satellite transmitters issued four fixes during daytime hours, with fixes being 2 h apart from each other.

All ten Red Kites (five females, five males) were territorial breeding birds, eight of them paired with an untagged mate and two of the tagged birds were a couple. The transmitters were attached during the breeding season. As some transmitters were still working in the year after capture, data was collected from 13 Red Kite seasons altogether. In total, 3381 fixes were analysed. The minimum number of fixes for an individual was 53 in nine days, and the maximum number was 412 fixes in 128 days (as shown in Table 1).

Table 1 Red Kites radio tagged for the study

Individual	Sex	Year	Tag	Days with fixes	Fixes
Barbarossa	m	2007	VHF	65	385
Barbarossa	m	2008	VHF	67	286
Arthur	m	2007	VHF	45	194
Ramona	f	2007	VHF	33	174
Karl	m	2007	VHF	54	297
Barbara	f	2007	VHF	64	398
Gishild	f	2008	VHF	33	177
Gishild	f	2009	VHF	15	115
Alte Dame	f	2009	VHF	9	53
Alte Dame	f	2010	VHF	44	193
Lui	m	2010	PTT	128	412
Erik	m	2010	PTT	111	388
Svenja	f	2010	PTT	90	309

The following potential effects on the movements of individual Red Kites were tested: the number of fixes per season; transmitter type (VHS vs. satellite); commencement of observations (before versus during breeding season); site; year; and sex. This was done by estimating the home ranges of Red Kites by determining 95% kernel home ranges (Harris et al. 1990; Kenward 1992, 2001).

To study the daily flight time, six individuals (three males, three females) were tracked continuously during the daylight period. Teams of two to four observers followed each target bird individually from dawn to dusk, and recorded its behaviour. Altogether 20 tracking days took place. The time period when individuals were not observable (on average 2 h 12 min per day), behaviour patterns were interpolated based on the periods of visual contact (on average 13 h 04 min per day).

From 2007 to 2008, data on flight altitude was collected by observing Red Kites flying within and through wind farms. Observers recorded all Red Kites passing through the wind farms. Flight altitude was recorded in the following height categories: 0–10 m, 11–25 m, 26–50 m, 51–100 m, 101–150 m, and above 150 m, making use of the known hub height and upper and lower height limits of the rotor diameter. Flight altitudes between 51 and 150 m were estimated to be within the rotor blades' swept zones.

The influence of distances between wind turbines and nest sites was addressed by modelling a collision risk index for wind turbines at various distances. Segments were formed around a Red Kite's nest. The segments were embedded by concentric circles around the Red Kite's nest with each ring separated by 250 m (250–2000 m). Risk of collision was assumed to be equal at all points within a segment (area between two concentric circular lines). A collision risk index was calculated for each segment. In order to calculate collision risk it was assumed that each segment contained one wind turbine.

Collision risk was calculated according to May et al. (2010). Collision risk is composed of three factors: (I) the frequency of passage through the rotor blades' spheres, (II) the probability of collision when passing through the rotor swept area (so called Band model, Band et al. 2007), and (III) avoidance rate (the relative rate when birds avoid flying through the rotor swept area at the last minute).

Frequency of passage through the rotor swept area (factor I) was estimated by calculating the total flight activity for the entire season. For this the average daily flight time was taken from the 20 tracking days multiplied by the length of stay of Red Kites in their breeding grounds (245 days, 1st March to 31st October). Then the total flight activity was allocated to each of the 250 m distance segments. This was done by interpolating the telemetry data, by multiplying the relative frequency of fixes per individual and season with the total duration of flight activity.

The time spent flying over the rotor projection area was then estimated by multiplying the time spent flying (per segment) with the share of the area of the rotor projection (of the total area of the segment). The latter was derived by dividing the ground projection of the rotor sphere (equal to the size of the rotor swept area) by the total area of the segment. We took an Enercon E-66, a wind turbine typical for the studied wind farms with a swept area of 3421 m^2.

The average number of flights over the rotor projection area was then estimated by dividing the total time spent flying in the rotor projection area by the average duration of a flight over the rotor projection area. This was taken to be 10 s, assuming an average flight speed by Red Kites of 6.9 m/s (see below). The procedure described so far assumes an even distribution of flight activity through space, which is obviously an approximation of the real situation. The number of flights of each bird/season through the rotor swept area was finally estimated. This was done by multiplying the number of flights over the rotor projection area by the relative frequency of flights in rotor height, as derived from the behavioural observation data (see above).

As the focus was on the potential spatial variation of collision risk, the following was calculated: means of daily flight activity; length of stay per season; flight altitude; and relative frequencies of fixes in different distance zones across all individuals and years.

Average duration of daily flight activity, length of season, flight height estimations and the relative frequencies of fixes in different segments were taken from different samples and different individuals. Therefore, individual variation in all parameters could not be studied at the same time. As the focus was on the spatial variation in collision risk, variation between individuals was only allowed for the relative frequencies of fixes in different segments. For all other factors the means were calculated, which were then applied for all individuals.

The probability of collision (factor II) was estimated for each single flight event through the rotor sphere, and calculated using Band et al. (2007). A Microsoft Excel spreadsheet was used, which was drafted by Band et al. (2007). Specific traits of Red Kites that were used in the calculations: body length 65 cm; wing span 160 cm; average flight speed 6.9 m per second; using the overall average for both sexes (Bauer et al. 2005; Mebs and Schmidt 2006).

The avoidance rate (factor III) for Red Kites is unknown. An avoidance rate of 0.98 was assumed, based on observations of the related Northern Hen Harrier *Circus cyaneus* (Whitfield and Madders 2006a; Urquhart 2010). Because the avoidance rate factor highly influences collision risk calculations, collision index figures for Red Kites have to be treated with care.

All factors (I, II, and III) were first calculated for each distance zone and finally multiplied over the whole study area. Analyses were done in the statistic package R 3.0.2 (R Development Core Team 2009).

Results

First, the number of fixes, transmitter type (VHS versus satellite), commencement of observations (before versus in the beginning of the breeding season), site, year and sex were tested, to determine whether these factors had an influence on the home range of individual Red Kites. A generalized linear model (GLM) and single factor analyses (Kruskal-Wallis tests) did not show any strong influence on home range (Table 2). Although sample sizes were small, the absence of significant factors and the relatively high p-values (probabilities of rejecting a true null hypothesis, in this case non-impact of factors) support pooling the data.

Red Kites had large home ranges (Fig. 1), and often wind farms were intersecting with these home ranges. Nests usually formed the activity centres of a home range, both during and after the breeding season (Mammen et al. 2013). The presence of Red Kites (as measured by the number of fixes of tagged individuals) was highest in the 250 m distance segment immediately surrounding the nest site. Overall, numbers of fixes were high in all distance segments up to a distance of 1250 m from the nest. Further away, fewer fixes were recorded (Fig. 2).

Table 2 Results of a GLM and Kruskal-Wallis tests on the effects of different factors on home range of Red Kites

Factor	Estimate	SE	t-value	p-value
GLM				
Intercept	3111.12	2255.77	1.379	0.181
No. fixes	−11.63	15.00	−0.776	0.446
Kruskal-Wallis				
Factor		chi^2	df	p-value
Tag	VHS/GPS	0.0162	1	0.8987
Start of observations	Before/during breeding season	2.8846	1	0.0894
Site	Querfurt/Speckberg/Queis/Druiberg	4.2874	3	0.2321
Year	2007/2008/2009/2010	2.4518	3	0.4841
Sex	m/f	1.1834	1	0.2767

SE standard error of estimate, t-value: test statistic of a t-test

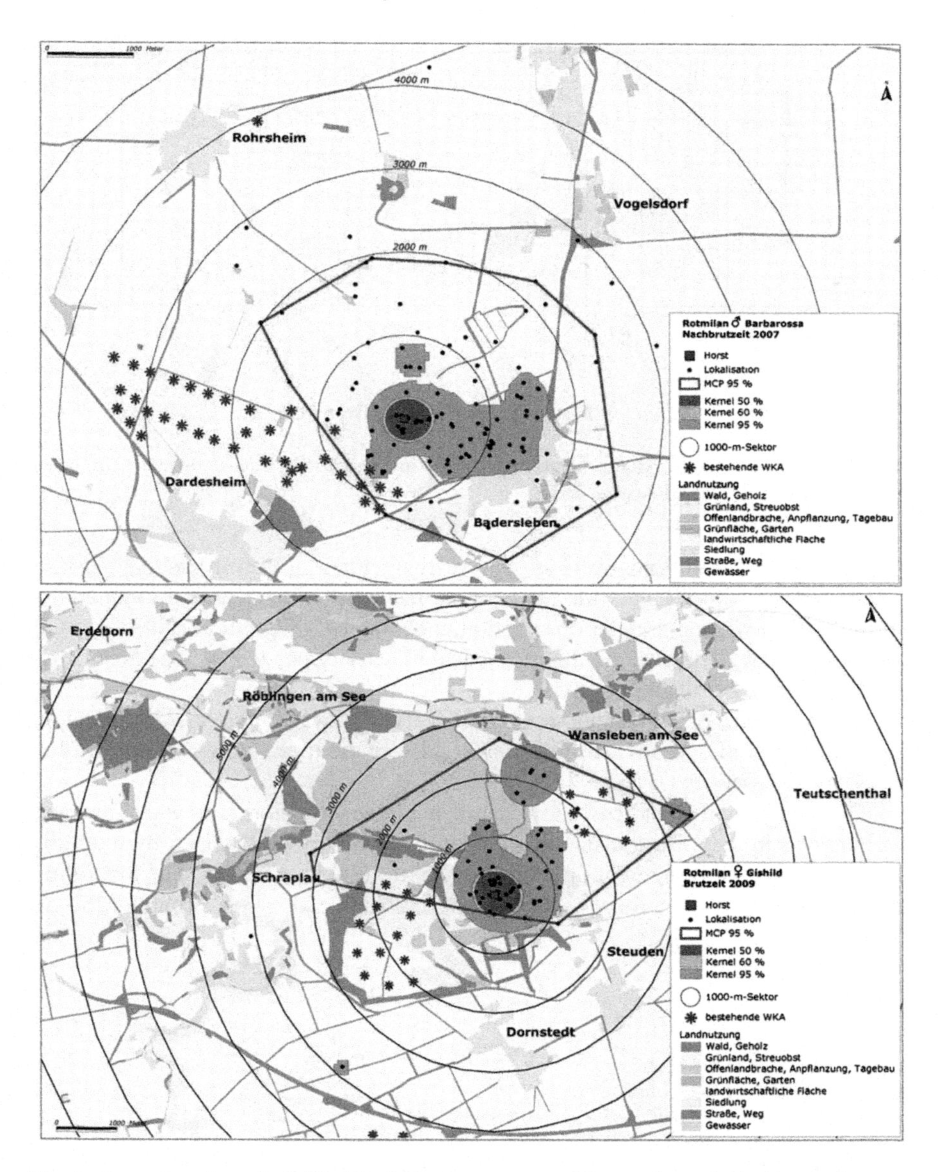

Fig. 1 Examples of two individual Red Kite home ranges. The nest is marked with a *red square*, and wind turbines are indicated with *blue stars*. The concentric circles indicate 1000 m sectors, the red polygon shows the 95% minimum convex polygon (MCP) home range, kernel home ranges are presented in brown (50%), yellow (60%) and green (95%). Other *colours* indicate land usage (forests, grassland, open area, settlement, water etc.)

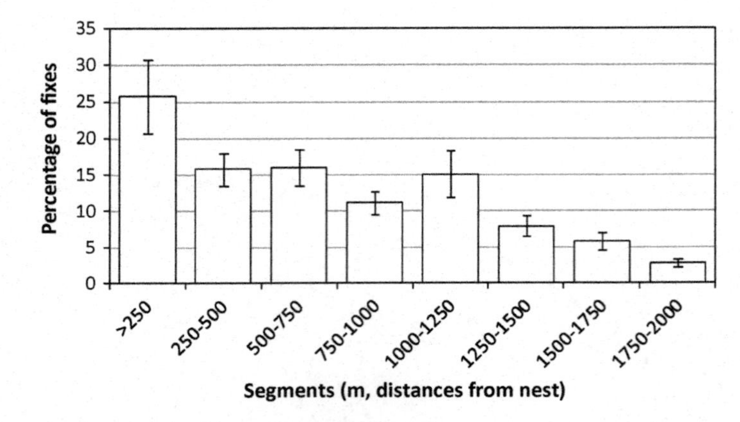

Fig. 2 Mean fixes (% ± SE) of tagged Red Kites in relation to distance segments from nest. n = 10 individuals and 13 seasons

On average, Red Kites were observed flying for 3 h 14 min per day (n = 20, SD = ±3 h 13 min). Under the assumption that Red Kites spend the same amount of time flying when not under observation, the total daily flight time added up to 3 h 47 min per day. The average duration of flight activity per season was 56,615 min (3 h 47 min * 245 days).

Flight altitude of Red Kites was recorded for 30,442 s altogether. When flying within wind farms, Red Kites spend about 25% of the time flying at a height of the rotor sphere (between 51 and 150 m, Fig. 3).

The collision risk indices show a much higher collision risk when the wind turbine was in the closest distance segment to the nest. This dropped by more than half at every 250 m segment up to a distance of 1000 ms (Fig. 4). The logarithmic scale reveals another strong decline in the collision risk index around 1250 and 1750 m.

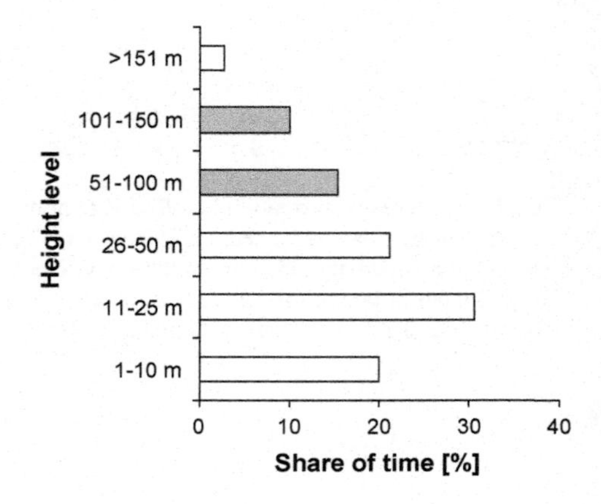

Fig. 3 Flight activity (percentage of the total flight time of 30,442 s) of Red Kites in different height segments. The *grey bars* show the altitude covered by rotor swept areas

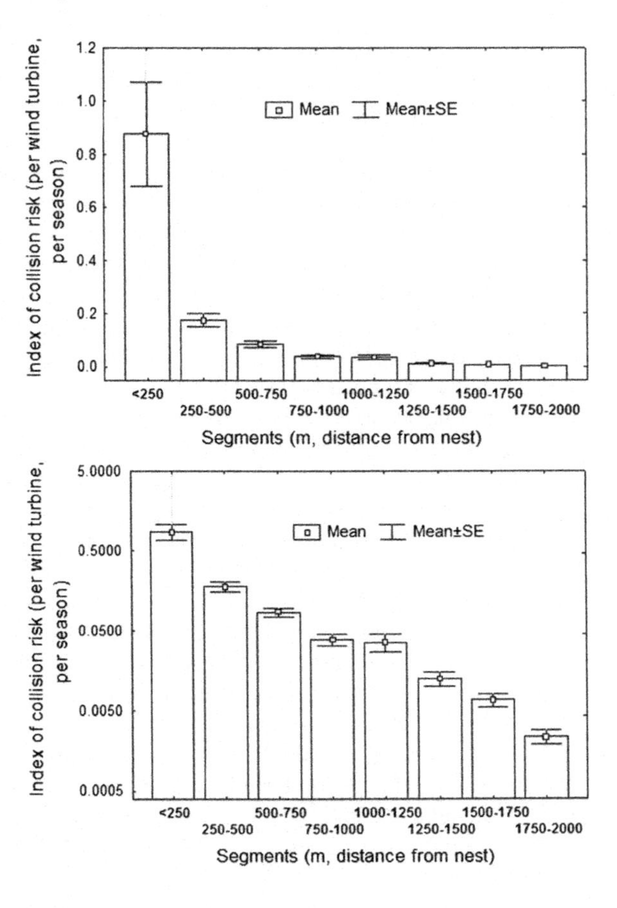

Fig. 4 Collision risk indices (mean ± SE) for Red Kites during the breeding season in different distance segments around the nest (see text for details). *Upper graph* normal scale, *lower graph* logarithmic scale. n = 13 seasons of 10 individuals

Discussion

Obviously, the collision risk index does not equal an actual collision risk. Unfortunately, the absence of field data for measuring avoidance rates does not allow exact collision rates to be determined. Also, a probable slight over-estimation of flight time within rotor swept areas increases the inaccuracy of measuring collision rates. A rather large range of critical (rotor swept area) flight heights was applied, in order to allow for extrapolation to other types of wind turbines present at the study sites. However, estimates of flight altitude were not precise enough to allow for better estimates. However, estimates of distances between nest sites and wind turbines were based on actual observations, and hence the collision risk index calculated can be regarded as a collision index relative to distance between nest and wind turbine.

The data presented here largely supports the assumption of Schaub (2012) who modelled the relationship of nest to wind-turbine distance and collision risk as a negative exponential function. Very clearly, our study also confirmed that Red Kites were at a much higher risk of collision when wind turbines are constructed close to their nest sites. The risk decreased rapidly with distance, and therefore the designation of 'no-go areas' for wind turbines (fixed radii) around Red Kite nesting sites is a useful measure for mitigating collision risk. The 'fixed radii' suggested by Länder-Arbeitsgemeinschaft der Vogelschutzwarten (2007) serve as a helpful guideline. Obviously, the larger the radius the more effective the mitigation is for avoiding and minimising impacts. On the other hand, large "no-go areas" of course have a large impact on the area available for future wind farms, a resource which is already scarce in parts of Germany (Eichhorn and Drechsler 2010).

Länder-Arbeitsgemeinschaft der Vogelschutzwarten (2014) recently suggested to increase of the "no-go areas" for wind turbines from 1000 to 1500 m in the vicinity of Red Kite nest locations (with reduced distances for some other species). Of course, there are shortcomings in the concept of 'no-go areas' for wind turbines at Red Kite nest locations: (1) Red Kites may change nest sites over the years. Although, such changes do not happen frequently enough to reject the concept of 'no-go areas', as most new nest sites are close to the old ones (Ortlieb 1989; Aebischer 2009); (2) Red Kites do not always distribute their flight activities evenly over their breeding territory, or nests may not be the centre but in one corner of their home range. This of course will have implications for the planning of wind farms. For example, even wind farms close to a nest site may pose a lower risk to the birds than expected by this model when the turbines are situated in an area not used by them. Whereas, wind turbines further away from nest sites can pose a high risk if they are located in a preferred feeding site. Red Kites are opportunistic in their choice of diet and, therefore, may change preferred foraging sites among seasons. Comprehensive, multi-seasonal and multi-annual studies are needed before building wind farms.

Relative to their population size (approximately 12,000–18,000 breeding pairs in Germany, Sudfeldt et al. 2013), fatalities of Red Kites have been reported frequently. Only White-tailed Eagles have a higher victim/population ratio among raptors in Germany (Hötker and Dürr, own calculations). Two obvious reasons may account for this fact. Firstly, Red Kites and White-tailed Eagles are rather large birds. Their size results in a higher probability to be hit by a rotor blade when passing through the rotor blade swept area (Band et al. 2007). Secondly, Red Kites fly relatively often at heights identical to the rotor swept area (Mammen et al. 2011), while other raptor species fly at lower altitudes, such as the Montagu's Harriers (Grajetzky and Nehls 2013) and the Hen Harrier (Whitfield and Madders 2006b).

According to their life history, Red Kite populations are expected to be very sensitive to an increasing adult mortality. Smart et al. (2010) and others have shown that mortality due to illegal persecution has a high impact on population growth. Schaub (2012) suggested that a further increase in wind turbine numbers may lead to population declines. Bellebaum et al. (2013) estimated that the local Red Kite population of Brandenburg is running the risk of decline as more wind farms are built. In addition, recent data has shown that the German Red Kite breeding population has been declining in recent years (Sudfeldt et al. 2013). At the same time, the German Federal Government has been supporting a further increase in the sector of renewables (Bundesregierung der Bundesrepublik Deutschland 2015).

Given the sensitivity of Red Kites to collisions with wind turbines, mitigation measures should be implemented immediately. Besides strictly applying "no-go areas" of wind turbines in certain distance to Red Kite nest sites, a land development plan concerning wind turbine locations should specify that wind turbines should be pooled in wind farms and not discretely scattered all over the landscape (Schaub 2012; Rasran and Dürr 2013). Additionally, land management issues within wind farm areas should be addressed. Red Kites strongly prefer freshly mown or farmed field as well as field margins. Moreover, Red Kites find ample food sources at the base of wind turbines, as these areas often hold dense rodent populations due to the lack of farming activity there (Mammen et al. 2013).

Given the above information regarding Red Kite behaviour, all activities that could potentially attract this species into wind farms should be ceased. For example, early season mowing of grass or other crops should be ceased, and areas around the base of the wind turbine should be made as unattractive to Red Kites as possible. This could be done by farming, paving or planting bushes at the base of the wind turbine. No other additional food sources that attract Red Kites, such as dung heaps should be present within the area of a wind farm (Hötker et al. 2013b). Instead, safe areas far away from wind farms but close to nest sites should be made more attractive for Red Kites (Martínez-Abraín et al. 2012).

Acknowledgements We would like to thank Alexander Resetaritz, Lukas Kratzsch and Ralf Siano for their help during the field work. We received valuable advice from Jutta Leyrer, who also improved the English of the manuscript.

We thank the Federal Ministry for the Environment, Nature Conservation, Building and Nuclear Safety for the financial support and we are grateful to the Project Management Jülich, in particular Tobias Verfuß, for their patience and support.

References

Aebischer A (2009) Der Rotmilan. Haupt Verlag, Bern

Band W, Madders M, Whitfield DP (2007) Developing field and analytical methods to assess avian collision risk at wind farms. In: de Lucas M, Janss G, Ferrer M (eds) Birds and wind power. Lynx Edicions, Barcelona

Knott J, Newbery P, Barov, B (2009) Action plan for the red kite *Milvus milvus* in the European Union. RSPB and BirdLife International, on behalf of the European Commission, Sandy

Bauer HG, Bezzel E, Fiedler W (2005) Das Kompendium der Vögel Mitteleuropas. Nonpasseriformes-Nichtsperlingsvögel, AULA, Wiebelsheim

Bellebaum J, Korner-Nievergelt F, Dürr T, Mammen U (2013) Wind turbine fatalities approach a level of concern in a raptor population. J Nat Conserv 21:394–400

BirdLife International (2015) European red list of birds. http://www.birdlife.org/europe-and-central-asia/european-red-list-birds-0. Accessed 7 Dec 2015

Bundesregierung der Bundesrepublik Deutschland (2015) Energiewende. http://www.bundesregierung.de/Content/DE/StatischeSeiten/Breg/Energiekonzept/1-EnergieErzeugen/23-11-11-wind.html. Accessed 19 Oct 2015

Carrete M, Sanchez-Zapata JA, Benitez JR, Lobon M, Donazar JA (2010) Large scale risk-assessment of wind-farms on population viability of a globally endangered long-lived raptor. Biol Conserv 142:2954–2961

Dahl EL, Bevanger K, Nygård T, Røskaft E, Stokke BG (2012) Reduced breeding success in white-tailed eagles at Smøla windfarm, western Norway, is caused by mortality and displacement. Biol Conserv 145:79–85

Drewitt AL, Langston RHW (2008) Collision effects of wind-power generators and other obstacles on birds. Ann NY Acad Sci 1134:233–266

Dürr T (2004) Vögel als Anflugopfer an Windenergieanlagen—ein Einblick in die bundesweite Fundkartei. Bremer Beitr Naturk Naturschutz 7221–228

Eichhorn M, Drechsler M (2010) Spatial trade-offs between wind power production and bird collision avoidance in agricultural landscapes. Ecol Soc 15:10 (online)

Garvin JC, Jenelle CS, Drake D, Grodsky SM (2011) Response of raptors to a windfarm. J Appl Ecol 48:199–209

Gedeon K, Grüneberg C, Mitschke A, Sudfeldt C (2015) Atlas Deutscher Brutvogelarten. Stiftung Vogelmonitoring Deutschland, Dachverband Deutscher Avifaunisten, Münster

Grajetzky B, Nehls G (2013) Wiesenweihentelemetrie. In: Hötker H, Krone O, Nehls G (eds) Greifvögel und Windkraftanlagen: Problemanalyse und Lösungsvorschläge. Schlussbericht für das Bundesministerium für Umwelt, Naturschutz und Reaktorsicherheit, FKZ 0327684, Michael-Otto-Institut im NABU, Leibniz-Institut für Zoo- und Wildtierforschung, BioConsult SH, Bergenhusen, Berlin, Husum

Harris S, Cresswell WJ, Forde PG, Trewhella WJ, Woollard T, Wray S (1990) Home-range analysis using radio-tracking data - a review of problems and techniques particularly as applied to the study of mammals. Mamm Rev 20:97–123

Hötker H, Krone O, Nehls G (eds) (2013a) Greifvögel und Windkraftanlagen: Problemanalyse und Lösungsvorschläge. Schlussbericht für das Bundesministerium für Umwelt, Naturschutz und Reaktorsicherheit, FKZ 0327684, Michael-Otto-Institut im NABU, Leibniz-Institut für Zoo- und Wildtierforschung, BioConsult SH, Bergenhusen, Berlin, Husum

Hötker H, Dürr T, Grajetzky B, Grünkorn T, Joest R, Krone O, Mammen K, Mammen U, Nehls G, Rasran L, Resetaritz A, Treu G (2013b) Fazit, Risikoeinschätzung, Minimierung von Konflikten, Empfehlungen für die Praxis, Forschungsbedarf. In: Hötker H, Krone O, Nehls G (eds) Greifvögel und Windkraftanlagen: Problemanalyse und Lösungsvorschläge. Schlussbericht für das Bundesministerium für Umwelt, Naturschutz und Reaktorsicherheit, FKZ 0327684, Mi chael-Otto-Institut im NABU, Leibniz-Institut für Zoo- und Wildtierforschung, BioConsult SH, Bergenhusen, Berlin, Husum

Kenward RE (1992) Quantity versus quality: programmed collection and analysis of radio-tacking data. In: Priede IG, Swift SM (eds) Wildlife Telemetry. Remote Monitoring and Tracking of Animals. Ellis Horwood, London, pp 231–246

Kenward RE (2001) A manual for wildlife radio tagging. Academic Press, London

Länder-Arbeitsgemeinschaft der Vogelschutzwarten (2007) Abstandsregelungen für Windenergieanlagen zu bedeutsamen Vogellebensräumen sowie Brutplätzen ausgewählter Vogelarten. Ber z Vogels 44:151–153

Länder-Arbeitsgemeinschaft der Vogelschutzwarten (2014) Abstandsempfehlungen für Windenergieanlagen zu bedeutsamen Vogellebensräumen sowie Brutplätzen ausgewählter Vogelarten. Ber z Vogels 51:15–42

Landesamt für Umwelt (2014) Zentrale Fundkartei über Anflugopfer an Windenergieanlagen (WEA). Download 11th Dec. 2014. Available via http://www.lugv.brandenburg.de/cms/detail.php/bb1.c.321381.de

Langston R (2002) Wind Energy and Birds: Results and Requirements. RSPB Research Report No. 2. RSPB, Sandy, pp 1–54

Mammen U, Mammen K, Heinrichs N, Resetaritz A (2011) Red Kite (*Milvus milvus*) fatalities at wind turbines—why do they occur and how are they to prevent? In: May R, Bevanger K (eds) Proceedings Conference on Wind energy and Wildlife impacts. 108. NINA Report 693, Trondheim, Norway

Mammen K, Mammen U, Resetaritz A (2013) Rotmilan. In: Hötker H, Krone O, Nehls G (eds) Greifvögel und Windkraftanlagen: Problemanalyse und Lösungsvorschläge. Schlussbericht für das Bundesministerium für Umwelt, Naturschutz und Reaktorsicherheit, FKZ 0327684, Michael-Otto-Institut im NABU, Leibniz-Institut für Zoo- und Wildtierforschung, BioConsult SH, Bergenhusen, Berlin, Husum

Martínez-Abraín A, Tavecchia G, Regan HM, Jiménez J, Surroca M, Oro D (2012) Effects of wind farms and food scarcity on a large scavenging bird species following an epidemic of bovine spongiform encephalopathy. J Appl Ecol 49:109–117

May R, Hoel PL, Langston R, Dahl E, Bevanger K, Reitan O, Nygård T, Pedersen HC, Røskaft E, Stokke BG (2010) Collision risk in white-tailed eagles. Modelling collision risk using vantage point observations in Smøla wind-power plant. 1–25. NINA Report 639, Trondheim, Norway

Mebs T, Schmidt D (2006) Die Greifvögel Europas. Nordafrikas und Vorderasiens, Franckh-Kosmos, Stuttgart

Ortlieb R (1989) Der Rotmilan. 3. Aufl. Ziemsen (Die neue Brehm Bücherei 532), Wittenberg-Lutherstadt

Rasran L, Dürr T (2013) Kollisionen von Greifvögeln an Windenergieanlagen—Analyse der Fundumstände. In: Hötker H, Krone O, Nehls G (eds) Greifvögel und Windkraftanlagen: Problemanalyse und Lösungsvorschläge. Schlussbericht für das Bundesministerium für Umwelt, Naturschutz und Reaktorsicherheit, FKZ 0327684, Michael-Otto-Institut im NABU, Leibniz-Institut für Zoo- und Wildtierforschung, BioConsult SH, Bergenhusen, Berlin, Husum

Sæther B-E Bakke Ø (2000) Avian life history variation and contribution of demographic traits to the population growth rate. Ecol 81:642–653

Schaub M (2012) Spatial distribution of wind turbines is crucial for the survival of red kite populations. Biol Conserv 155:111–118

Smallwood KS (2007) Estimating wind turbine-caused bird mortality. J Wildl Manage 71:2781–2791

Smallwood KS, Thelander C (2008) Bird Mortality in the Altamont Pass Wind Resource Area, California. J Wildl Manage 72:215–223

Smart J, Amar A, Sim IMW, Etheridge B, Cameron D, Christie G, Wilson JD (2010) Illegal killing slows population recovery of a re-introduced raptor of high conservation concern—the red kite *Milvus milvus*. Biol Conserv 143:1278–1286

Sudfeldt C, Dröschmeister R, Frederking W, Gedeon K, Gerlach B, Grüneberg C, Karthäuser J, Langgemach T, Schuster B, Trautmann S, Wahl J (2013) Vögel in Deutschland—2013. DDA, BfN, LAG VSW, Münster

R Development Core Team (2009) R: A language and environment for statistical computing. R Foundation for Statistical Computing, ISBN 3-900051-07-0. Wien. http://www.R-project.org

Urquhart B (2010) Use of avoidance rates in the SNH wind farm collision risk model. SNH Avoidance Rate Information & Guidance Note, Scottish Natural Heritage 10

Whitfield DP, Madders M. (2006a) A review of the impacts of wind farms on Hen Harriers *Circus cyaneus* and an estimation of collision avoidance rates. Natural Research LTD, Banchory

Whitfield DP, Madders, M (2006b) Flight height in the Hen Harrier *Circus cyaneus* and its incorporation in wind turbine collision risk modelling. Natural Research Information, Note 2, Banchory

Part II
Collision Risk and Fatality Estimation

Unforeseen Responses of a Breeding Seabird to the Construction of an Offshore Wind Farm

Andrew J.P. Harwood, Martin R. Perrow, Richard J. Berridge, Mark L. Tomlinson and Eleanor R. Skeate

Abstract Sheringham Shoal Offshore Wind Farm (OWF), comprised of 88 3.6 MW turbines, was built within foraging range of Sandwich Tern *Thalasseus sandvicensis* breeding at a European designated site. Boat-based surveys (n = 43) were used to investigate changes in tern abundance within the site and within 0–2 and 2–4 km buffer areas before and throughout the construction of the OWF, over a study period between 2009 and 2012. Visual tracking of individual birds (n = 840) was also undertaken to document any changes in behaviour. This study is amongst the few to detail the response of a breeding seabird to the construction of an OWF. Navigational buoys in the 0–2 km buffer were used extensively by resting and socialising birds, especially early in the breeding season. Visual tracking illustrated avoidance of areas of construction activity and birds surprisingly kept their distance from installed monopiles. Avoidance was strengthened during turbine assembly, with around 30% fewer birds entering the wind farm, relative to the pre-construction baseline. Flight lines of birds that entered the site were generally along the centre of rows between turbines. A focus on transit flight meant that feeding activity was lower in the site than the buffer areas. As the site remained permeable to terns flying to and from foraging grounds further offshore, the overall abundance within the site was not significantly reduced. Although a number of the responses observed were unforeseen by Environmental Impact Assessment, the overall conclusion of only minor adverse effects was upheld. Analysis of further data from the operational site is now planned.

Keywords Sandwich Tern · Offshore wind farm · Visual tracking · Boat-based survey · Avoidance behaviour

---

A.J.P. Harwood (✉) · M.R. Perrow · R.J. Berridge · M.L. Tomlinson · E.R. Skeate
ECON Ecological Consultancy Limited, Unit 7, The Octagon Business Park, Little Plumstead, Norwich NR13 5FH, UK
e-mail: a.harwood@econ-ecology.com

J. Köppel (ed.), *Wind Energy and Wildlife Interactions*,
DOI 10.1007/978-3-319-51272-3_2

Introduction

Offshore wind energy is a rapidly developing industry, particularly in countries bordering the North Sea in north-western Europe, but increasingly across the globe including China and the USA (Breton and Moe 2009; Da et al. 2011). The associated risks of offshore wind farm (OWF) development for seabirds are well documented (e.g. Garthe and Hüppop 2004; Furness et al. 2013; Gove et al. 2013) and the following effects are typically assessed during Environmental Impact Assessment (EIA): mortality through collision with rotating blades, disturbance due to construction and maintenance activities, displacement leading to direct habitat loss, and barriers to movement resulting in changes in energy expenditure during commuting and foraging flights (see DECC 2011 in relation to National Policy in the UK). In the case of breeding birds, changes in energy budget may impact upon dependent chicks and thus breeding productivity, although this has not been quantified as yet (Masden et al. 2010).

Recent evidence also suggests the potential for indirect effects of construction upon seabirds, in particular the effect of piling noise on sensitive fish species such as clupeids, with consequent effects on prey availability (Perrow et al. 2011a). However, in the long-term, indirect effects could benefit seabirds through improved prey resources associated with reef and sanctuary effects (Linley et al. 2007). Such benefits may be countered by increased collision risk. For example, Thelander and Smallwood (2007) reported increased mortality of Red-Tailed Hawk *Buteo jamaicensis* at onshore turbines due to increased prey (rodents) around the turbine bases.

Despite the large number of offshore wind farms currently in operation or under construction, there are few detailed published studies on the real impacts upon birds (Desholm and Kahlert 2005; Petersen et al. 2006, 2014; Masden et al. 2009, 2010; Plonczkier and Simms 2012; Lindeboom et al. 2011; Skov et al. 2012; Leopold et al. 2013; BSH and BMU 2014; Vanermen et al. 2012, 2015a). This is partly because of the significant technical challenges and costs associated with monitoring and quantifying the response of birds to OWFs.

To determine changes in the distribution and abundance that indicate displacement, surveys of large areas around or away from the development are required to allow investigation of natural variation or gradient effects. However, appropriate spatial and temporal resolution must be maintained to provide sufficient data and statistical power to detect changes associated with the development (Vanermen et al. 2015b). Digital aerial surveys are increasingly being used (Buckland et al. 2012) to efficiently cover large study areas, although intensive boat-based surveys may allow rapid changes in the distribution and abundance of birds, for example due to tidal cycles, to be more effectively sampled (Embling et al. 2012). Sophisticated modelling techniques have also been developed to discriminate the effects of development from natural background variation (see Petersen et al. 2011, 2014).

Visual survey techniques using standard visual aids (e.g. binoculars) and laser rangefinders (Pettersson 2005; Skov et al. 2012), have been used to monitor the response of birds to structures. Technical equipment, including a variety of radar, video and thermal imaging systems (Desholm et al. 2006; Krijgsveld et al. 2011; Plonczkier and Simms 2012; Skov et al. 2012; BSH and BMU 2014) have also been employed to attempt to quantify avoidance and collision risk from the movements of individuals and/or flocks of both seabirds and migrating land birds. However, the observation of actual collisions remains an extremely rare event and risk is typically assessed through modelling of passage rates (Skov et al. 2012; Brabant et al. 2015).

Monitoring the behavioural response of birds is more readily achieved and, for some breeding species in particular, individual-based tracking with radio and GPS devices, to determine general patterns of use, has recently been employed for wind farms (Perrow et al. 2006, 2015; Wade et al. 2014; Thaxter et al. 2015). However, remote monitoring tools may not be suitable for all species and the sample sizes and behavioural detail (e.g. foraging activity and subtle responsive changes in flight height and direction) that can be achieved may be limiting.

Defining a behavioural reaction to the construction and operation of a wind farm is also complicated by the fact that it may illicit a gross response in sensitive species, with avoidance beginning several kilometers from the potential risk (Desholm and Kahlert 2005; Plonczkier and Simms 2012). Even where a species is less sensitive, the response can vary according to environmental conditions (Skov et al. 2012) or be subject to considerable inter-annual, seasonal and individual variation (Thaxter et al. 2015). Establishing a baseline prior to construction is likely to be essential to help separate cause and effect of behavioural responses, but this is rarely accommodated in studies. Furthermore, most studies to date have been conducted on non-breeding birds during passage. For seabirds, the energetic constraints imposed by provisioning chicks seem likely to modify the risks that adult birds may take. Thus, observations derived from birds during dispersal should only be applied to breeding birds with extreme caution, if at all.

Here, we present findings from monitoring work specifically targeting Sandwich Tern at the Sheringham Shoal OWF. This study contains a number of important elements that further the understanding of interactions between birds and wind farms: (1) a breeding seabird is monitored when the population is most sensitive to impacts, (2) the initial response to wind farm construction is investigated—such studies are generally inhibited by restricted access to survey vessels (e.g. Lindeboom et al. 2011; Leopold et al. 2013; Vanermen et al. 2015a, b), (3) gross changes in the use of the study area during construction are explored using a gradient analysis applied to boat-based survey data, (4) complementary visual tracking data is used to evaluate individual responses to the development, and (5) the study provides an opportunity to test the EIA predictions.

Methods

Study Site

The 317 MW Sheringham Shoal OWF consists of 88 3.6 MW Siemens wind turbine generators (rotor diameter of 107 m) and two substations. It was the first of the OWFs within the Crown Estate's Greater Wash (UK) Round Two development area to be consented (Fig. 1). Construction began in February 2010 with the installation of eight navigation buoys to delimit the site for marine vessels. To protect against scour, large quantities of rock were installed at 75 turbine locations and the two substations in March 2010. Monopile installation began on 24 June, and by the end of November 2010 22 monopiles and the two substations had been

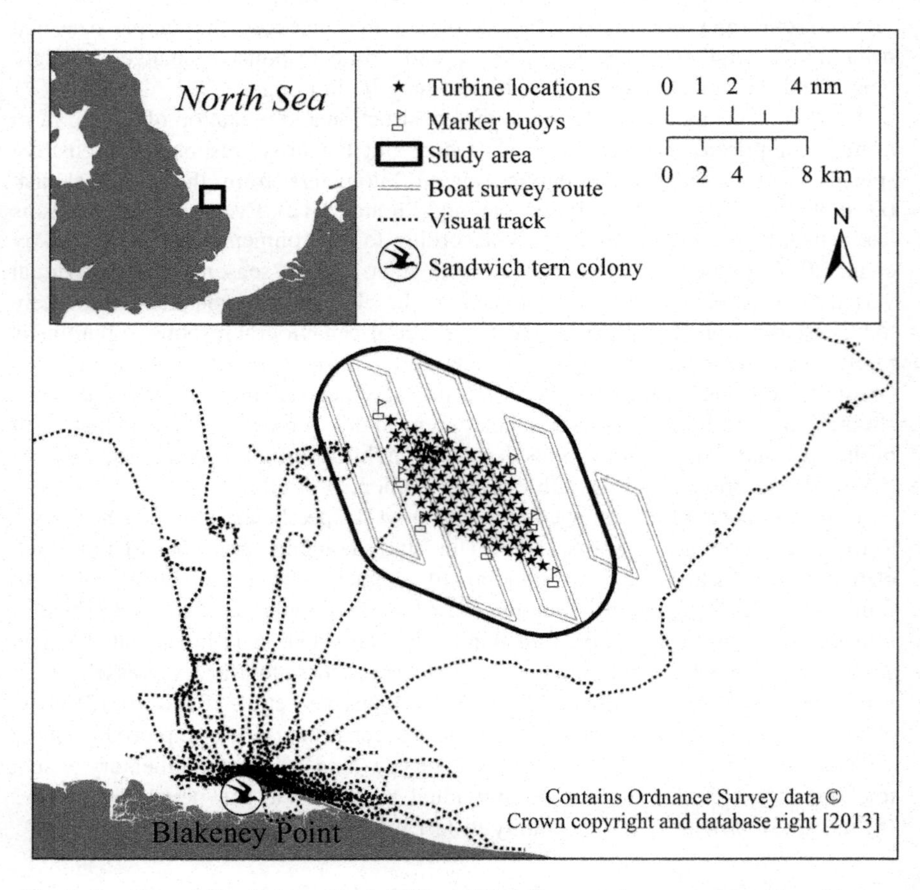

Fig. 1 Location of Sheringham Shoal Offshore Wind Farm, study area and boat-based transect route relative to visual tracking from the Blakeney Point colony conducted prior to the development of the site in 2007 and 2008 (Perrow et al. 2010)

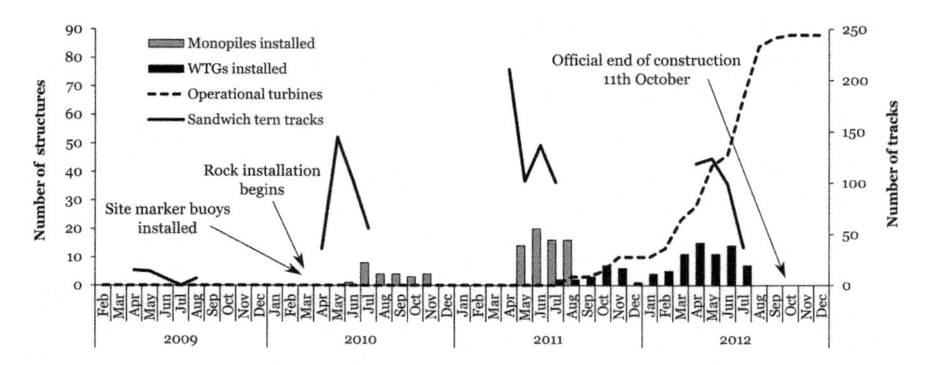

Fig. 2 Sheringham Shoal OWF construction progress relative to the number of visual tracks achieved in the study area during the Sandwich Tern breeding season from April to July inclusive in each year

installed (Fig. 2). Pile driving of individual monopiles was brief, taking between 30 and 40 min. Construction works were more frequent and sustained during 2011, including regular pile driving and cable laying during the installation of the final 66 monopiles between May and August. Assembly of the tower, nacelle and blades for each turbine started in July 2011, with the first power produced in August 2011. Turbine assembly was completed by the end of July 2012, by which time 46 turbines were connected and able to deliver power. The OWF was officially opened on 27 September 2012.

The OWF is located between 18 and 24 km from Blakeney Point, which, in combination with Scolt Head, contains the largest concentration of breeding Sandwich Terns in the UK and is designated as a qualifying feature of the internationally important North Norfolk Coast Special Protection Area (SPA). The SPA can support over 4000 pairs and is designated as containing 24.7% of the UK population (Mitchell et al. 2004). Both colonies lie within the mean maximum foraging range of the birds from the OWF (49 km—Thaxter et al. 2012). However, Blakeney Point is closer to the OWF and previous tracking work (Fig. 1) has suggested this is likely to supply the majority of birds encountered (Perrow et al. 2010).

Although both colonies may be active in the same year, the typical pattern is for the highly colonial Sandwich Tern to favour one or the other. Previous data suggests a periodicity of four or more years between colony switches (NNNS 2007), although Blakeney Point was the dominant colony throughout the duration of this study, with 2500–3753 pairs (Table 1). Boat-based surveys of Sheringham Shoal, as part of the EIA in 2004–2006, confirmed the presence of Sandwich Terns throughout the breeding season (April to July inclusive), with maximum densities of 0.3 and 0.72 ind. km^{-2} in the site and study area respectively (SCIRA Offshore Energy Ltd 2006). It was assumed that most, if not all, of these birds originated from Blakeney Point.

The relatively low density of Sandwich Terns recorded during site characterisation resulted in low numbers of annual predicted collisions (12 at 99% avoidance).

Table 1 Estimated number of pairs, fledged chicks and chicks fledged per pair of Sandwich Terns nesting at Blakeney Point over the study period of 2009–2012 inclusive

Parameter	2009	2010	2011	2012
Number of pairs	3100	2500	3562	3753
Chicks fledged	1300	900	1700–2000	2200
Chicks pr^{-1} $year^{-1}$	0.43	0.36	0.48–0.56	0.59

Thus, the effect was deemed to be '*minor adverse*' in EIA terms; that is, undesirable but of limited concern (SCIRA Offshore Energy Ltd 2006). A similar effect was predicted in relation to temporary disturbance during construction, through increased boat traffic. The potential for minor adverse barrier effects to occur was thought likely to be offset by the orientation and layout of the turbines, which incorporated the preferred northeast-southwest flight lines of Sandwich Terns. No disruption of flight lines leading to increased energy expenditure of the birds was anticipated. Indirect effects upon the available prey base and effects of increased noise and vibration, as well as cable laying activities, were all predicted to be of negligible significance (SCIRA Offshore Energy Ltd 2006).

Use of the Wind Farm Area

In order to determine any changes in the abundance and distribution of birds according to the construction and operation of the wind farm, a gradient design (Strickland et al. 2007) was employed. This incorporated the wind farm site and two sequential buffer areas at 0–2 and 2–4 km from the site. These areas were surveyed by boat-based line transects (300 m either side of the vessel) for birds 'on the water' (perching on surface floating objects) and using radial snapshots (180° scan centred on the bow of the vessel out to 300 m) for birds in flight (Fig. 1). Two experienced ornithologists (one on each side of the vessel) carried out observations at all times whilst a third recorded data. Survey intensity varied slightly between months, with two surveys completed in April each year and three in the following months. However, the monitoring schedule set with the statutory authorities only incorporated two surveys in May 2010. A total of $n = 43$ surveys were therefore available for analysis.

Density estimates were calculated for each of the three areas by combining separate densities of Sandwich Tern 'on the water' and in flight, derived by dividing the numbers of observed birds by the respective areas sampled. Distance sampling corrections were not employed as there were insufficient observations of birds on the water to generate a viable detection function. It was assumed that all birds in flight were detected to a distance of 300 m, according to standard practice (Camphuysen et al. 2004). Population estimates for each of the areas were estimated by scaling the densities to each respective area for later analyses.

Variations in the abundance of birds over the breeding season and between areas were investigated using Generalised Additive Models (GAMs). GAMs were chosen as they allow for data which is non-normally distributed and could potentially better describe complex seasonal trends in the abundance of Sandwich Terns in the area (Hastie and Tibshirani 1990). Thus, the model framework included 'year day' (i.e. day 1–365 or 366 in a leap year) as a continuous variable (using a smooth function with degrees of freedom limited to 4 or less), 'site' (i.e. wind farm, 0–2 km buffer and 0–4 km buffer) and 'year' (2009–2012 inclusive) as factors. Year was used as a factor, rather than discreet development periods, as it provided a balanced dataset, which also accounted for inter-annual variability in abundance. An interaction between 'site' and 'year' was tested first to determine whether there were significant variations between combinations of sites and years. 'Monitoring year' and 'site' were investigated independently if the interaction was not significant. To account for the variability in the size of each of the areas, for which populations were derived from survey densities, an offset (log area) was also included in the model.

A negative binomial distribution (including log-link function) was used as it outperformed others trialled, due to its ability to deal with over dispersion in the data (Zuur et al. 2009). The optimal model was chosen as the one with the lowest Akaike information criterion (AIC) value and in which all remaining explanatory variables presented significant effects. The deviance explained by the model was used to evaluate the fit with Pearsons correlation (r) and Spearman rank correlation (ρ) coefficients as measures of model accuracy. A non-parametric Runs test was used to determine whether there was significant autocorrelation within the model residuals. All analyses were carried out using R 3.1.2 software (R Core Team 2014), stats (R Core Team 2014), mgcv (Wood 2011) and lawstat (Noguchi et al. 2009) packages.

Visual Tracking in the Wind Farm Study Area

Visual tracking of Sandwich Terns applied the methods established by Perrow et al. (2011b), and later adopted by Robertson et al. (2014) and Wilson et al. (2014). Birds were followed at a distance, so that they are not influenced by the vessel (generally upward of 50 m), whilst continually recording positions and behaviour. The resultant tracks aim to closely represent the path taken by the birds, albeit undertaken a few seconds later.

The movements of Sandwich Terns were tracked within a study area, defined by a 4 km buffer around the site (Fig. 1), throughout the breeding seasons in 2009 to 2012 inclusive. In 2009, tracking effort was limited (four days) during trialling of the method. Tracking was undertaken from a high-powered rigid-hulled inflatable boat (RIB) in a range of weather conditions (with at least reasonable visibility) up to sea state four (Fig. 3). Birds apparently heading toward the wind farm site were generally detected and tracked from a distance of greater than 2 km from the site, a

Fig. 3 Example of RIB (10 m) typically used for visual tracking (*left*) and representation of the view of a tracked bird (*right*)

distance at which Sandwich Terns are unlikely to exhibit any avoidance behaviour in relation to structures or activity (Everaert and Stienen 2007). One ornithologist continuously observed the bird whilst a second took notes. Tracking ended when the bird left the study area, was lost from view (due to speed or weather) or landed on an object.

Tracks were plotted and analysed in ArcGIS v.10.1 (ESRI 2011: ArcGIS Desktop, Release 10, Redlands, CA, USA) and using Geospatial Modelling Environment (GME) software (Beyer 2012). Data were processed to remove tracks not entering the study area, that were short (arbitrarily selected as <1 km) and where birds did not fly toward the wind farm or the track was not long enough to reach it. Processed tracks were plotted by month against structures (monopile foundations or wind turbines) installed prior to, or during the month in 2011 and 2012. Tracks were also assigned to three discreet periods associated with particular site activity for further analysis: pre-construction (20th April 2009–23rd June 2010), initial construction (monopile installation between 24th June 2010 and 2nd July 2011) and final construction (turbine installation between 3rd July 2011 and 10th July 2012).

Cumulative proximity distributions (see Petersen et al. 2006) were calculated based on the numbers of birds flying within binned distances (50 m intervals truncated at 2 km—the main zone of interest) of the nearest structure present on the day of tracking. Pre-construction tracks were used as an indicative baseline, where distributions were calculated as if all structures had been present. As both monopiles and turbines were present in the final construction phase, the cumulative proximity distributions were calculated for each structure type separately to investigate any differences in response. All samples were non-normally distributed or did not show homogeneity of variance (Shapiro-Wilk and Fligner-Killeen tests respectively). Thus, non-parametric Kruskal-Wallis tests were carried out to determine if the phase had a significant effect on the distribution of the data. Multiple Kolmogorov-Smirnov tests were used to determine if there were significant differences between the cumulative distributions for each phase.

Results

Use of the Wind Farm Area

Boat-based surveys suggested that the area supported relatively low densities of Sandwich Terns of <1 ind. km^{-2}, with a few exceptions, during the breeding season (Fig. 4). The abundance of birds and seasonal trends in the 2–4 km buffer area were consistent throughout. In this outer buffer, densities peaked in May before falling in subsequent months. In the 0–2 km buffer this trend was seen in 2009, but in the subsequent years densities peaked in April. Although not as clear, the use of the wind farm site was also generally greatest in April. Abundance in the 0–2 km buffer increased from 2010 (when construction began) onwards, largely due to increased densities in April and May. Densities in the wind farm site also increased in 2011 when the use of the 0–2 km buffer peaked. The observed trends in the 0–2 km buffer and wind farm site suggest some attraction to these areas. Estimates of birds perched on floating objects, especially navigation buoys or turbine structures, appeared to contribute greatly to these observed trends (Fig. 4).

The gradient analysis resulted in 'year day' (edf = 1.001, $p < 0.001$) and 'site' (df = 2, $p = 0.003$) being included in the most parsimonious model. Although the interaction between 'site' and 'year' was significant in the full model (also including 'year day'), there was no improvement based on AIC values (ΔAIC = 8.07). The selected model had a deviance explained of 27.8% and predictions appeared to fit the data well (Pearsons correlation coefficient of 0.58 and Spearman rank correlation of 0.64). The Runs test found no autocorrelation in the model residuals (Runs statistic = −0.97, $p = 0.331$). The modelled relationship between abundance and 'year day' was approximately linear, decreasing from a peak at the start of April to a minimum at the end of July, with larger confidence intervals at the start and end of the breeding season. This is consistent with the trends seen in the mean density data illustrated in Fig. 4, although there was a lag in the peak abundance in the 2–4 km buffer. Birds were significantly more abundant in both the 0–2 km ($p < 0.001$) and the 2–4 km ($p = 0.043$) buffers relative to the wind farm site.

The flight directions of birds using the study area varied between years, but north-eastern (overall mean of 20.56%) and south-western (overall mean of 18.61%) trajectories were generally preferred (Table 2). In the wind farm site, there was a clear switch in preference from a northerly flight trajectory in 2009 and 2010 to north-easterly in 2011 (main piling period). In 2012 (installation of turbines), almost 50% of birds observed were heading in this direction (Table 2). Within the 0–2 km buffer no such trend was observed, although far fewer birds (almost half) were seen heading on a north-easterly trajectory in 2012 when the turbines were being installed. This reduction was balanced by more birds heading back toward the colony on a south-westerly course in 2012. In contrast, there was no apparent drop-off in the proportions of birds heading north-easterly in the 2–4 km buffer in 2012 (Table 2).

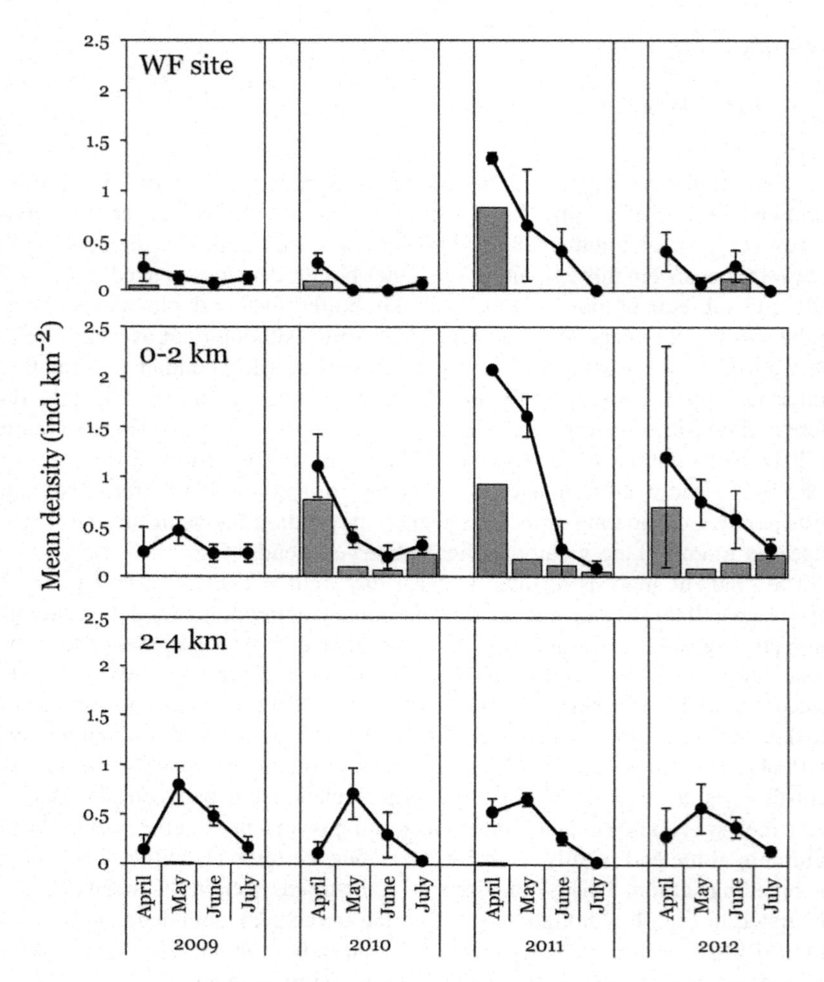

Fig. 4 Mean monthly overall Sandwich Tern density estimates (*solid lines* with associated standard errors) and densities of birds perched on objects (*grey bars*) derived from boat-based surveys of the wind farm site (WF), 0–2 km buffer and 2–4 km buffer during the breeding seasons in 2009–2012 inclusive

Visual Tracking Across the Wind Farm

A total of 154 days of tracking were achieved across the three study phases, comprised of 28 days during the pre-construction phase, 59 days during initial construction and 67 days during the final construction phase. A total of 1256 tracks

Table 2 Percentages of birds observed heading in different flight directions during boat-based surveys within the Sandwich Tern breeding season in 2009–2012 inclusive

Area	Year	Flight direction (%)								
		N	NE	E	SE	S	SW	W	NW	No direction
WF site	2009	***23.19***	10.14	5.80	0.00	13.04	20.29	13.04	1.45	13.04
	2010	***27.27***	4.55	13.64	6.82	6.82	11.36	9.09	4.55	15.91
	2011	4.55	***26.52***	12.88	10.61	9.09	18.18	2.27	0.76	15.15
	2012	10.91	***47.27***	1.82	1.82	9.09	16.36	1.82	3.64	7.27
0–2 km buffer	2009	12.30	***17.21***	12.30	11.48	10.66	16.39	9.84	3.28	6.56
	2010	13.38	***24.84***	13.38	8.92	7.64	10.19	6.37	0.00	15.29
	2011	5.84	19.84	4.28	5.84	11.28	20.23	7.78	3.11	***21.79***
	2012	12.98	12.50	9.13	4.33	12.98	***29.81***	8.65	3.37	6.25
0–4 km buffer	2009	14.98	12.15	10.93	4.86	13.36	***27.53***	9.72	2.02	4.45
	2010	8.00	***26.67***	16.67	2.67	14.67	9.33	6.67	2.00	13.33
	2011	5.41	***24.84***	10.51	10.83	10.83	21.34	5.10	5.10	6.05
	2012	10.09	20.18	6.42	14.98	8.56	***22.32***	4.89	3.06	9.48
Overall mean (±se)		12.41 (2.00)	***20.56*** (3.18)	9.81 (1.27)	6.93 (1.30)	10.67 (0.72)	18.61 (1.85)	7.10 (0.94)	2.69 (0.43)	11.21 (1.54)

The predominant flight directions in each year are highlighted in bold italics

were recorded in this time. Post processing removed 33% of these, leaving 840 tracks (covering a total distance of almost 9700 km) for further analysis. The mean track distance and durations were 11 km (1–39.6 km) and 17.6 min (1.2–82.5 min) respectively. The average estimated flight speed was 40.2 km h^{-1}, although on occasion birds outpaced the RIB at full speed (>70 km h^{-1}). Tracks were also cut short as a result of poor weather, exclusions around operational vessels or by birds landing on buoys. Otherwise, 65% tracks were completed by a bird leaving the study area.

All tracks from each monitoring phase are shown in Fig. 5, illustrating the dominant north-east to south-west flyway, with passages across a broad front through the study area. During pre-construction, some of the tracks clearly reflect transits to and from site marker buoys (installed in March 2010), particularly to the south-west, west and south of the wind farm. The main flyway appeared to split during the initial construction phase, with many tracks heading east-northeast/west-southwest or north/south. When turbine installation began in

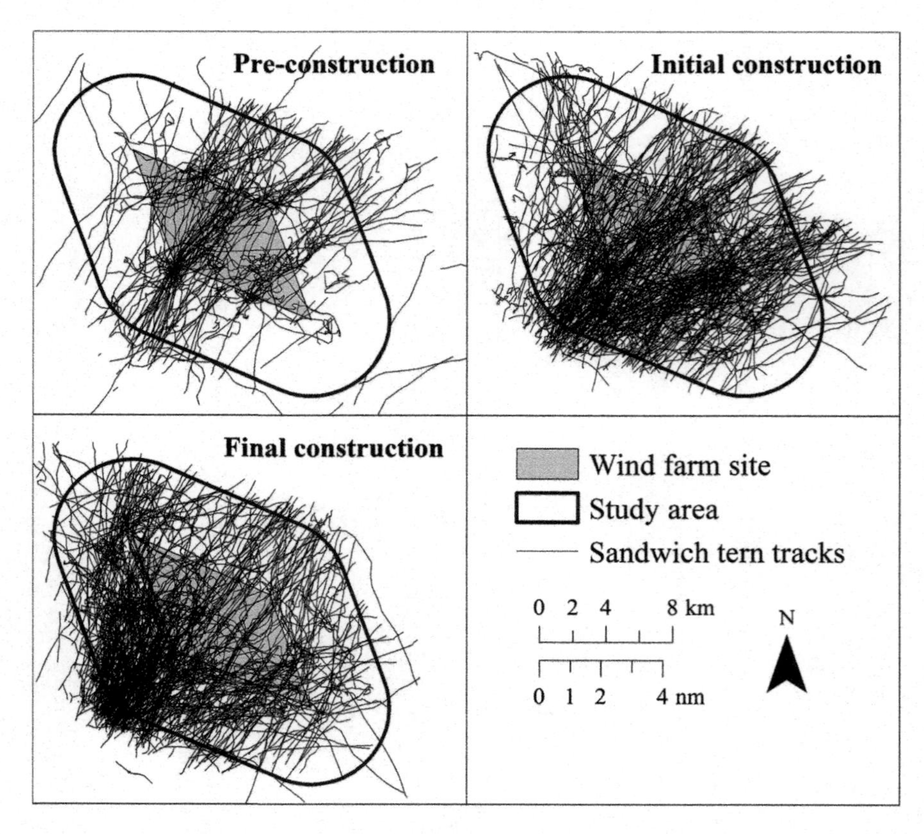

Fig. 5 All Sandwich Tern tracks recorded during the pre-construction, (n = 277), initial construction (n = 530) and final construction (n = 449) phases

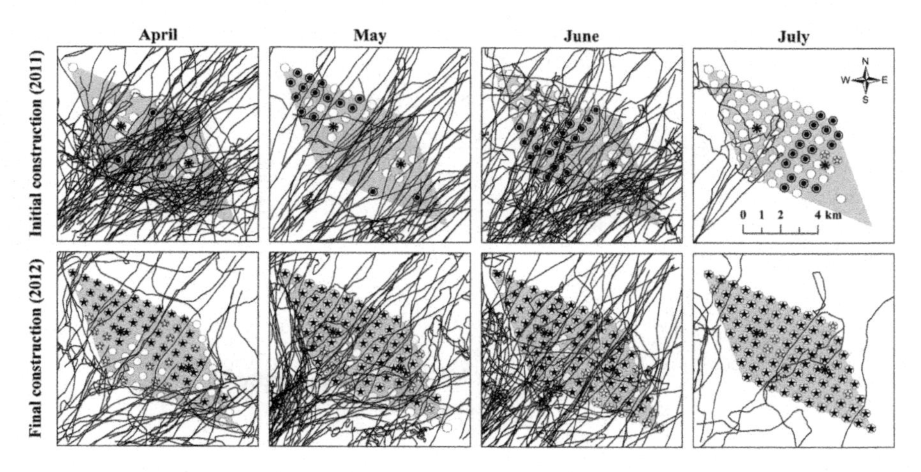

Fig. 6 Filtered tracks in each month during initial construction in 2011 (n = 340) and final construction in 2012 (n = 256), relative to monopiles installed (*solid black circles*) or already present (*solid white circles*) and turbines installed (*white stars*) or already present (*solid black stars*) in respective months. *Asterisks* denote sub-stations

the final construction phase, the tracks showed much clearer diversions around the periphery of the wind farm site. However, birds still used the wind farm throughout construction and well-defined passage routes through the site started to emerge.

Further segregation of tracks into individual months in 2011 and 2012, when the bulk of the construction was carried out, demonstrated that Sandwich Terns were avoiding areas where structures were being installed in 2011, and areas where turbines were being installed or were already present in 2012 (Fig. 6). For example, in April 2011 the areas around the substations, where much of the work was taking place, were used much less than other part of the site. In May, June and July the main areas where piling was taking place were also avoided. In 2012, fewer birds penetrated the site and those that did avoided areas where construction was taking place or turbines were already present (Fig. 6). When individuals did enter the site, the tracks tended to be linear and followed corridors aligned south-west to north-east within the array. Many individuals that seemingly diverted around the site were observed cutting the corners of the array, where the chance of encountering a structure is lowest. Indeed, the overall proportions of tracks which entered the site changed dramatically from 95.0% during pre-construction, to 82.5% when the monopiles were being installed and to only 65.1% during the installation of turbines. The proportions of observed foraging attempts within the wind farm also suggested a coincident decline in the use of the area from 48% during the pre-construction phase, to 30% during initial construction and only 19% in the final construction phase (Table 3). Conversely, the area around the wind farm became proportionally more important in relation to foraging; particularly in the 0–2 km buffer (Table 3).

Table 3 The percentages of foraging attempts (n = 3342) by all tracked birds in the wind farm site and two buffers during the three monitoring phases

Year	% of foraging attempts in each area		
	WF site	0–2 km buffer	2–4 km buffer
Pre-construction	47.9	38.8	13.3
Initial construction	30.1	39.0	30.9
Final construction	19.0	46.5	34.5
Overall mean	32.3	41.4	26.2

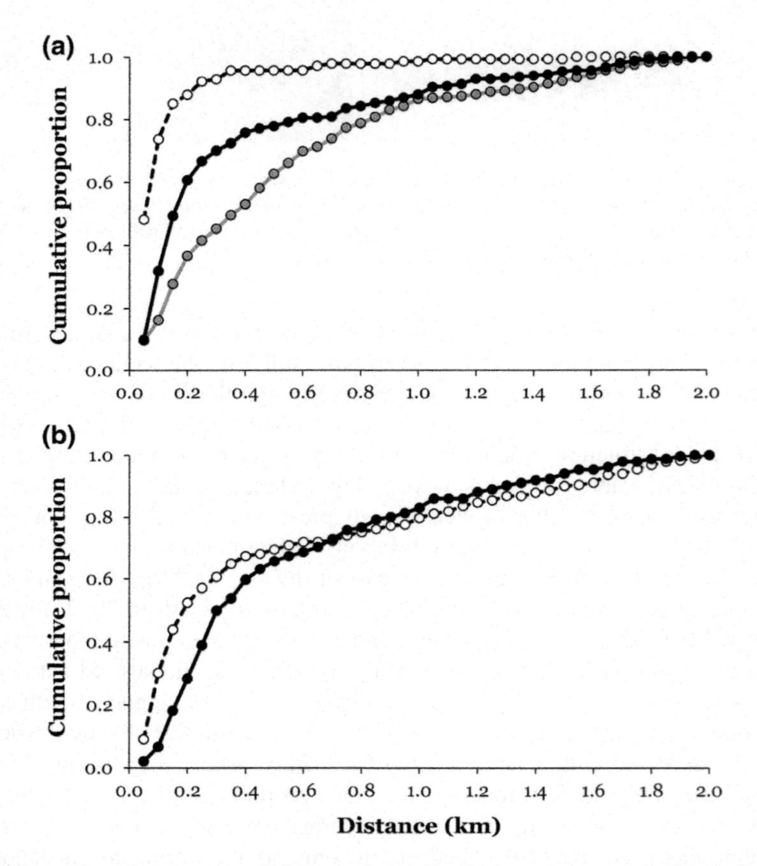

Fig. 7 Comparison between cumulative proximity distributions (truncated to 2 km) for: **a** the pre-construction phase assuming all structures were present (*open circles*, n = 141), initial construction phase (*grey circles*, n = 349) and final construction phase (*black circles*, n = 298), and **b** the final construction phase in relation to monopiles (*open circles*, n = 282) and turbines (*black circles*, n = 236) present at the time of tracking

Figure 7a illustrates the cumulative proximity distributions for tracked birds during the three phases. A high proportion of the birds (48%) flew within 50 m of a future turbine during pre-construction, whilst considerably fewer came as close to a structure during the initial and final construction phases (both at 10%). Furthermore, 88% of the tracks during the pre-construction phase would have passed within 250 m of a future turbine, compared to 42 and 67% during the initial and final construction phases respectively. The Kruskal-Wallis test indicated a significant effect ($x^2 = 158.64$, $p < 0.001$) of the study phase on the proximity of tracks to structures. Subsequent Kolmogorov-Smirnov tests between different combinations of phases suggested highly significant differences ($p < 0.001$) in the distributions of the data in each case. However, these results should be treated with caution as relatively few monopiles were present within the site for much of the breeding season during the initial construction phase, and therefore the chance of birds coming close to them would have been proportionally lower. In the final construction phase, birds were more sensitive to the presence of turbines than to the monopiles present at the same time ($D = 0.243$, $p < 0.001$), with almost no birds coming within 50 m of a turbine (Fig. 7b).

Discussion

The general preference for north-eastern and south-western flight trajectories through the study area, shown by both boat-based surveys and visual tracking, mirrored the pattern for the two breeding seasons (2004 and 2005) monitored for the Environmental Statement (SCIRA Offshore Energy Ltd 2006). These trajectories are consistent with outbound and inbound flights from the expected origin of birds from Blakeney Point. The general decline in use of the study area over the breeding season is consistent with the abandonment of more distant foraging areas with increasing pressure to provision chicks (Ojowski et al. 2001). This further reinforces the previous assumption that breeding birds comprise the majority, if not all, of the birds observed in the study area. Moreover, there is no evidence for the alternative view of a sizeable pool of non-breeding birds in the region, which would be manifested as nightly roosts of birds that could not be attributed to colonies. Such a phenomenon has not been recorded in the extensive local literature (e.g. Taylor and Marchant 2011).

A low proportion of feeding activity was recorded during boat-based surveys with just 120 (4.6%) of the 2602 Sandwich Terns recorded noted as fishing, closely matching the 3.5% of 624 records in 2004–2005 (SCIRA Offshore Energy Ltd 2006). This tends to support the theory that Sandwich Terns are principally transiting through the Sheringham Shoal OWF study area to foraging grounds around Inner Cromer Knoll and Cromer Knoll, as well as Haddock Bank, a large linear sandbank.

These features are largely within mean maximum foraging range from the colony at between 40 and 50 km from Blakeney Point and are potentially attractive to shoaling clupeids (Atlantic herring *Clupea harengus* and European sprat *Sprattus sprattus*) and sandeels (*Hyperoplus lanceolatus* and *Ammodytes* spp.) that dominate the diet of Sandwich Terns (Stienen et al. 2000).

The modelling did not identify a significant change in the birds' use of the different components of the study area between years, despite the obvious difference in mean (±1SE) breeding season population size in the study area. The population size ranged from 53 ± 15.1 in 2010 to 114 ± 30.7 in 2011. The abundance of birds in the study area broadly reflected changes in size of the Blakeney Point colony, which supported the lowest number of pairs in 2010. However, the magnitude of any inter-annual variability was insufficient to detect a statistically significant change in the abundance of birds. Thus, a strong link to construction events in different years could not be established. However, the modelling did confirm that both the 0–2 km buffer and 2–4 km buffer areas supported higher populations of birds than the wind farm. The accumulation of Sandwich Terns around buoys, present only in the 0–2 km buffer, was thought to be responsible for the higher densities in this area, especially early in the season (Fig. 4). Up to 18 individuals were noted on a single buoy, with birds recorded on at least one buoy during 56% of surveys. The numbers of birds on buoys was sufficient to contribute an approximately equivalent density of 'birds on the water' to that of 'birds in flight'. In fact, as only three of the eight (37.5%) buoys fell within the transect route, the true numbers of Sandwich Terns using buoys at any one time seems likely to have been considerably underestimated. The buoys were attractive as a platform on which to rest (Fig. 8), although a variety of social interactions, including courtship feeding and mating, were also recorded. Buoys thus operated as a social hub from which foraging flights were initiated or interrupted, as also revealed during tracking. This was particularly evident in April and May, early in the breeding season when pair bonds were being reinforced. It is plausible that the lack of significant difference in population size between the 0–2 and 2–4 km buffer partly reflected an attraction of Sandwich Terns to buoys from a wider area, with birds having to cross the 2–4 km buffer to ultimately reach the buoys. Alternatively, the maintenance of the population size in the 2–4 km buffer may be linked to the fact that at least the southern part of it was the closest point to the source of birds from the colony.

Lindeboom et al. (2011) observed a similar attraction to OWF infrastructure at Egmond aan Zee. Here, large numbers of Great Cormorant *Phalacrocorax carbo* utilised the actual turbine bases to rest and dry their wings between foraging trips mainly within the wind farm. When foraging, Great Cormorant dive from the surface and pursue fish underwater and, if flight heights in the wind farm are below the sweep of the rotor blades, they could use habitat within wind farms with little risk. In fact, use of turbine bases at Egmond aan Zee allowed Great Cormorant to

Fig. 8 Sandwich Terns using one of the Sheringham Shoal OWF site marker buoys

move further offshore than they could otherwise. At the Blighbank OWF, Lesser Black-backed Gull *Larus fuscus* and Herring Gull *Larus argentatus* were generally attracted to the site and were seen roosting on structures. Lesser Black-backed Gulls were also seen feeding on pelagic prey within the array and around the turbine bases (Vanermen et al. 2015a). Unlike gulls and cormorants, terns have not been seen using the handrails around turbine bases, but were recorded resting on the transition pieces of the monopiles before turbine assembly in 2011. It may be speculated that individuals that were already familiar with buoys were responsible for this rare behaviour.

Otherwise, there was evidence of Sandwich Terns avoiding areas under construction and keeping their distance from standing structures in a similar fashion to Common Eider *Somateria mollissima* in the study of Larsen and Guillemette (2007). Seaduck and waterfowl in general appear to be wary of novel objects, although they can quickly habituate where there is a reason to do so. At Horns Rev, Common Scoter *Melanitta nigra* overcame what appeared to be a particularly strong fear of the wind farm after their bivalve mollusc prey colonised turbine bases (Petersen and Fox 2007). It was not anticipated that the fast-flying and agile Sandwich Tern would show wariness of structures that actually posed no risk. However, during the initial construction phase a small proportion of flights (8% relative to the pre-construction baseline conditions) that were initially heading for the site deviated away from it.

One obvious alternative explanation for the avoidance of areas of construction activity by Sandwich Terns is that construction activity affected the distribution of important prey, particularly hearing-specialist clupeids that are especially sensitive to pile driving noise (Thomsen et al. 2006). Sandeels, the other prey species of choice for Sandwich Terns, are thought to be relatively insensitive as they have no swim bladder. However, piling of individual turbines at Sheringham Shoal OWF was of very short duration (30–40 min), with the time between events generally being at least one day. This potentially provided ample time for fish to quickly recolonise and maintain abundance. In fact, there was some evidence that Sandwich Terns were occasionally attracted to, rather than repelled by, construction activity due to prey abundance, with a few records of Sandwich Terns aggregating immediately after piling events. For example, during tracking on July 3rd 2011 more than 200 terns and several hundred gulls were observed feeding on fish, mostly clupeids, that may have been affected by piling noise.

Furthermore, apart from these isolated events, tracking revealed a general decline in feeding activity in the site relative to the buffers during both the initial and final construction phases. With no obvious source of noise in the final construction phase there would appear to be no reason why the prey would be affected. However, prey distribution may be influenced by the increased use of the site by larger predators such as Harbour Seal *Phoca vitulina* that is now known to forage at Sheringham Shoal OWF (Russell et al. 2014), as well as large predatory fish such as Atlantic Cod *Gadus morhua* that also favour OWFs (Reubens et al. 2013). But perhaps a more tenable explanation is simply that Sandwich Terns were less inclined to forage within the site as they focussed on maximising the distance from each turbine and thus tended to pass through the centre of the rows within the array.

During the final construction phase, when testing of turbines presented some collision risk, 30% fewer of the tracked birds, relative to the pre-construction baseline, entered the wind farm site and instead appeared to deviate around it. This closely aligns with the macro-avoidance rate of 28% reported by Krijgsveld et al. (2011) from radar studies of the operational Egmond aan Zee. Petersen et al. (2006) had previously demonstrated that Sandwich Terns were significantly more likely to enter the Horns Rev OWF where one or both of the turbines either side of the point of entry were not in operation. Taken together, these results imply that Sandwich Terns have a good perception of danger and modify their actions accordingly. However, as some other authors (e.g. Leopold et al. 2013) have not detected a clear response, this may vary on a case-by-case basis. At this stage, there is no particular evidence that the response of breeding Sandwich Terns using the Sheringham Shoal OWF was radically different to that of migrating birds at Egmond aan Zee and Horns Rev, despite the potential difference in energetic costs for birds in different stages of their reproductive cycle.

According to Masden et al. (2010), in their comparison of a range of common seabirds, terns would have the lowest additional energy cost associated with increased foraging distance as a result of any deviation around wind farms. Put simply, the Sandwich Terns from the Blakeney Point colony may be able to undertake noticeable modifications to flight patterns, or accommodate the loss of some foraging habitat, without incurring a significant energetic cost.

Conclusions

This study is one of the very few to shed light on the response of a breeding seabird to the construction of an OWF. Although no statistically significant changes in the use of the study area over time were detected, the boat-based survey results suggested an increased use of the 0–2 km buffer area consistent with the installation of navigation buoys prior to construction. Unforeseen by the EIA, these buoys became the focus of courtship and social activity early in the breeding season. The visual tracking revealed Sandwich Terns avoided areas where piling was taking place and, also unforeseen by the EIA, were initially wary of the installed monopiles, with birds maintaining distance from them despite relatively little associated risk. As construction advanced and turbines were installed, an increasing number of tracked birds appeared to deviate around the wind farm rather than entering it. This is consistent with the predictions of the EIA that minor adverse barrier effects could occur. Indeed, the general conclusion of the EIA of only a minor adverse effect during the construction phase was upheld.

Although a relatively high proportion of tracked birds appeared to be displaced from the OWF in the presence of fully assembled turbines, boat-based survey densities did not decline significantly. In part, this may be because even when all turbines were constructed, the spacing of turbines at 650–720 m meant the site was still highly permeable to transiting birds that tended to select the centre of rows between turbines. In turn, such flight behaviour was favoured by the general layout of the wind farm array providing corridors with a northeast-southwest alignment. Reductions in foraging observations within the wind farm were also noted as construction advanced, likely reflecting the reduced time spent in the site and the more direct flights of individuals through the array.

Analysis of further data from the operational site, when there is a greater risk of collision, is ongoing. Considering the response of Sandwich Terns observed to date, further modification of behaviour is anticipated. At this stage, it is also important to note that the general use of the site and number of birds at potential risk in the operational site would be likely to decline considerably if Sandwich Terns resume periodic switching of breeding between Blakeney Point and the more distant colony at Scolt Head.

Acknowledgements We are indebted to SCIRA Offshore Energy Ltd, a joint venture company owned by Statoil and Statkraft, who have funded and supported the work alongside the development of their wind farm. We are also grateful for the logistical support of all staff at Statkraft who currently operates the site on behalf of SCIRA. We would like to thank all who have taken part in the tracking studies, including Seafari Marine Services led by Iain Hill, Hovercraft Services and Safety Boat Services. Finally, we thank Johann Köppel and his team at Technische Universität Berlin for inviting us to contribute to this volume and appreciate the helpful comments and suggestions provided by the reviewers.

References

Beyer HL (2012) Geospatial modelling environment (Version 0.7.2.1). url:http://www.spatialecology.com/gme/

BSH & BMU (2014) Ecological research at the offshore windfarm *alpha ventus*—challenges, results and perspectives. Federal Maritime and Hydrographic Agency (BSH), Federal Ministry for the Environment, Nature Conservation and Nuclear Safety (BMU). Springer Spektrum, Berlin, 201

Brabant R, Vanermen N, Stienen EWM, Degraer S (2015) Towards a cumulative collision risk assessment of local and migrating birds in North Sea offshore wind farms. Hydrobiologia 756:63–74

Breton SPh, Moe G (2009) Status, plans and technologies for offshore wind turbines in Europe and North America. Renew Energy 34:646–654

Buckland ST, Burt ML, Rexstad EA, Mellor M, Williams AE, Woodward R (2012) Aerial surveys of seabirds: the advent of digital methods. J Appl Ecol 49:960–967

Camphuysen CJ, Fox AD, Leopold MF (2004) Towards standardised seabirds at sea census techniques in connection with environmental impact assessments for offshore wind farms in the U.K.: a comparison of ship and aerial sampling for marine birds, and their applicability to offshore wind farm assessments. Report commissioned by COWRIE (Collaborative Offshore Wind Research into the Environment). Available via: www.offshorewindfarms.co.uk

Da Z, Xiliang Z, Jiankun H, Qimin C (2011) Offshore wind energy development in China: current status and future perspective. Renew Sust Energ Rev 15:4673–4684

Department of Energy and Climate Change (DECC) (2011) National policy statement for renewable energy infrastructure (EN-3). The Stationary Office, London

Desholm M, Kahlert J (2005) Avian collision risk at an offshore wind farm. Biol Lett 1:296–298

Desholm M, Fox AD, Beasley PDL, Kahlert J (2006) Remote techniques for counting and estimating the number of bird–wind turbine collisions at sea: a review. Ibis 148:76–89

Embling CB, Illian J, Armstrong E, van der Kooij J, Sharples J, Camphuysen CJ, Scott BE (2012) Investigating fine-scale spatio-temporal predator-prey patterns in dynamic marine ecosystems: a functional data analysis approach. J Appl Ecol 49:481–492

Everaert J, Stienen EWM (2007) Impact of wind turbines on birds in Zeebrugge (Belgium): significant effect on breeding tern colony due to collisions. Biodivers Conserv 16:3345–3359

Furness RW, Wade HM, Masden EA (2013) Assessing vulnerability of marine bird populations to offshore wind farms. J Environ Manage 119:56–66

Garthe S, Hüppop O (2004) Scaling possible adverse effects of marine wind farms on seabirds: developing and applying a vulnerability index. J Appl Ecol 41:724–734

Gove B, Langston RHW, McCluskie A, Pullan JD, Scrase I (2013) Wind farms and birds: an updated analysis of the effects of wind farms on birds and best practice guidance of integrated planning and impact assessment. RSPB/BirdLife in the UK for BirdLife International on behalf of the Bern Convention, Bern Convention Bureau meeting, 17 Sept 2013, Strasbourg, p 89

Hastie T, Tibshirani R (1990) Generalized additive models. Chapman and Hall, New York

Krijgsveld KL, Fijn RC, Japink M, van Horssen PW, Heunks C, Collier MP, Poot MJM, Beukers D, Dirksen S (2011) Effect studies offshore wind farm Egmond aan Zee. Flux, flight altitude and behaviour of flying birds. Bureau Waardenburg report 10-219. Bureau Waardenburg, Culemborg, p 334

Larsen JK, Guillemette M (2007) Effects of wind turbines on flight behaviour of wintering common eiders: implications for habitat use and collision risk. J Appl Ecol 44:516–522

Leopold MF, van Bemmelen RSA, Zuur AF (2013) Responses of local birds to offshore wind farms PAWP and OWEZ off the Dutch mainland coast. IMARES Report C151/12 Wageningen, The Netherlands, p 108

Lindeboom HJ, Kouwenhoven HJ, Bergman MJN, Bouma S, Brasseur S, Daan R, Fijn RC, de Haan D, Dirksen S, van Hal R, Hille Ris Lambers R, ter Hofstede R, Krijgsveld KL, Leopold M, Scheidat M (2011) Short-term ecological effects of an offshore wind farm in the Dutch coastal zone, a compilation. Environ Res Lett 6(3):13

Linley EAS, Wilding TA, Black K, Hawkins AJS, Mangi S (2007) Review of the reef effects of offshore wind farm structures and their potential for enhancement and mitigation. Report from PML Applications Ltd. and the Scottish Association for Marine Science to the Department for Business, Enterprise and Regulatory Reform (BERR), Contract No: RFCA/005/0029P. BERR/DEFRA, London, UK, p 132

Masden EA, Haydon DT, Fox AD, Furness RW, Bullman R, Desholm M (2009) Barriers to movement: impacts of wind farms on migrating birds. ICES J Mar Sci 66:746–753

Masden EA, Haydon DT, Fox AD, Furness RW (2010) Barriers to movement: modelling energetic costs of avoiding marine wind farms amongst breeding seabirds. Mar Pollut Bull 60:1085–1091

Mitchell PI, Newton S, Ratcliffe N, Dunn TE (2004) Seabird populations of Britain and Ireland (results of the seabird 2000 census 1998–2000). T&D Poyser, London, p 511

Noguchi K, Hui WLW, Gel YR, Gastwirth JL, Miao W (2009) lawstat: an R package for biostatistics, public policy, and law. R package version 2.3

Norfolk and Norwich Naturalists' Society (NNNS) (2007) Norfolk bird and mammal report 2006. Trans Norfolk Norwich Naturalists' Soc 40(2):145–372. ISSN 0375 7226

Ojowski U, Eidtmann C, Furness RW, Garthe S (2001) Diet and nest attendance of incubating and chick-rearing Northern Fulmars (*Fulmarus glacialis*) in Shetland. Mar Biol 139:1193–1200

Perrow MR, Skeate ER, Lines P, Brown D, Tomlinson ML (2006) Radio telemetry as a tool for impact assessment of wind farms: the case of Little Terns Sterna albifrons at Scroby Sands, Norfolk, UK. Ibis 148(Suppl. 1):57–75

Perrow MR, Gilroy JJ, Skeate ER, Mackenzie A (2010) Quantifying the relative use of coastal waters by breeding terns: towards effective tools for planning and assessing the ornithological impacts of offshore wind farms. ECON Ecological Consultancy Ltd. Report to COWRIE Ltd. ISBN 978-0-9565843-3-5

Perrow MR, Gilroy JJ, Skeate ER (2011a) Effects of the construction of Scroby Sands offshore wind farm on the prey base of little tern *Sternula albifrons* at its most important UK colony. Mar Pollut Bull 62:1661–1670

Perrow MR, Skeate ER, Gilroy JJ (2011b) Visual tracking from a rigid-hulled inflatable boat to determine foraging movements of breeding terns. J Field Ornithol 82:68–79

Perrow MR, Harwood AJP, Skeate ER, Praca E, Eglington SM (2015) Use of multiple data sources and analytical approaches to derive a marine protected area for a breeding seabird. Biol Conserv 191:729–738

Petersen IK, Fox AD (2007) Changes in bird habitat utilisation around the Horns Rev 1 offshore wind farm, with particular emphasis on common scoter. NERI Report commissioned by Vattenfall A/S, National Environmental Research Institute, Ministry of the Environment, Denmark

Petersen IK, Christensen TK, Kahlert J, Desholm M, Fox AD (2006) Final results of bird studies at the offshore wind farms at Nysted and Horns Rev, Denmark. NERI Report commissioned by DONG Energy and Vattenfall A/S. National Environmental Research Institute, Ministry of the Environment, Rønde, Denmark

Petersen IK, MacKenzie ML, Rexstad E, Wisz MS, Fox AD (2011) Comparing pre- and post-construction distributions of long-tailed ducks *Clangula hyemalis* in and around the Nysted offshore wind farm, Denmark: a quasi-designed experiment accounting for imperfect detection, local surface features and autocorrelation. CREEM technical report no. 2011-1, University of St Andrews

Petersen IK, Nielsen RD, Mackenzie ML (2014) Post-construction evaluation of bird abundances and distributions in the Horns Rev 2 offshore wind farm area, 2011 and 2012. Report commissioned by DONG Energy. Aarhus University, DCE—Danish Centre for Environment and Energy, p 51

Pettersson J (2005) The impact of offshore wind farms on bird life in Southern Kalmar Sound, Sweden. Report to Swedish Energy Agency, p 128

Plonczkier P, Simms C (2012) Radar monitoring of migrating pink-footed geese: behavioural responses to offshore wind farm development. J Appl Ecol 49:1187–1194

R Core Team (2014) R: a language and environment for statistical computing. R Foundation for Statistical Computing, Vienna, Austria. url:http://www.R-project.org/

Reubens JT, Pasotti F, Degraer S, Vincx M (2013) Residency, site fidelity and habitat use of Atlantic cod (*Gadus morhua*) at an offshore wind farm using acoustic telemetry. Mar Environ Res 90:128–135

Russell DJF, Brasseur Sophie MJM, Thompson D, Hastie GD, Janik VM, Aarts G, McClintock BT, Matthiopoulos J, Moss SEW, McConnell B (2014) Marine mammals trace anthropogenic structures at sea. Curr Biol 24(14):R638–R239

Robertson GS, Bolton M, Grecian WJ, Wilson LJ, Davies W, Monaghan P (2014) Resource partitioning in three congeneric sympatrically breeding seabirds: foraging areas and prey utilization. Auk 131:434–446

SCIRA Offshore Energy Ltd (2006) Sheringham Shoal Offshore Wind Farm Environmental Statement. SCIRA Offshore Energy Ltd, London, p 722

Skov H, Leonhard SB, Heinänen S, Zydelis R, Jensen NE, Durinck J, Johansen TW, Jensen BP, Hansen BL, Piper W, Grøn PN (2012) Horns Rev 2 Monitoring 2010–2012. Migrating Birds. Orbicon, DHI, Marine Observers and Biola. Report commissioned by DONG Energy

Stienen EWM, van Beers PWM, Brenninkmeijer A, Habraken JMPM, Raaijmakers MHJE, van Tienen PGM (2000) Reflections of a specialist: patterns in food provisioning and foraging conditions in Sandwich Terns *Sterna sandvicensis*. Ardea 88:33–49

Strickland D, Erickson W, Young D, Johnson G (2007) Selecting study designs to evaluate the effect of windpower on birds. In: de Lucas M, Janss GFE, Ferrer M (eds) Birds and wind farms: risk assessment and mitigation, Chap 6. Quercus/Servicios Informativos Ambientales, Madrid. ISBN 978-84-87610-18-9

Taylor M, Marchant JH (2011) The Norfolk Bird Atlas: summer and winter distributions 1999-2007. British Trust for Ornithology, Thetford, 528 pp. ISBN 978-1-906204-82-2

Thaxter CB, Lascelles B, Sugar K, Cook ASCP, Roos S, Bolton M, Langston RHW, Burton NHK (2012) Seabird foraging ranges as a tool for identifying candidate marine protected areas. Biol Conserv 156:53–61

Thaxter CB, Ross-Smith VH, Bouten W, Clark NA, Conway GJ, Rehfisch MM, Burton NHK (2015) Seabird–wind farm interactions during the breeding season vary within and between years: a case study of lesser black-backed gull *Larus fuscus* in the UK. Biol Conserv 186:347–358

Thelander CG, Smallwood KS (2007) The Altamont Pass Wind Resource Area's effects on birds: a case history. In de Lucas M, Janss GFE, Ferrer M (eds) Birds and wind farms: risk assessment and mitigation. Servicios Informativos Ambientales/Quercus, Madrid, Spain, pp 25–46

Thomsen F, Lüdemann K, Kafemann R, Piper W (2006) Effects of offshore wind farm noise on marine mammals and fish. Biola, Hamburg, Germany on behalf of COWRIE Ltd, p 62

Vanermen N, Stienen EWM, Onkelinx T, Courtens W, Van de walle M, Verschelde P, Verstraete H (2012) Seabirds and offshore wind farms monitoring results 2011. Research Institute for Nature and Forest, Brussels. INBO.R.2012.25

Vanermen N, Onkelinx T, Courtens W, Van de walle M, Verstraete H, Stienen EWM (2015a) Seabird avoidance and attraction at an offshore wind farm in the Belgian part of the North Sea. Hydrobiologia 756:51–61

Vanermen N, Onkelinx T, Verschelde P, Courtens W, Van de walle M, Verstraete H, Stienen EWM (2015b) Assessing seabird displacement at offshore wind farms: power ranges of a monitoring and data handling protocol. Hydrobiologia 756:155–167

Wade HM, Masden EA, Jackson AC, Thaxter CB, Burton NHK, Bouten W, Furness RW (2014) Great skua (*Stercorarius skua*) movements at sea in relation to marine renewable energy developments. Mar Environ Res 101:69–80

Wilson LJ, Black J, Brewer MJ, Potts JM, Kuepfer A, Win I, Kober K, Bingham C, Mavor R, Webb A (2014) Quantifying usage of the marine environment by terns *Sterna* sp. around their breeding colony SPAs. JNCC report no. 500, Peterborough, p 125. Available via: http://jncc.defra.gov.uk/pdf/JNCC_Report_500_web.pdf

Wood SN (2011) Fast stable restricted maximum likelihood and marginal likelihood estimation of semiparametric generalized linear models. J Roy Stat Soc B 73:3–36

Zuur AF, Ieno EN, Walker NJ, Saveliev AA, Smith GM (2009) Mixed effects models and extensions in ecology with R. Springer, New York

A Large-Scale, Multispecies Assessment of Avian Mortality Rates at Land-Based Wind Turbines in Northern Germany

Thomas Grünkorn, Jan Blew, Oliver Krüger, Astrid Potiek, Marc Reichenbach, Jan von Rönn, Hanna Timmermann, Sabrina Weitekamp and Georg Nehls

Abstract Collisions of birds with wind turbines are a focal point when discussing the implications of renewable energies on nature conservation. The project "Prognosis and assessment of collision risks of birds at wind turbines in northern Germany" (PROGRESS) focused on the extent and consequences of bird mortality at wind turbines.

Collision victims were searched in five search efforts from spring 2012 to spring 2014 (three spring and two autumn field searches of 12 weeks). 46 different wind farms were examined. The total searched transect length amounted to 7672 km. With a total of 291 birds found, an average of one bird was found every 27 km. Common bird species with habitat use (feeding and staging) of the wind farm area prevail the list of fatalities found. Birds of prey did not dominate the list. Nocturnal broad front migratory songbirds (especially thrush species) were hardly represented among the fatalities. The total number of fatalities was estimated incorporating search efficiency and carcass removal experiments.

An extrapolation of the results for the entire project area leads to an annual mortality of around 8500 Common buzzards, 11,300 Wood pigeons and 13,000 Mallards. Based on the breeding population in the project area this translates to 0.5% of Wood pigeons, 5.0% of Mallards and 7% of Common buzzards (assuming 50% floaters).

Results of vantage point watches indicate that the species-specific collision risk with wind turbines largely differs between species as a result of clear behavioural differences.

For the vast majority of wind farms the numbers of collision victims predicted by the BAND-model were clearly below the number of collision victims estimated from

T. Grünkorn (✉) · J. Blew · J. von Rönn · G. Nehls
BioConsult SH GmbH & Co KG, Husum, Germany
e-mail: t.gruenkorn@bioconsult-sh.de

O. Krüger · A. Potiek
Department of Animal Behaviour, University of Bielefeld, Bielefeld, Germany

M. Reichenbach · H. Timmermann · S. Weitekamp
ARSU GmbH, Oldenburg, Germany

J. Köppel (ed.), *Wind Energy and Wildlife Interactions*,
DOI 10.1007/978-3-319-51272-3_3

carcass searches. The suitability of the BAND-model for the evaluation of an anticipated collision risk for a planned wind farm at an 'average' onshore site is limited.

Four modelled populations of Common buzzard in northern Germany are predicted to decline when incorporating the median estimates of additional mortality derived from fatality estimates.

Statutory species protection conflicts might not always be adequately solvable for an individual project. Therefore, overarching solutions are required to accompany the further expansion of wind farms, which ensure that this does not lead to a severe decline of certain bird species that are particularly affected by collisions. Specifically, the following strategies need to be addressed:

- Large-scale wildlife conservation programs e.g. for Red kites and Common buzzards that improve habitats, particularly in terms of food availability
- Identification of species-specific density centres that are of particular importance as source populations, and assessing targeted measures to protect and promote them,
- Development of concepts and practical testing of a post-construction species protection support in terms of their effectiveness and their economic effects.
- Increased research efforts in terms of scale and addressing cumulative effects.

Keywords Wind turbine · Collision · Search efficiency · Carcass removal · Vantage point watch · BAND-Model · Population projections · Guidelines · Site planning

Introduction

Collisions of birds (and bats) with wind turbines are a focal point when discussing the implications of renewable energies on nature conservation. Although a large number of studies have already addressed this topic, there are only a few systematic studies that have quantified the collision rates of birds and judged the importance at population level. This complicates the assessment of possible conflicts. In regard to the EU's strict species protection legislation affecting permission for the construction of wind farms, a lack of knowledge on the actual impacts on wildlife presents obstacles for the expansion of wind energy in Germany and other countries.

The project "Prognosis and assessment of collision risks of birds at wind turbines in northern Germany" (PROGRESS) focused on the extent and consequences of bird mortality at wind turbines.

Even though wind energy use has greatly expanded in Germany over the last two decades and a total number of approximately 25,000 turbines have been installed, only a few small-scale studies have so far tried to quantify bird collisions at wind turbines. It was thus the aim to collect a comprehensive and representative dataset on collision rates of birds at land-based wind farms by a systematic field study across several federal states of Northern Germany. Based on this, conclusions and recommendations for the conflict assessment and resolution were drawn. In the context of PROGRESS, the North German lowland was investigated, as it is a focus area for current and future wind energy development in Germany. Emphasis was placed on the federal states of Schleswig-Holstein, Lower Saxony, Mecklenburg-Western Pomerania and Brandenburg. The target species were: birds of prey, large birds (e.g. Black and white stork or Common crane with often small populations) and breeding and staging bird species (e.g. Lapwing or Golden Plover with utilization of the wind farm area).

Project Modules

Search for Fatalities

Methods, Fieldwork of Fatality Searches

Collision victims were searched on a weekly basis in five search campaigns from spring 2012 to spring 2014 (three spring and two autumn field searches of 12 weeks). 46 different wind farms were examined (8–18 turbines, average 11). Since some wind farms were examined more than once (two to three times) in different seasons, overall 55 search campaigns were conducted (hereafter wind-farm-seasons). Searches were carried out with a transect design where mostly two observers searched for carcasses along predefined parallel transects that were 20 m apart. All retrievals within a search plot—defined by the radius of the total height of the wind turbine—were classified as collision victims (Fig. 1).

Fig. 1 Example of logged search transects. *Red dot* wind turbine, *red circle* search plot with radius of turbine height, *yellow lines* 12 consecutively logged transect lines (Color figure online)

Results of Fatality Searches

A total of 291 birds were found during the study period. The two most frequently species were the common species Wood Pigeon and Mallard. Among the 15 most frequent species were five target species of the project: Common Buzzard, Lapwing, Golden Plover, Red Kite and Kestrel. Waterfowl (ducks, geese/waders/gulls) represent half of the fatalities. The group of other non-passerines is the largest group of fatalities, as doves were found frequently. Birds of prey did not dominate the list. Nocturnal broad front migratory songbirds (especially thrush species) were hardly represented among the fatalities.

The total searched transect length amounted to 7672 km. With a total of 291 birds found, an average of one bird was found every 27 km (Fig. 2).

Estimation of Total Number of Fatalities

Methods of Fatality Estimation

The total number of fatalities was extrapolated from the number of birds that were actually found by considering several correction factors. The rendered surface

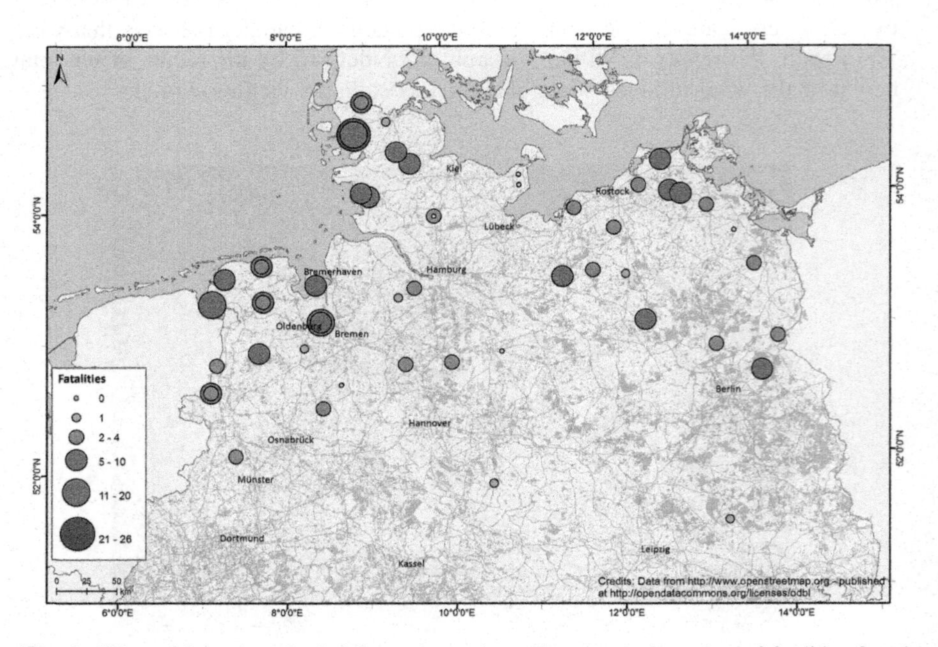

Fig. 2 Siting of investigated wind farms in northern Germany and number of fatalities found (n = 291)

extent was determined by buffering the actual transect line in the search circle with 10 m wide strips on both sides.

The daily carcass removal rate (caused e.g. by predators) was determined by conducting 81 experiments in which 1208 birds were laid out in 46 wind farms. The calculated daily removal rate was usually below 10%.

The search efficiency was determined by placing birds (two classes of conspicuity) in different areas covering five vegetation classes. Under good search conditions approximately 50% of the inconspicuous birds and 72% of the more noticeable birds were found. The good conformance between the observers (high observer reliability) justifies the general applicability of the survey results (Fig. 3).

The expected distribution of collision victims was derived by placing the individually measured distances of the retrievals from the wind turbine in relation to the total rendered surface in that particular distance ring.

The proportion of collision victims outside the search circle was between 7 and 20%. However, for the collision victim estimate, only birds that were found within the search circle were taken into account, as the search effort can only be calculated for these retrievals. In order to avoid an underestimation of the number of actual collision victims, it is necessary to correct for the proportion of collision victims outside the search circle.

The total bird fatality estimates within the investigated wind farms over the 12 week period were achieved in two key steps. First, the following figures were considered: the detection probability of a given fatality in a specific vegetation class; the distribution of the fatalities around the wind turbines in relation to the total height of the wind turbine and conspicuous class of the fatality species; the coverage of searched area; the search efficiency in different vegetation classes; the removal rate (daily "survival rate of carcasses"); and the proportion of fatalities outside the search plot.

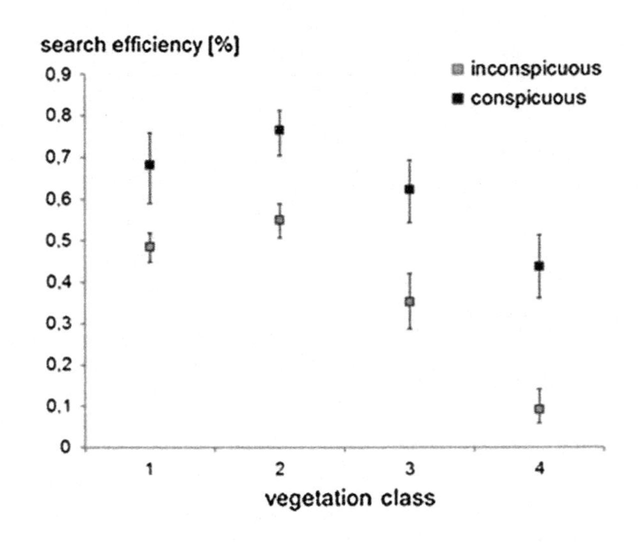

Fig. 3 Search efficiency of two different classes of conspicuity (product of classes of bird size and colour) in relation to vegetation class (1 = good search condition with low vegetation coverage and low vegetation height)

Second, the detection probabilities were averaged per search effort with the help of mixture-models (Brinkmann et al. 2011; Korner-Nievergelt et al. 2013).

Results of Fatality Estimations

For the following species and species groups an extrapolation to each surveyed wind farm was carried out: Common buzzard, Golden plover, gulls, Kestrel, Mallard, Lapwing, Red kite, waders, Wood pigeon, Skylark and Starling.

The relative uncertainty of these projections decreased only with more than ten fatalities. Therefore, an extrapolation for five species/species groups for the non-examined wind turbines and wind farms in the entire study area of PROGRESS (Lower-Saxony, Schleswig-Holstein, Mecklenburg-Western Pomerania und Brandenburg) was performed. These were waders, Common buzzard, gulls, Wood pigeon and Mallard.

Table 1 lists the number of fatalities actually found under each search effort, the mean estimated number of fatalities per wind turbine, and the number of covered wind turbine days (number of wind turbines x number of days with potential presence of the bird species).

An extrapolation of the results for the entire project area leads to an annual mortality of around 8500 Common buzzards, 11,300 Wood pigeons and 13,000 Mallards. Based on the breeding population in the project area this translates to 0.5% of Wood pigeons, 5.0% of Mallards and 7% of Common buzzards (assuming 50% floaters).

For comparison, the annual harvest by hunting is 11 times higher for the Mallard and 15 times higher for the Wood pigeon (hunting statistics of the four federal states of the covering the project area.

Table 1 Estimation of the number of fatalities of different bird species or species classes in the investigated wind farms in 11 weeks

Species	Fatalities (n)	Estimate (n)	Lower CRI (n)	Upper CRI (n)	Turbine days
Skylark	9	**291**	124	554	43.736
Starling	9	**184**	88	324	43.736
Mallard	22	**116**	72	180	43.736
Gulls	15	**102**	60	158	43.736
Wood Pigeon	14	**101**	58	157	43.736
Waders	16	**99**	54	167	33.649
Common buzzard	12	**76**	42	124	43.736
Lapwing	9	**70**	34	122	31.416
Golden plover	5	**34**	15	70	14.553
Red kite	3	**26**	7	61	21.868
Kestrel	3	**25**	7	54	43.736

Vantage Point Watches—Flight Behaviour

Methods, Fieldwork of Vantage Point Watches

In parallel to bird collision searches, behavioural observations were carried out from two observation points following a specified temporal sequence to assess the spatial distribution of birds flying near or within wind farms and their avoidance behaviour to wind turbines. This typically consisted of three units of one hour each, resulting in a total of 36 observation hours per wind farm. Different to standard SNH approach (SNH 2010a, b) we observed a full 360° circle and a 1-km maximum observation distance was set.

The behavioural observations distinguished between target species (birds of prey, waders, geese, cranes and other large birds) and secondary species (all other species) that were recorded with a varying degree of intensity as recording of secondary species was subsidiary to recording of target species. The target species were observed from the moment of sighting until they had left the study area. The species, number and the length of flight sequences were documented within and outside the wind farm in three-different height classes (under, in and above rotor swept zone), allowing for observer error. Observers have also been trained in the recording of different heights from a distance (i.e. by using a quadrocopter). For the relevant height band of the BAND-model a slight overestimation of 12% was registered with these quadrocopter experiments. Comparing observer estimates with already known heights (in relation to features such as existing turbines, trees and pylons) were used in training and observation.

In order to record avoidance behaviour of birds to the wind turbines, the flight behaviour was categorized. If a change in flight trajectory was observed it was classified as recognisable avoidance behaviour (e.g. horizontal/vertical avoidance of the whole wind farm, horizontal/vertical avoidance of a single turbine). However it might be more difficult to assess vertical responses and VP watches might not give other possible responses, such as anticipatory evasion (c.f. May et al. 2015) sufficient consideration.

Results of Observed Flight Behaviour

The most common target species were Lapwing with 29,671 individuals recorded and Golden plover with 26,789 individuals, followed by White-fronted geese and Barnacle geese with 22,901 and 20,769 individuals, respectively. Other common species with more than 10,000 individuals recorded were the Common crane, Greylag goose and Bean goose. The most frequent bird of prey was the Common buzzard (2403 individuals), followed by Red kite (869 individuals), Common kestrel and Marsh harrier (753 and 639 individuals). Other raptor species only occurred in much lower numbers.

The most abundant secondary species were Wood Pigeon and Starling. The most abundant individuals observed at rotor height were pigeons and swifts. Songbirds, gulls and ducks were predominantly observed below the rotor swept zone. The largest share of sightings of target species consisted of raptors, while the most frequently observed group of birds were geese.

From these results, the following conclusions concerning the flight behaviour at presence of wind turbines, as well as the consequences for the collision risk were derived:

- Geese, Golden plovers, and amongst the birds of prey, Buzzard, Red and Black kite exhibited the highest proportions of observed flight activity within rotor height at the study sites.
- Cranes and geese showed a high percentage of avoidance behaviour (42% of the registered flight events of geese within the wind farms showed a visible avoidance reaction).
- Birds of prey and shorebirds occurred much more frequently within the wind farms compared to geese and cranes.
- Birds of prey showed only very small amounts of recognisable avoidance behaviour towards wind turbines (3%), especially in contrast to geese and cranes.
- The most common recognisable behavioural avoidance reaction in all species was a horizontal circumnavigation of the entire wind farm or individual installations.
- Flights through the rotor with a corresponding risk of collision as a whole were rarely observed in birds of prey, however, they are observed more often than in other species.
- Birds of prey faced a significantly higher proportion of hazardous situations than other species.

Overall, these results indicate that the species-specific collision risk with wind turbines, which shows large differences between species, is the result of clear behavioural differences. Accordingly, only two collision fatalities of Common crane were recorded in the PROGRESS project, despite more than 12,000 sighted individuals. In contrast, 2400 sighted individuals of Common buzzard relate to 12 collision victims (absolute values without inclusion of correction factors).

Validation of the Band-Model

Methods of Validation

Across 55 search efforts, systematic collision victim searches were carried out and data on flight activity was collected for target species in parallel. This approach allowed an analysis of the extent to which the number of estimated collision victims, based on the collision victim searches, is dependent on the determined flight activity. Furthermore, it was assessed whether the expected number of collision victims, based on the flight activity data and projected by the BAND-model (SNH 2010a, b) was consistent with the numbers determined by collision victim searches.

In this comparison, only those target species were included which showed a sufficient amount of fatalities and flight activity totaled over all 55 wind-farm-seasons (Common buzzard n = 12, Kestrel n = 3, Red kite n = 3, Golden plover n = 5 and Lapwing n = 9, cf. Table 1). However, the estimates for species with few recorded collision fatalities (particularly Kestrel and Red kite) are associated with an increased uncertainty. The estimated variance between different wind farms is also very high for some species (e.g. Red kite). Similarly, the data on flight activity underlying the results of the test of the BAND-model are characterized by a high variability, which should be considered when comparing the two data sets.

Results of Validation

For the vast majority of wind farms the numbers of collision victims predicted by the BAND-model were clearly below the number of collision victims estimated from carcass searches.

On the basis of the recorded flight activity data of the Common buzzard, for example, the prognosis of collision victims via the BAND-model leads to a dramatic underestimation of the number of collision victims (calculated with the assumption of no avoidance) compared with the adjusted estimates based on carcass searching. A total of 12 Common buzzard collision victims were found (Table 1), however the corresponding median estimate was 76 collision victims (credibility interval: 42–124). In contrast, the BAND-model led to an estimation of 35 collision victims (without application of an avoidance rate) across all 55 wind farm-seasons.

Step 3 of the BAND-model normally includes the application of the avoidance rate, which considers the fact that in most cases birds prevent a collision. The avoidance rate is generally derived by comparing the number of birds that actually collide—calculated from carcass search studies—with the number of collisions predicted by the BAND-model before allowing for avoidance. It is important to realize that the correction factor may in fact encompass different sources of error in the model (i.e. stage 1 and 2). The correction factor likely represents the total effect resulting from many unknown factors (c.f. May et al. 2010).

$$\text{Avoidance rate} = 1 - \frac{\text{No. of observed collisions}}{\text{No. of predicted collisions with no avoidance}}$$

Considering the prognosis for the buzzards, in most wind farms it is not possible to calculate such an avoidance rate, as it mainly results in a negative avoidance. In comparison, SNH (2010a, b) proposes avoidance rates of 95–99% for many species. Assuming a conservative avoidance rate of 95% for Common buzzards, this would lead to only 1.75 predicted collision victims in total according the our Band-model—i.e. only 2.3% of the number of estimated victims from the carcass searching and only 15% of the numbers actually found. So the calculations according to the Band-model based on the flight activity data led to a drastic underestimation of fatality numbers.

These results are in line with the critical discussion of the Band-model in the literature (Chamberlain et al. 2006; May et al. 2010, 2011). The observed discrepancy between the estimates is only partially due to methodological problems of the observation of spatial use, which fail to predict the quantity or quality (small sample size, high impact of individual events). The standard way of estimating the correction factor (Avoidance rate), as done above, does not render any information on the uncertainty involved in the modelling. Here, we have also modelled collision risk incorporating the (observed) variability during the 36 observation hours (see error bars in Fig. 4 according to Douglas et al. 2012), as VP watches are likely to be inherently variable, and collision rate predictions should assess the potential error associated with such results.

In particular, the potential error due to the high variability of flight activity in the danger zone and a combination of other factors caused this mismatch. A number of inherent weaknesses of the Band-model approach also contribute to this discrepancy. In addition, it is sensitive when changes are made to various input data. These calculated values of the Band-model e.g. assume that the wind turbines operated continuously with the respective RPMs.

As the Band-model should have been tested considering the every-day planning process we didn't make any efforts to derive an overall correction factor for each site for all considered input parameters. Although some technical parameters such

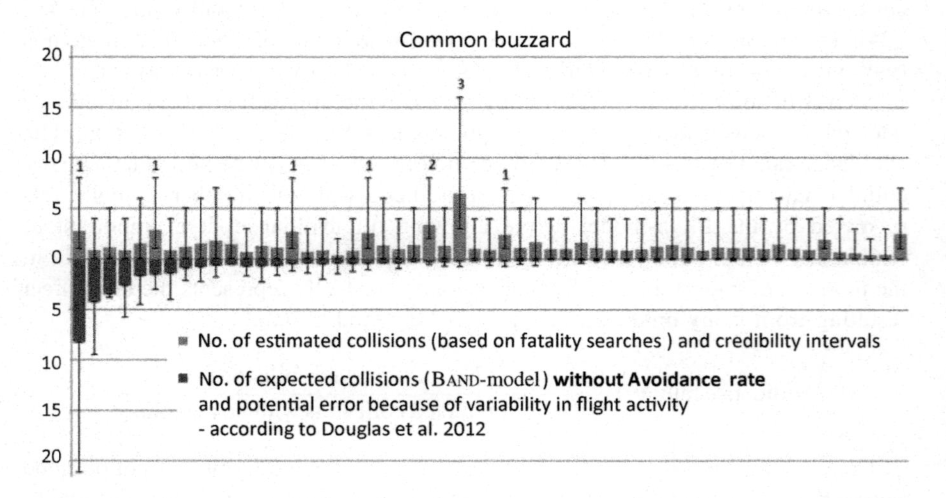

Fig. 4 Prognosis of common buzzard fatality numbers per wind farm-season as a result of the Band-model (*red columns*) plotted against the adjusted estimates based on fatality searches (*blue columns*). Actual recorded fatalities for each site are given in *red numbers* (Color figure online)

as rotor diameter and total height are fixed input parameters in the model, other parameters such as the mentioned rotor speed or the angle of the tilt of blades are difficult to include in the model precisely. In relation to bird behaviour, it was shown that next to the proportion of activity in the danger zone, flight speed and flight height are also subject to a high level of variability.

Overall, the main problem with the calculations using the BAND-model seems to be the vague relationship between the observable flight activity and the collision rate. This is because collisions are rare events. Even for the most frequent collision victims, the results of this study indicate a frequency of less than 1 collision per turbine and year. The model assumes a linear relationship between the length of stay in the danger zone and the resulting collision risk, which could not be confirmed with the PROGRESS data. For Common buzzard and Golden plover, no significant influence of flight activity on the number of estimated collision victims was found. This corresponds to the results of De Lucas et al. (2008) and Ferrer et al. (2012).

Given these results, the suitability of the BAND-model for the evaluation of an anticipated collision risk for a planned wind farm at an 'average' onshore site is limited. The projections do not predict absolute collision victims sufficiently. However, the model allows standardized comparisons of relative risks, such as for the assessment of various repowering scenarios (Dahl et al. 2015) or to illustrate the influence of various distances to a breeding site (Rasran and Thomsen 2013).

On the other hand, it appears that the model can only be used sensibly as a possible planning tool when the variability of flight activity is low. Low variability means that there is a good predictability of course, altitude, direction and intensity of flight paths. For example, this might be the case for flights between breeding colonies of gulls, terns or herons and their foraging grounds. In addition, the actual or missing breeding success, changes in availability of prey and the presence of possible neighbours and their breeding success (Meyburg et al. 2006; Langgemach and Meyburg 2011; MELUR and LLUR 2013) can play a crucial role in the anticipated flights through the area of a planned wind farm.

Population Projections

Methods

The effect of the estimated additional mortality (Table 1) on the population level was modelled for target species. As Common buzzard and Red kite populations are expected to be strongly affected because of their long lifespans and relatively low reproduction rates, those species were defined as target species. Additionally, Lapwing was included, as many collision victims of this already declining species were found (Hötker 2015). Matrix models were developed for each species, in

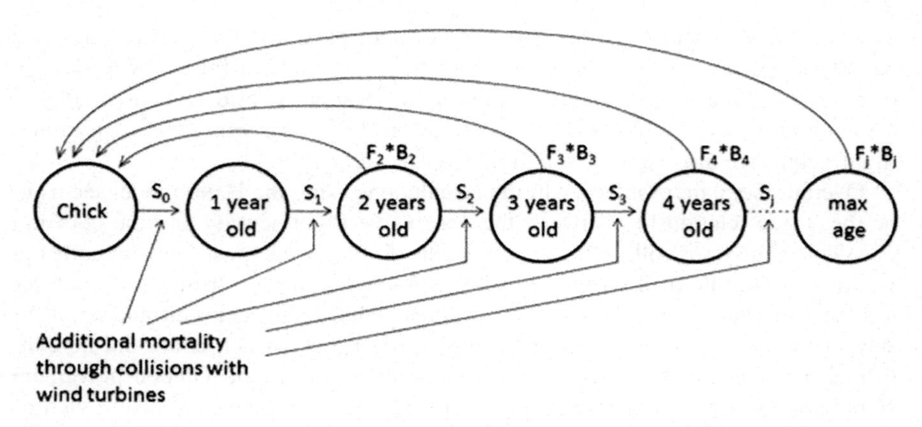

Fig. 5 Example of a matrix model, with age at first breeding equal to two as in most Common buzzards. *Arrows* indicate transition probabilities. Survival of age class j in a situation without additional mortality through collisions is indicated with S_j. Fertility is indicated with F_j, and breeding probability with B_j. Additional mortality through collisions is integrated as a factor affecting the survival, assuming equal additional mortality across the age classes. The model is single-sex, so fecundity is halved, assuming a 1:1 sex ratio

which demographic data and the population structure were used to predict the future population trend (Fig. 5). Due to limited space, the following section only focuses on the Common buzzard, with results for the red kite being qualitatively very similar.

Long-term data from the population around Bielefeld (Krüger and Lindström 2001) was used to construct a matrix model, with age-specific fertility and breeding probability. Survival estimates were calculated for five age classes, based on mark-recapture using re-sightings of wing tags. By calculating the dominant eigenvalue from the resulting matrix, the population growth rate (λ) was estimated (Caswell 2001). The effect of additional mortality may differ between populations. Therefore, the matrix was adapted from Bielefeld to fit the reproduction and growth rate of three other populations in northern Germany (Dänischer Wohld, Altenpleen, Rathenow). This estimated the survival, as specific data on survival was not available. We chose populations from areas with low turbine density, or from before the strong development of wind energy.

For this study, the additional mortality estimates shown in Table 1 were extrapolated to two different scenarios of turbine densities. The first scenario represents the current average density in the northern federal states of Germany (12 turbines per 100 km^2). In the second scenario, an increasing turbine density (observed development over the last 15 years in the federal state of the particular population) was used, in order to see how well the additional mortality predicts the observed population trends. For this extrapolation, the estimates from Table 1 (per turbine) were multiplied by the number of turbines per 100 km^2, to give the expected number of collision victims per 100 km^2. Differences in area suitability and usage are not taken into account, and the additional mortality is assumed to be a constant fraction of population size, i.e. not density-dependent. Due to space limitations, we will focus on the results for the first scenario over the next 30 years.

Results

The predicted population trend of the Common buzzard in the area around Bielefeld is an annual increase of 4.0%, which fits to the observed population increase. In the other areas, a smaller population increase is observed (0.7% and 1.4% in Dänischer Wohld and Altenpleen), or even a population decline (−3.1% in Rathenow). According to the additional mortality estimates (Table 1), the median collision rate for Common buzzard is 0.433 individuals per wind turbine per year, with a credibility interval between 0.131 and 0.836. As the models are for females only, we halved the collision rates, using 0.217 as the median, with a credibility interval between 0.065 and 0.418.

For the population in Bielefeld, the scenario with the current turbine density predicts the trends shown in Fig. 6. The median estimate for the additional mortality predicts an annual decline of 5.7%, with a credibility interval between an increase of 1.2% and a decrease of 15.7%. Qualitatively, the results for the scenario with a changing turbine density over the last 15 years shows similar results, with a predicted annual decrease of 5.3% for 2015 (CI between +1.3 and −14.9%).

For the other populations, similar results were found. For the scenario with the current turbine density, the relatively stable population around the study area of Dänischer Wohld (+0.7%) is predicted to decrease with additional mortality (median: −2.6%, CI between −0.3 and −5.8%). In the study area of Altenpleen, the increasing population (+1.5%) is predicted to decrease as well, with a median of −2.7% (CI between +0.2 and −6.7%). As the population in the study area of

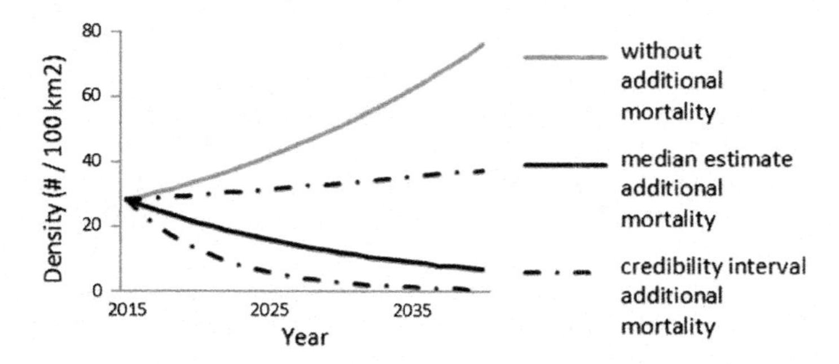

Fig. 6 Trend of the common buzzard population around Bielefeld/Germany without additional mortality (*grey*), and with additional mortality through collisions with turbines (*black* with credibility interval) (Color figure online)

Rathenow already shows a decline (−3.1%), the predicted decline will be only stronger (median: −27.4%, CI between −9.9 and 57%). However, the effect on the population in Rathenow is likely to be an overestimation, as the density of Common buzzards in this area is rather low, and hence the number of collision victims is expected to be lower than the average for the whole area. The effect of environmental stochasticity and the proportion of floaters on the modelled qualitative effect of additional mortality was relatively small.

In conclusion, it is shown that all four modelled populations of Common buzzard are predicted to decline when incorporating the median estimates of additional mortality, as calculated in Chap. 3.2. This coincides with the population decline observed in some areas in Germany (e.g. Grünkorn 2014). With the further expansion of wind energy, the effects could be even stronger.

Consequences for Planning Procedures

The systematic collision victim searches for PROGRESS have demonstrated that collision victims are to be expected at any wind farm site. In addition, it is apparent from the PROGRESS list of species and from the German and European collision victim data base[1] that in principle any species might collide with a wind turbine. However, there are marked and specific differences how different species are affected. In absolute terms, common species that stay within wind farms without a pronounced avoidance behaviour collide the most (e.g. Skylark, Starling, Wood pigeon, Mallard, Common buzzard, gulls). In relation to population size, birds of prey and large birds collide disproportionately frequently.

Generalized statements for the occurrence of situations or locations with increased collision risk are limited. For breeding birds the distance to the nesting site can be used as a first approximation, within which an increased flight activity or collision prone behaviours (e.g. courtship and territorial flights) have to be expected for certain species (Langgemach and Meyburg 2011; Eichhorn et al. 2012; Hötker et al. 2013; Dahl et al. 2015; LAG VSW 2015). A tangible assessment of the collision risk is only possible for individual cases, for which a qualitative behavioural ecological assessment based on a specific bird utilisation study is proposed. The species-specific spatio-temporal variability of land use has to be considered. It must also be considered whether the collected data represents only a snapshot and if it constitutes a reliable basis for assessing the operational span of the planned wind farm.

Although the results of PROGRESS and some other studies (De Lucas et al. 2008; Ferrer et al. 2012) show that—at least for some species—there is no direct quantitative relationship between the detectable flight activity and the risk of collision, it is still reasonable to assume that certain species will exhibit a higher

---

[1]Access via: http://www.lugv.brandenburg.de/cms/detail.php/bb1.c.312579.de (07.01.2016).

collision rate. These particular species include those with possible high flight activity within wind farms (showing no displacement) in combination with specific collision prone behaviour. Therefore, data collection in the pre-construction phase should focus on behaviour and habitat use to identify ecologically and functionally important habitat areas for these species, and assess whether a planned wind farm is located in such areas. Such functionally important areas might only temporarily play an important role, such as immediately after mowing or harvesting (as for White stork and Red kite) or dependent of the annual cycle (courtship period, fledging period or winter feeding areas).

Possible factors influencing the assessment of collision risk are outlined below:

- The distance of a planned wind farm to nearby breeding areas of species with high collision risk is, in general, a first indication of the risk of collision. The shorter this distance, the sooner a flight activity can be expected which is associated with a high risk of collision.
- Attractive foraging grounds result in an increased flight activity, where the risk of collision while searching for food depends on the flight altitude and the distance from the ground to the lower rotor tip and this risk may be very species-specific.
- Flight paths: Perhaps some species are less at risk of collision when flying in a directed manner than in other situations, when they might be distracted by looking for prey, conspecifics or harassing crows. Nevertheless, it is likely that the fundamental inability of a realistic assessment of the speed of the rotor blades and the lack of pronounced avoidance behaviour towards wind turbines during regular flights at rotor height through a wind farm lead to an increased risk of collision.
- Intra- and interspecific interactions can lead to an increase in flight activity at rotor height. The extent of territory defence behaviour close to breeding sites depends on the intra- and interspecific nearest neighbour distances. The birds are distracted in such situations, for example the mobbing of crows can lead to a situation with a high collision risk (Hötker et al. 2013).
- Changes in food supply may exert a strong influence on the habitat use of collision-endangered species. Thus, a temporary lack of food availability in the vicinity (e.g. as in high-standing crops) can lead to increased foraging within the wind farm (foraging along the access roads, Hötker et al. 2013).

Quantitative collision models, which are still widely used and, in the meantime, are not limited to the Band-model (Masden and Cook 2016), can be used as an additional auxiliary method to the above-stated behavioural and ecological assessment, but cannot replace these.

As shown, the assessment of collision risk is often associated with a high uncertainty. However, uncertainty is a fundamental feature when dealing with the planning of wind farms and their consequences for wildlife and this will remain, despite increased research efforts (Masden et al. 2015). Usually, prediction uncertainties are dealt with under the principle of precaution (be sure to stay in safe distance to the birds or quit the whole project), which, however, would lead to an exclusion of large areas for wind energy use, and thus would lead to conflicts with climate objectives (Köppel et al. 2015). Given these problems, it appears that solutions should not just target the planning phase, but also the operational phase of wind turbines. This approach, the so-called "adaptive management", is already practiced internationally in various contexts and can be characterized as follows (Köppel et al. 2015):

- Clear definition of objectives
- Accompanying research (monitoring)
- Link between the monitoring results and unequivocal management implications.

Thus the established hierarchy of conflict management—prevention, minimization, compensation—should be enhanced by a mandatory monitoring in combination with any necessary adjustment measures. A caveat of this approach is that the wind farm operators face a more or less incalculable economic risk for the duration of the operation phase. Therefore, solutions must be found which, on the one hand, ensure that operators can minimise the collision risk for certain species and, on the other hand, do not create too large economic uncertainties during the operation of the wind turbines. This approach might make it feasible to find locations that would otherwise not be available for precautionary reasons.

Thus, it is proposed to combine this approach with a stronger species protection support in the operation phase. This may be particularly necessary for species that are already adversely affected on the population level by collision mortality. According to current knowledge this is relevant for the Common buzzard and the Red kite. However, it is to be expected that with continued expansion, these cumulative effects will also occur for other species.

Species protection would be constituted in monitoring (inventory control), protective measures (e.g. habitat improvement, increasing breeding success) and possibly temporary operating restrictions. Each of these depend on the target species and the local population trend. In regards to the additional costs for the individual wind farm project, at least some costs could be absorbed by reducing the extensive bird utilisation studies during the pre-construction phase. A realistic assessment of the individual case would show that their value has already to be considered limited due to spatial and temporal variability.

It can be assumed that cumulative effects will become more significant with an increasing number of wind farms. Accordingly, the demands on conflict resolution will increase from a species protection perspective. It will also have to be expected that statutory species protection conflicts might not always be adequately solvable for an individual project.

Therefore, overarching solutions are required to accompany the further expansion of wind farms, which ensure that this does not lead to a severe decline of certain bird species that are particularly affected by collisions. Specifically, the following strategies need to be addressed:

- Large-scale wildlife conservation programs e.g. for Red kites and Common buzzards that improve habitats, particularly in terms of food availability and lead to a compensation of collision losses at the population level (increase in reproduction rate, reduction of other anthropogenic mortalities). This cannot be imposed on an individual wind farm project, but should be seen as a political task to accompany the large-scale implementation of renewable energy.
- Identification of some species-specific core areas of high density that are of particular importance as source populations, and assessing targeted measures to protect and promote them, e.g. by appropriate species relief measures, protection against collisions by having wind turbine free areas or by increasing requirements on the avoidance of losses (unless already protected by legal reserve categories).
- Development of concepts and practical testing of a post-construction species protection support in terms of their effectiveness and their economic effects.
- Increased research efforts in terms of scale and addressing cumulative effects.

Conclusions

The project PROGRESS investigated, for the first time on a large scale and quantitatively, the collision rates of birds with wind turbines accompanied by visual flight activity surveys. The North German lowland was chosen as the study area, because of its particular importance for the use of wind energy in Germany and because approximately half of the wind turbines currently operating in Germany are located here. As of 2014, 12,841 turbines (out of a total of 24,867 in Germany[2]) were located in the project area of PROGRESS. Thus, representative statements on the collision risk of birds of northern Germany are possible for all species. However, due to the relatively small numbers of fatalities found, the extent of collisions can only be quantified for few species.

The project is based on collision victim searches that were conducted with considerable effort, and a simultaneous determination of detection errors, such as search efficiency and carcass removal of the collision victims as well as an accurate determination of the searched area. The determination of these factors allows an estimation of the actual collision victims for the investigated wind farms and the investigation period. Importantly, PROGRESS's determined correction factors are relatively small, such as the search efficiency within the selected transects (20 m wide) with a search efficiency of 50–70% under good conditions. The findings of carcasses of collision victims was quite high with a daily removal rate of <10%. These two correction factors only contribute little to the uncertainty in the estimated number of fatalities.

[2]https://www.wind-energie.de/themen/statistiken/deutschland

As searches along linear transects results in a decreased area coverage as the distance to the wind turbine increases, a larger correction for area coverage is necessary. The method developed for PROGRESS is thus considered as very suitable for the estimation of collision victims. However, a simple transfer of the determined correction factors to other studies is not recommended, as they were derived for local conditions and with the particular methodology used. Given these assumptions and based on of this methodology for the determination of collision victims, it is emphasised that:

1. The required effort is very high. For PROGRESS and under good search conditions (flat agricultural land with low vegetation) in a 20 m wide search strip, one collision victim was found for every 27 km of transect line searched. With a total area of 7500 km of covered transect lines, 291 fatalities were found which were distributed among 57 species. The effort required to obtain robust data on species-specific collision rates is thus very high.
2. The possible investigation effort is limited by geographical features and vegetation structure. The search for collision victims in areas with a higher and/or denser vegetation than what has been accepted for PROGRESS would severely restrict search efficiency. It would significantly increase the necessary effort to obtain a sufficient sample size. Search effort in fully grown cereal or maize fields as well as in forests was therefore deemed unacceptable during PROGRESS. This limits the applicability of the method, both seasonally and spatially. However, since no more efficient methods are currently available, this is regarded as a tolerable restriction for the determination of collision rates, but it has to be considered when actual collision rates are projected.
3. the low finding rates—in accordance with the low collision rates of most species —impedes the formulation of quantitative statements, especially for rare species, because the necessary effort for these species cannot be rendered. However, since some particularly relevant species, e.g. birds of prey, have relatively low abundances, the necessity arises that other methodological approaches for the determination of collision rates have to be developed, too.

The overall low number of fatalities found allows for a projection of collision numbers for the investigated wind farms for eleven species/groups and a projection for the entire study area of PROGRESS for five species/groups. Among the eleven frequent collision victims 71% are accounted for in five species/groups: Skylark, Starling, Mallard, gulls and Wood pigeon. It is noteworthy that these species account for only 28% in the national reference database.[3] Birds of prey, which are represented in the national reference database with 35%, account for only 11% of fatalities according to PROGRESS data. This highlights the necessity for systematic studies, taking into account the investigation effort and correction factors. Findings of fatalities under unsystematic controls automatically lead to a bias for more

[3]http://www.lugv.brandenburg.de/cms/detail.php/bb1.c.312579.de

noticeable species and species with higher public interest. This hampers the assessment of the actual degree to which the various species are affected.

In accordance with the accompanying visual observations, the majority of collisions happen to abundant and non-endangered species in the agricultural landscape, which are resting or foraging in the wind farm. The collision risk is species specific, but a high similarity seems to exist for related species. This allows, at least within certain boundaries, a transfer of the assessment of the collision risk to species for which little data exists so far. Those species that collide frequently, such as birds of prey, were found disproportionately often in the vicinity of wind turbines and displayed hardly recognizable avoidance behaviour. Geese and cranes, on the other hand, exhibit both macro and micro avoidance behaviour around wind turbines.

Among the fatalities, nocturnally migrating species are significantly underrepresented and a threat by wind turbines to species of the nocturnal broad front migration of northern songbirds can be ruled out. Given the high investigation effort, it can be assumed that those species that are common in the study area, but were only detected in small numbers as collision victims, are not significantly affected by the recent expansion of wind energy in the North German lowland. For rare species, however, this conclusion cannot be drawn due to a limited sample size even in this investigation.

Overall it is noteworthy that collisions with wind turbines predominantly occur during the day and affect species with good flying capabilities. On the other hand, species with poor manoeuvrability, such as geese or cranes, and nocturnal migrant species collide significantly less frequently with wind turbines. Reflecting the results of the flight activity observations, the species composition of the collision victims indicates that the collision risk is largely determined by the behaviour of the birds towards the wind turbines. While some species apparently perceive wind turbines as disturbing structures, other species approach them without showing any avoidance behaviour at all and are thereby endangered by the rotors. Specific behaviours (courtship, territorial fights, foraging, etc.) can affect the perception of wind turbines. The flight activity observations of birds within the investigated wind farms, as well as a habitat analysis, did not clearly determine the circumstances under which collisions occur. This indicates that the risk of collision significantly arises from the situational behaviour of birds towards wind turbines, which currently cannot be generalized. Therefore, the existing projection models cannot predict collision rates of birds based on their flight behaviour with satisfactory accuracy.

The population models indicate a likely negative effect on the population level for the Common buzzard. This is a new and surprising result, since the Common buzzard—the most common raptor in Germany—has so far not been considered in the planning process of wind farms. Considering the nationwide distribution and a generally increased collision risk for raptors, a population level effect for this species due to the expansion of wind farms is plausible. The models indicate that similar effects are possible for the Red kite, although in the state of Brandenburg the population was not predicted to be negatively affected by the current extent of wind farms.

Only a limited number of relevant variables for raptor populations can be incorporated into models and the population projections have rather large confidence intervals. However, other less abundant raptor species lacking sufficient data could be affected on the population level by the current number of installed wind turbines in northern Germany. Relevant factors at the local level cannot be ruled out for other species such as Lapwing.

The outcome of PROGRESS gives an all-clear signal of no concern for the majority of bird species of northern Germany. For other species, especially Common buzzard, the results indicate that estimated fatality rates based on the current state of wind farm development could already lead to a population decline. It is worth noting that the predicted impacts on raptor populations are very sensitive to current population trends. Wind farms are therefore likely to cause a decrease in certain populations, where populations are already negatively affected by other factors.

The outcomes of PROGRESS reveal difficulties in identifying mitigation measures to reduce the risk of collision of vulnerable species in the planning process of wind farms. Previous approaches addressed primarily minimum distances from breeding sites of endangered species to reduce the collision risk. This is justified as the breeding site is an activity centre—at least in the breeding season. The effectiveness of standard distance radii is however countered by the fact that flight activity of species is not evenly distributed across different habitats and that habitat use is rather variable throughout the year and over the years.

For all species showing frequent fatalities, the fatalities also occur outside the breeding season in northern Germany, and some of the collision victims only occur as staging birds. The number of fatalities was comparable between the spring and autumn seasons. For some species—e.g. Skylark—the collision risk is influenced by specific flight activity pattern in the breeding season, but for most other species, there is no such evidence.

For all species with frequent fatalities, their populations depend on the actual type of land use which changes over seasons and years. Changes in land use result in changes of the breeding site and feeding and resting areas. This limits the possibilities for mitigation and avoidance at the project level to a great extent. As a consequence, the total number of fatalities depends on the total number of wind turbines installed across a larger area, which cannot be addressed in the planning process of single wind farm projects or even wind turbine projects.

In the context of the proposed increase of wind farms it is recommended to (1) examine the consequences of collisions for bird populations of conservation concern in more detail, and to identify methods, (2) how to avoid conflicts and (3) support populations of conservation concern. The following points provide further detail:

1. Comprehensive population studies on Common buzzard, Red kite and other potentially endangered species are strongly recommended. Models should incorporate individual based modelling (IBM), which account for density-dependent processes, resources and other causes of mortality. Further

studies to quantify collision rates of birds at wind turbines applying the PROGRESS methodology are recommended, as collision rates may differ in other landscapes.

2. As options to avoid or mitigate impacts of wind turbines are apparently limited for some species, a further increase of wind farms should be accompanied by protection measures that aim to reduce other anthropogenic impacts on these species. Compensatory measures must be established not only at the project level, but within the framework of regional planning or based on nationwide species management plans. Further, crucial core areas for breeding or staging of endangered species should have no or fewer additional wind turbines. Repowering should be considered as a potential mechanism to constrain the growth of, or even reduce, the number of wind turbines per unit area, especially in core areas of species of conservation concern.
3. Since most species affected by wind farms inhabit farmland, further intensification of agriculture should be avoided within particularly important habitat areas. Re-establishing the structural diversity of the agricultural landscape would be a powerful tool to counterbalance negative impacts from wind farms.

References

Brinkmann R, Behr O, Niermann I, Reich M (eds) (2011) Entwicklung von Methoden zur Untersuchung und Reduktion des Kollisionsrisikos von Fledermäusen an Onshore-Windenergieanlage. Umwelt und Raum Bd. 4. Civillier Verlag, Göttingen, p 457

Caswell H (2001) Matrix population models, 2nd edn. Sinauer Associates, Sunderland

Chamberlain DE, Rehfisch MR, Fox AD, Desholm M, Anthony SJ (2006) The effect of avoidance rates on bird mortality predictions made by wind turbine collision risk models. Ibis 148:198–202

Dahl EL, May R, Nygård T, Aström J, Diserud O (2015) Repowering Smøla wind-power plant. An assessment of avian conflicts. NINA Report, 41

De Lucas M, Janss GFE, Whitfield DP, Ferrer M (2008) Collision fatality of raptors in wind farms does not depend on raptor abundance. J Appl Ecol 45:1695–1703

Douglas DJT, Follestad A, Langston RHW, Pearce-Higgins JW, Lehikoinen A (2012) Modelled sensitivity of avian collision rate at wind turbines varies with number of hours of flight activity input data. Ibis 154(4):858–861

Eichhorn M, Johst K, Seppelt R, Drechsler M (2012) Model-based estimation of collision risks of predatory birds with wind turbines. Ecol Soc 17(2):12

Ferrer M, de Lucas M, Janss GFE, Casado E, Muñoz AR, Bechard MJ, Calabuig CP (2012) Weak relationship between risk assessment studies and recorded mortality in wind farms. J Appl Ecol 49(1):38–46

Grünkorn T (2014) Rückgang des Mäusebussards im Landesteil Schleswig. Jahresbericht Jagd und Artenschutz 2014. MELUR Schleswig-Holstein, pp 106–109

Hötker H (2015) Überlebensrate und Reproduktion von Wiesenvögeln in Mitteleuropa. Vogelwarte 53:93–98

Hötker H, Krone O, Nehls G (2013) Greifvögel und Windkraftanlagen: Problemanalyse und Lösungsvorschläge. Schlussbericht für das Bundesministerium für Umwelt, Naturschutz und Reaktorsicherheit., Michael-Otto-Institut im NABU, Leitnitz-Institut für Zoo- und Wildtierforschung, BioConsult SH, Bergenhusen, Berlin, Husum

Köppel J, Dahmen M, Helfrich J, Schuster E, Bulling L (2015) Cautious but committed: moving toward adaptive planning and operation strategies for renewable energy's wildlife implications. Environ Manage 54:744–755

Korner-Nievergelt F, Brinkmann R, Niermann I, Behr O (2013) Estimating bat and bird mortality occurring at wind energy turbines from covariates and carcass searches using mixture models. PlosOne 8:e67997

Krüger O, Lindström J (2001) Lifetime reproductive success in Common Buzzard Buteo buteo: from individual variation to population demography. Oikos 93:260–273

LAG VSW (Länderarbeitsgemeinschaft der Vogelschutzwarten) (2015) Abstandsempfehlungen für Windenergieanlagen zu bedeutsamen Vogellebensräumen sowie Brutplätzen ausgewählter Vogelarten in der Überarbeitung vom 15. April 2015, 29 S

Langgemach T, Meyburg BU (2011) Funktionsraumanalyse - ein Zauberwort der Landschaftsplanung mit Auswirkung auf den Schutz von Schreiadlern (Aquila pomarina) und anderen Großvögeln. Berichte zum Vogelschutz 47(48):167–181

Masden EA, Cook ASCP (2016) Avian collision risk models for wind energy impact assessments. Environ Impact Assess Rev 56:43–49

Masden EA, McCluskie A, Owen E, Langston RHW (2015) Renewable energy developments in an uncertain world: the case of offshore wind and birds in the UK. Mar Policy 51:169–172

May R, Hoel PL, Langston RH, Dahl EL, Bevanger K, Reitan O, Nygård T, Pedersen HC, Røskaft E, Stokke BG (2010) Collision risk in white-tailed eagles. Modelling collision risk using vantage point observations in Smøla wind-power plant. NINA Report 639, Trondheim, p 25

May R, Nygård T, Dahl EL, Reitan O, Bevanger K (2011) Collision risk in white-tailed eagles. Modelling kernel-based collision risk using satellite telemetry data in Smøla wind-power plant. Tagungsband der Fachtagung: "May, 2011", Trondheim

May R, Reitan O, Bevanger K, Lorentsen SH, Nygård T (2015) Mitigating wind-turbine induced avian mortality: sensory, aerodynamic and cognitive constraints and options. Renew Sustain Energy Rev 42:170–181

MELUR & LLUR (Ministerium für Energiewende, Landwirtschaft, Umwelt und ländliche Räume des Landes Schleswig-Holstein & Landesamt für Landwirtschaft, Umwelt und ländliche Räume des Landes Schleswig-Holstein) (2013) Errichtung von Windenergieanlagen (WEA) innerhalb der Abstandsgrenzen der sogenannten potenziellen Beeinträchtigungsbereiche bei einigen sensiblen Großvogelarten - Empfehlungen für artenschutzfachliche Beiträge im Rahmen der Errichtung von WEA in Windeignungsräumen mit entsprechenden artenschutzrechtlichen Vorbehalten

Meyburg BU, Meyburg C, Matthes J, Matthes H (2006) GPS-Satelliten-Telemetrie beim Schreiadler Aqulia pomarina: Aktionsraum und Territorialverhalten. Vogelwelt 127:127–144

Rasran L, Thomsen KM (2013) Auswirkungen von Windenergieanlagen auf den Bestand und die Nistplatzwahl der Wiesenweihe Circus pygargus in Nordfriesland In: Hötker H, Krone O, Nehls G (eds) Greifvögel und Windkraftanlagen: Problemanalyse und Lösungsvorschläge. Schlussbericht für das Bundesministerium für Umwelt, Naturschutz und Reaktorsicherheit, Michael-Otto-Institut im NABU, Leibniz-Institut für Zoo- und Wildtierforschung, BioConsult SH, Bergenhusen, Berlin, Husum

SNH (Scottish Natural Heritage) (2010) Guidance: survey methods for use in assessing the impacts of onshore windfarms on bird communities. Scottish Natural Heritage, November 2005 (revised December 2010), 50 p

SNH (Scottish Natural Heritage) (2010) Use of avoidance rates in the SNH wind farm collision risk model, 10 p

A Method to Assess the Population-Level Consequences of Wind Energy Facilities on Bird and Bat Species

Jay E. Diffendorfer, Julie A. Beston, Matthew D. Merrill, Jessica C. Stanton, Margo D. Corum, Scott R. Loss, Wayne E. Thogmartin, Douglas H. Johnson, Richard A. Erickson and Kevin W. Heist

Abstract For this study, a methodology was developed for assessing impacts of wind energy generation on populations of birds and bats at regional to national scales. The approach combines existing methods in applied ecology for prioritizing species in terms of their potential risk from wind energy facilities and estimating impacts of fatalities on population status and trend caused by collisions with wind energy infrastructure. Methods include a qualitative prioritization approach, demographic models, and potential biological removal. The approach can be used to prioritize species in need of more thorough study as well as to identify species with minimal risk. However, the components of this methodology require simplifying assumptions and the data required may be unavailable or of poor quality for some species. These issues should be carefully considered before using the methodology. The approach will increase in value as more data become available and will broaden the understanding of anthropogenic sources of mortality on bird and bat populations.

J.E. Diffendorfer (✉) · J.A. Beston · M.D. Merrill · J.C. Stanton
M.D. Corum · W.E. Thogmartin · D.H. Johnson · R.A. Erickson
U.S. Geological Survey, Reston, USA
e-mail: jediffendorfer@usgs.gov

S.R. Loss
Department of Natural Resource Ecology and Management,
Oklahoma State University, Stillwater, USA

K.W. Heist
Department of Fisheries, Wildlife, and Conservation Biology,
University of Minnesota, Minneapolis, USA

J.E. Diffendorfer · J.A. Beston
Geosciences and Environmental Change Science Center, Lakewood, CO, USA

M.D. Merrill · M.D. Corum
Eastern Energy Resources Science Center, Reston, VA, USA

J.C. Stanton · W.E. Thogmartin · R.A. Erickson
Upper Midwest Environmental Sciences Center, La Crosse, WI, USA

J. Köppel (ed.), *Wind Energy and Wildlife Interactions*,
DOI 10.1007/978-3-319-51272-3_4

Keywords Birds · Bats · Wind farms · Population level · Collision · Population models · Data · Mortality

Introduction

While many studies and multiple reviews exist regarding wind energy-wildlife interactions, we know little about the population-level consequences of direct and indirect effects of wind energy generation on species (Schuster et al. 2015). To date, only a few studies address the issue of broad-scale, population-level impacts from wind turbines. For example, two studies used modeling to examine how fatalities from collision with turbines will impact populations of Egyptian vultures (*Neophron percnopterus*) across Spain (Carrete et al. 2009) and Red Kites (*Milvus milvus*) across Germany (Schaub 2012). Bellebaum et al. (2013) compared an estimate of annual Red Kite fatalities from collisions with turbines to the species' estimated potential biological removal (described below) to assess population-level risk across Germany.

This chapter describes a methodology designed to address questions about the direct and indirect impacts of wind energy generation on populations of birds and bats. It focuses on regional- to national-scale population impacts, not impacts to local populations interacting with a single facility. As such, the methodology evaluates the cumulative impacts of many turbines within a species' range on its overall population status or trend.

The U.S. Geological Survey (USGS) began developing this population-level methodology in 2013. During a series of meetings with organizations ranging from industry to federal agencies and conservation groups, three main ideas emerged. First, stakeholders generally agreed that a rapid method for prioritizing species was needed to help identify species with different levels of potential risk from wind energy development. Second, groups consistently echoed the need for understanding the population-level effects of wind energy facilities. Third, stakeholders expressed an interest in a methodology that could be used to forecast the impacts of future wind energy facilities on species.

Over the two years that followed, the USGS developed a methodology designed to meet these three goals. This chapter summarizes the approach, discusses how these results might be used, and points out areas of future research that would improve the methodology and help refine the understanding of wind energy impacts on species. The methodology is considered to be one of potentially many ways to estimate population-level impacts of wind energy on wildlife at regional to national spatial scales. It is recognized that the approach has shortcomings and may be limited by data availability; however, it should help focus more research on population-level effects of wind, and other forms of energy development, on wildlife.

A Summary of the Methodology

The methodology has four steps: (1) data collection, (2) a qualitative species prioritization approach, (3) demographic modeling, and (4) the potential biological removal (PBR) approach developed to manage incidental by catch of marine mammals in fisheries (Fig. 1). These steps were based on tools and techniques previously developed and commonly used in other arenas of applied ecology. After data are collected (Step 1), species are prioritized based on four risk metrics: estimates of mortality at wind turbines, impact from habitat loss, life history characteristics, and conservation status (Step 2). Those species receiving high-priority ranks are then run through two demographic components (Steps 3 and 4). The first estimates a change in the population growth rate (lambda, λ) given an estimate of the mortality rate from collisions with wind energy turbines. The second calculates the PBR and compares this, via a ratio, to the estimated number of animals killed each year by turbines.

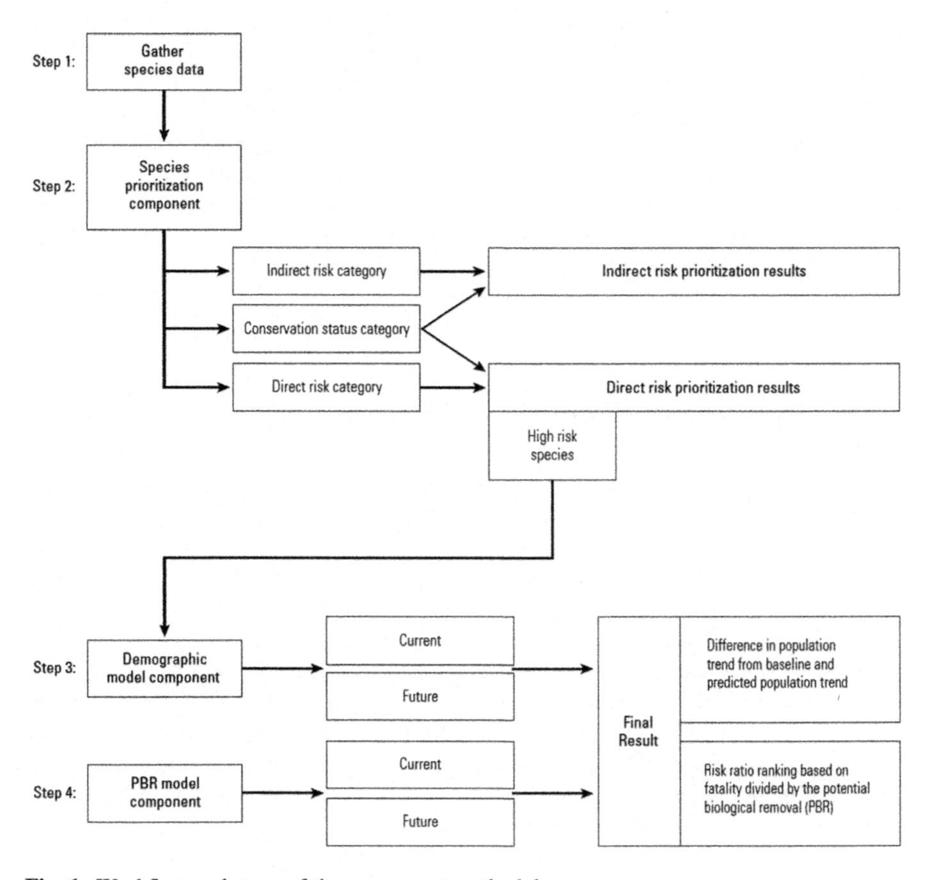

Fig. 1 Workflow and steps of the assessment methodology

Potential Biological Removal is defined as the maximum number of animals, not including natural mortalities that may be removed from a population while allowing that population to reach or maintain its optimum sustainable population (Wade 1998). The demographic components are then repeated using forecasts of future wind energy development to estimate expected changes under alternative future wind energy development forecasts.

Overall the methodology produces (1) a list of species ranked by their average priority score (Step 2) as well as the influence of each risk metric on the rank; and for those species with a high priority, (2) an estimate of the predicted change in probability that the population growth rate, given energy development induced mortality, will be less than 1 (Step 3); and (3) a risk ratio of PBR relative to the number of animals killed each year for both current and future levels of installed wind energy (Step 4).

The methodology was originally designed for species present in the United States during any part of their life cycle. However, the approach could be applied to species with sufficient data in any region of the world. The goal of this chapter is to present an overview of the approach so that other scientists will consider using or modifying it to study the population-level consequences of wind energy, or perhaps other anthropogenic sources of mortality on species. A detailed report of the approach, including the data sources used to parameterize the methodology and the way these are used in the United States has been published (Diffendorfer et al. 2015). This chapter provides a broad overview of the approach by describing each step and its key parameters, but does not include details on sources of data and parameterization methods. These details will vary by country, region, or taxon considered and depend on the types of data available. Data collection (Step 1) is not discussed, as the nuances of how data are generated to estimate parameters in the methodology vary by species and region. As such, our description begins with Step 2.

Step 2. Species Prioritization

Species are prioritized by combining information from four sources: conservation status and three "turbine risk" metrics (defined below). For most species, current conservation status is a consequence of existing population status and trend. At-risk species are generally less capable of handling additional negative impacts including those caused by power generation. Turbine risk is assessed as the direct and indirect risk to a species caused by wind energy facilities.

Three turbine risk metrics were designed to estimate different ways wind turbines could potentially impact a species. The metrics follow a risk framework that generally includes (1) the proportion of the population potentially impacted (exposure) and (2) a measure of the way a species might respond to the hazard. This exposure-response approach has been used in other ecological prioritization schemes (Andow and Hilbeck 2004; Parker et al. 1999; Regan et al. 2008).

Step 2 generates average risk scores, for each species by combining information from conservation status and turbine risk using a Monte Carlo simulation. The average risk scores can range from 1 to 9, with 9 being the highest risk. For each species, the approach simulates 10,000 randomly generated cutoff values for defining high, medium, or low risk scores for each metric, combines these into a single score, then estimates an average across all simulations (Beston et al. 2016; Diffendorfer et al. 2015). This approach was designed to avoid using arbitrary cut-off values for risk scores. The three turbine risk metrics are described below.

Proportion of Fatalities Due to Turbines (FT).

$$\mathrm{FT} = n/(1 - s)\mathrm{N},$$

where n is the number of individuals killed by turbines annually, N is the total population size, and $1 - s$ is the adult mortality rate because s is adult survival.

The denominator, $(1 - s)$N, represents the predicted number of adult fatalities each year for a species and survival, s should be calculated in the absence of fatalities from wind energy. A key assumption of FT is that species with long lifespans and low adult mortality will be more likely to experience additive mortality from human-caused factors than will species with short lifespans and high adult mortality (Péron et al. 2013; Wilson et al. 2010). Thus, species with a high proportion of their fatalities caused by turbines are more likely to be at higher risk.

Fatality Risk Index (FRI)

$$\mathrm{FRI} = p/(m/a),$$

where p is the proportion of the population exposed to turbines, m is maternity (the number of female offspring per female per year) and a is the age at first reproduction.

This metric was based on a similar approach developed by Desholm (2009) for use at individual wind farms. The ratio of maternity to age at first reproduction is a measure of a species' life history speed and is associated with the elasticity of the population growth rate. When $m{:}a$ is high, elasticity in population growth is driven primarily by changes in reproduction, but when $m{:}a$ is low, survival drives change in population growth (Stahl and Oli 2006). Because wind energy impacts survival, risk should vary across species depending, in part, on their life history speed. Species with either a higher exposure to turbines, or a lower life history speed, are likely to be at more risk.

Indirect Risk Index (IRI)

$$\mathrm{IRI} = p/h,$$

where p is, again, the proportion of the population exposed and h is the number of habitats a species uses.

This metric was developed to assess risk caused by the indirect effects of habitat loss and fragmentation from wind facilities. Studies indicate that across many taxa, species with lower niche breadth (i.e., habitat specialists) are more sensitive to habitat loss and fragmentation than habitat generalists (Carrascal et al. 2013; Swihart et al. 2003, 2006; Watling and Donnelly 2007). The number of habitats, *h*, is an index of niche breadth. For IRI, risk is assumed to be greater when either a higher proportion of the population is exposed to turbines or the species uses fewer habitats.

Step 3. Estimating Change in a Species Population Growth Rate

For species with high-risk prioritization scores, the next step estimates the probability of change in the population growth rate. To do this, an estimate of turbine mortality rate (*c*, the annual chance an individual will die from a collision, or barotrauma, with a wind turbine), is combined with either an empirical or modeled estimate of population growth rate in the absence of wind energy facilities. The approach assumes density-independent population growth and that all stage/age classes and sexes are equally likely to die from wind turbines. With these assumptions, simplified population dynamics can be described as:

$$N_{t+1} = \lambda N_t - c\lambda N_t,$$

where N_t is the population size at year *t*, λ is the population growth rate, *c* is the annual chance an individual will die from a collision with a turbine, and $c\lambda N_t$ is the number of individuals killed each year by turbines. A $\lambda < 1$ indicates a declining population.

This can be rewritten as $N_{t+1} = (1 - c)\lambda N_t$, and $N_{t+1}/N_t = (1 - c)\lambda$. If N_{t+1}/N_t is the population growth with wind, λ_w, then, $\lambda_w = (1 - c)\lambda_b$ where λ_b is the population growth rate in the absence of wind. Thus, the change in population growth rate caused by wind energy is simply $\lambda_w = \lambda_b - c\lambda_b$ and $\lambda_b - \lambda_w = c\lambda_b$.

To estimate the predicted change in population growth rate due to fatalities from wind turbines, this approach requires an estimate of the turbine mortality rate and an estimate of population growth rate in the absence of wind energy. Turbine mortality rate, *c*, can be estimated as the number of animals killed by turbines/total population size, while λ_b could be directly observed from monitoring data or estimated as the dominant eigenvalue from a matrix population model (Diffendorfer et al. 2015). The sources of monitoring data could be, for example, the Breeding Bird Survey in the United States and southern Canada. This approach generates a point estimate of change in population growth rate and the uncertainty around the point estimate.

Because uncertainty exists around *c* and λ (actually, estimates of these values, c-hat, and λ-hat), this step generates a distribution—rather than a single value—of the predicted change in population growth rate due to wind energy fatalities. For each distribution of population growth rates, one with fatalities from wind turbines, and one without, the probability of $\lambda < 1$ can be estimated. The difference between the two probabilities represents the estimated change in the probability that the population growth rate is less than 1. So, change in $\text{Prob}(\lambda < 1) = \text{Prob}(\lambda_w < 1) - \text{Prob}(\lambda_b < 1)$.

Step 4. Potential Biological Removal and the Risk Ratio

Potential Biological Removal was developed to manage human-caused deaths to marine mammals (Taylor et al. 2000). It is one of several 'reference point' approaches used to assess, and set limits on, anthropogenic sources of mortality to species (Moore et al. 2013). PBR estimates the number of individuals that could be killed before a population will fall below a size considered sustainable, typically half a population's carrying capacity (Wade 1998). The approach has been extended to issues of harvest in birds (Johnson et al. 2012; Runge et al. 2009) and applied to at least one species in relation to wind energy (Bellebaum et al. 2013).

The PBR is calculated as:

$$\text{PBR} = \text{F}(r_{\text{max}}/2)\text{N}_{\text{min}},$$

where F is a 'recovery factor' set by regulatory agencies or existing law.

'F' adjusts the value of PBR to change the rate of recovery depending on a species' population size or status, or to account for uncertainties in the input parameter estimates and assure the estimated value of PBR is not too high. The 'r_{max}' is the maximum annual population growth rate under ideal conditions. There are a variety of approaches to estimate r_{max} (Slade et al. 1998; Millar and Meyer 2000; Niel and Lebreton 2005; Dillingham et al. 2016). Finally, N_{min} is a lower bound on the total population size. Under the Marine Mammal Protection Act in the U.S., N_{min} is set at the 20th percentile of observed abundance assuming a log-normal distribution. The 20th percentile level was based on results from a computer simulation (Wade 1998), whereas F is set by regulatory agencies at 0.1, 0.5, or 1 depending on population status.

The methodology presented here uses the PBR approach to calculate a risk ratio (RR), which is the number of annual fatalities due to wind energy divided by the PBR (Richard et al. 2011). This represents the proportion of the PBR caused by collisions with wind turbines. Risk from wind energy is low when RR is near zero. At RR equal to 1, fatalities from wind turbines are equal to the PBR. At RR greater than 1, the population is expected to decline below the level considered sustainable due to fatalities from wind energy.

Final Output, Use of the Methodology, and Areas for Future Research

Ultimately, the methodology produces (1) a list of species ranked according to their average risk score calculated in the prioritization step. In addition, for those species further evaluated in Steps 3 and 4, the methodology produces (2) an estimate of the predicted change in probability the population will decline in the face of mortality from wind energy (i.e. the population growth rate will be less than 1), and (3) an estimate of the risk ratio based on PBR. It is also possible to use Steps 2 and 3 to forecast the effects of future growth in wind energy generation on species (Diffendorfer et al. 2015). To do so, the values of variables *c* (used in Step 3, the demographic model), and the number of animals killed annually (used in Step 4, PBR) must be updated based on scenarios of future wind facility development. Table 1 shows an example of output for a case study of six species.

To model future growth in wind energy, the average and 95% upper limit of 11 projections of installed wind energy capacity were used as the medium and high growth scenarios (Diffendorfer et al. 2016). Installed capacity in gigawatts (GW) is current (2014) scenario, 62.3 GW; medium scenario for 2025, 94 GW; high scenario for 2025, 121 GW. Risk-ratio values are presented as an average, and values in parentheses represent the 95% confidence interval. NA is not applicable; PBR is potential biological removal; and λ is the population growth rate.

Three of the species (bird 4, 5 and 6) had relatively low direct risk during the prioritization process and were not further analyzed. Of the remaining three species, bird 1 showed low risk, with predicted changes in $\lambda < 1$ smaller than 0.5% and a risk ratio near zero. Bird 2, however, had a much larger predicted change in population growth and the confidence interval around the risk ratio included 1 in all three scenarios, with the mean risk ratio greater than 1 in the high scenario. Thus, for Bird 2, PBR may be currently exceeded by mortality caused by collisions and is likely to be exceeded if future levels of wind energy development follow our scenarios. Bird 3's results were intermediate in comparison to the bird 1 and bird 2. Small changes in population growth for bird 3 were predicted while mean risk ratios were <1 for all scenarios, and the 95% Confidence intervals overlapped 1 for medium and high scenarios.

What sorts of decision making can the methodology support? We first distinguish between absolute and relative risk. Absolute risk is the actual risk to population status and trend from the impacts of wind energy generation on a species. Relative risk is a comparison of risk level between species. The prioritization component of the methodology is a first attempt at estimating relative risk. Species with higher average risk scores are expected to be at more relative risk than species with lower average risk scores. Furthermore, the approach produces additional information about the possible cause of the risk by identifying which turbine risk metric has the most influence on the risk ranking. For example, a species might rank highly because the Indirect Risk Index is high but the remaining metrics that include collision fatalities are low. In this case, the species may be a habitat

Table 1 Example main output from the methodology for six bird species in the United States

Species	Scenario	Species prioritization		Demographic model	PBR model
		Average rank for direct risk (range 1–9)	Average rank for indirect risk (range 1–9)	Projected increase in percentage of $\lambda < 1$	Risk ratio (95% confidence interval)
Bird 1	Current	6.16	2.46	0.2	0.002 (0.001–0.003)
	Medium	NA	NA	0.3	0.003 (0.001–0.004)
	High	NA	NA	0.4	0.004 (0.001–0.005)
Bird 2	Current	7.05	3.13	18.2	0.570 (0.345–1.568)
	Medium	NA	NA	21.1	0.861 (0.527–2.390)
	High	NA	NA	25.0	1.106 (0.674–3.102)
Bird 3	Current	5.80	4.52	3.1	0.339 (0.129–0.709)
	Medium	NA	NA	4.1	0.512 (0.194–1.084)
	High	NA	NA	6.0	0.650 (0.253–1.397)
Bird 4	Current	4.35	2.97	NA	NA
Bird 5	Current	2.57	7.75	NA	NA
Bird 6	Current	4.08	5.54	NA	NA

specialist whose range overlaps turbines, but is rarely recorded as colliding with turbine blades. If studies show this species is not displaced from areas with turbines, or that breeding success is not affected by facility infrastructure, then impacts to the species may be limited to direct habitat loss caused by the turbine. On the other hand, if FT (proportion of fatalities from turbines) is high and heavily influences a species risk rank, then additional research on population trends and demography may be required.

The approach may also perform well at identifying species at low risk. Species with low average risk scores, very small or negligible estimated changes in population growth, and small risk ratios are less likely to be impacted by fatalities from turbines. In these cases, observed mortality will likely be small relative to population size and productivity.

More species must be included and additional analyses performed to understand how finely the prioritization approach can separate species. For example, given uncertainties in the input data, the methodology may only be able to distinguish five distinct levels of risk, despite ranking hundreds of bird species. Higher uncertainty associated with the input data will likely result in higher uncertainty around the average risk scores, more overlap between species in the distributions of average risk score, and thus lower resolution of relative risk across species. Ultimately, additional research, and simulation of the prioritization approach, is required to address this issue.

Both the change in population growth rate and the risk ratio will produce accurate estimates of absolute risk for a species, but only if the input data are accurate and if the assumptions of the approaches are met. It is suspected that the data quality for at least some parameters might be low for many bird and bat species. This data limitation, coupled with the generalized structure of the population growth and risk ratio approaches, likely means Steps 3 and 4 will produce only approximate estimates of absolute risk. In these cases, it would be prudent to consider the results to be working hypotheses about the impacts of wind energy facilities on a species that require additional research. Such research could include more realistic models that better match a species' life history and behavior in relation to wind turbines, and field studies to better estimate model parameters. Thus, in cases with high uncertainty, the approach may be best used to identify species potentially at risk and to highlight them for additional work.

One approach to validate the population growth rate and risk ratio components is to compare them to alternative approaches for estimating population-level impacts. For example, relatively complex models have been produced for Red Kites (Schaub 2012) and Egyptian vultures (Carrete et al. 2009), linking population growth to fatalities from wind turbines. These species could be evaluated using the methodology described here, and the results from the more simplistic models could be compared to the more complex models previously used.

The assessment methodology is considered a work in progress. The approach has been purposely based on existing methods successfully used in other applications. However, for many species, the methodology is limited in use by missing or poor-quality information on species-specific turbine collision mortality, population size, and demographic rates. We also know little about the distribution, local movements, or long-distance migratory patterns of many species killed by wind turbines, especially bats. This hampers our ability to estimate the proportion of the population exposed to wind turbines.

We currently recommend using this approach to identify species at high and low risk of experiencing population-level impacts of wind energy, and then performing more detailed investigation of the species identified as high risk. However, this recommendation should be investigated with simulations that explore how robust the outputs of the methodology are to parameter uncertainty. In some cases, such as the steps to calculate population growth rate and PBR, much of this uncertainty can be transparently carried forward and expressed as uncertainty in the modelled outputs. Development, use, and refinement of methods such as these will be crucial in efforts to understand and mitigate the risks posed to vertebrate populations by the development of wind and other energy sources.

Acknowledgements We thank V. Bennett, P. Cryan, D. Houseknect, T. Katzner, and M. Runge for reviewing early versions of the methodology. We particularly thank M. Runge for asking us to think harder about the utility of PBR. We are extremely grateful for external review panel members, T. Allison, W. Erickson, A. Hale, and F. Bennet. J. Havens developed nearly all the complex graphics and we appreciate his help. Finally, we thank the organizers of the CWW conference in Berlin and the many individuals who gave us excellent feedback on the methodology.

References

Andow DA, Hilbeck A (2004) Science-based risk assessment for nontarget effects of transgenic crops. Bioscience 54:637–649

Bellebaum J, Korner-Nievergelt F, Dürr T, Mammen U (2013) Wind turbine fatalities approach a level of concern in a raptor population. J Nat Conserv 21:394–400

Beston JA, Diffendorfer JE, Loss S, Johnson DH (2016) Prioritizing avian species for their risk of population-level consequences from wind energy development. PLoS ONE 11:Article e0150813

Carrascal LM, Galván I, Sánchez-Oliver JS, Benayas JMR (2013) Regional distribution patterns predict bird occurrence in Mediterranean cropland afforestations. Ecol Res 29:203–211

Carrete M, Sánchez-Zapata JA, Benítez JR, Lobón M, Donázar JA (2009) Large scale risk-assessment of wind-farms on population viability of a globally endangered long-lived raptor. Biol Conserv 142:2954–2961

Desholm M (2009) Avian sensitivity to mortality; prioritizing migratory bird species for assessment at proposed wind farms. J Environ Manage 90:2672–2679

Diffendorfer JE, Beston JA, Merrill MD, Stanton JC, Corum MD, Loss SR, Thogmartin WE, Johnson DH, Erickson RA, Heist KW (2015) Preliminary methodology to assess the national and regional impact of U.S. wind energy development on birds and bats: U.S. Geological Survey Scientific Investigations Report 2015–5066, p 40. http://dx.doi.org/10.3133/sir20155066

Dillingham DW, Moore JE, Fletcher D, Cortés E, Curtis KA, James KC, and Lewison, RL (2016) Improved estimation of intrinsic growth rmax for long-lived species: integrating matrix models and allometry. Ecol Appl 26:322–333. http://dx.doi.org/10.1890/14-1990

Johnson FA, Walters MAH, Boomer GS (2012) Allowable levels of take for the trade in Nearctic songbirds. Ecol Appl 22:1114–1130

Millar RB, Meyer R (2000) Non-linear state space modelling of fisheries biomass dynamics by using Metropolis-Hastings within-Gibbs sampling. J Roy Stat Soc: Ser C (Appl Stat) 49: 327–342

Moore JE, Curtis KA, Lewison RL, Dillingham PW, Cope JM, Fordham SV, Heppell SS, Pardo SA, Simpfendorfer CA, Tuck GN, Zhou S (2013) Evaluating sustainability of fisheries by catch mortality for marine megafauna; A review of conservation reference points for data-limited populations. Environ Conserv 40:329–344

Niel C, Lebreton JD (2005) Using demographic invariants to detect overhear vested bird populations from incomplete data. Conserv Biol 19:826–835

Parker IM, Simberloff D, Lonsdale WM, Goodell K, Wonham M, Kareiva PM, Williamson MH, Von Holle B, Moyle PB, Byers JE, Goldwasser L (1999) Impact: toward a framework for understanding the ecological effects of invaders. Biol Invasions 1:3–19

Péron G, Hines JE, Nichols JD, Kendall WL, Peters KA, Mizrahi DS (2013) Estimation of bird and bat mortality at wind-power farms with super population models. J Appl Ecol 50:902–911

Regan HM, Hierl LA, Franklin J, Deutschman DH, Schmalbach HL, Winchell CS, Johnson BS (2008) Species prioritization for monitoring and management in regional multiple species conservation plans. Divers Distrib 14:462–471

Richard Y, Edward R, Filippi A, Filippi D (2011) Assessment of the risk to seabird populations from New Zealand commercial fisheries. Final Research Report for New Zealand Ministry of Fisheries projects IPA2009/19 and IPA2009/20 (Unpublished report held by the Ministry of Fisheries, Wellington), p 66. Available via http://www.dragonfly.co.nz/publications/pdf/Richardetal_2011a_IPA2009-20.pdf. Accessed on 28 Sept 2015

Runge MC, Sauer JR, Avery ML, Blackwell BF, Koneff MD (2009) Assessing allowable take of migratory birds. J Wildlife Manage 73:556–565

Schaub M (2012) Spatial distribution of wind turbines is crucial for the survival of red kite populations. Biol Conserv 155:111–118

Schuster E, Bulling L, Koppel J (2015) Consolidating the state of knowledge: a syntopic review of wind energy's wildlife effects. Environ Manage 56:300–331

Slade NA, Gomulkiewicz R, Alexander HM (1998) Alternatives to Robinson and Redford's method of assessing overharvest from incomplete demographic data. Conserv Biol 12:148–155

Stahl JT, Oli MK (2006) Relative importance of avian life-history variables to population growth rate. Ecol Model 198:183–194

Swihart RK, Gehring TM, Kolozsvary MB, Nupp TE (2003) Responses of "resistant" vertebrates to habitat loss and fragmentation: the importance of niche breadth and range boundaries. Divers Distrib 9:1–18

Swihart RK, Lusk JJ, Duchamp JE, Rizkalla CE, Moore JE (2006) The roles of landscape context, niche breadth, and range boundaries in predicting species responses to habitat alteration. Divers Distrib 12:277–287

Taylor BL, Wade PR, De Master DP, Barlow J (2000) Incorporating uncertainty into management models for marine mammals. Conserv Biol 14:1243–1252

Wade PR (1998) Calculating limits to the allowable human-caused mortality of cetaceans and pinnipeds. Mar Mammal Sci 14:1–37

Watling JI, Donnelly MA (2007) Multivariate correlates of extinction proneness in a naturally fragmented landscape. Divers Distrib 13:372–378

Wilson JC, Elliott M, Cutts ND, Mander L, Mendão V, Perez-Dominguez R, Phelps A (2010) Coastal and offshore wind energy generation: is it environmentally benign? Energies 3:1383–1422

Part III
Landscape Features and Gradients

Bat Activity at Nacelle Height Over Forest

Hendrik Reers, Stefanie Hartmann, Johanna Hurst and Robert Brinkmann

Abstract The number of wind power facilities (wind farms) has rapidly increased in Germany, with a number of these constructed in forested areas. As most bat species use forests to forage and roost, concerns have been raised in relation to the potentially higher collision risk with wind turbines in forests than in open landscapes. In addition, the standard curtailment algorithms used in open landscapes might not be appropriate in forests. An ample acoustic dataset derived from 193 nacelle height surveys of 130 individual turbines was used to investigate whether bat activity, phenology or species composition differ between forests and open landscapes. The data showed no significant differences between bats in forests and open landscape habitats, but revealed strong regional differences. Overall bat activity increases towards the east of Germany, which is mirrored by an increase of the dominant group of Nyctaloids, whereas the activity of common pipistrelles increases towards the south. These findings suggest that acoustic surveys must be interpreted on a regional and species-specific level. In summary, wind farms within forested areas do not seem not to inherently show higher bat activity at nacelle height, suggesting no increased collision risk for bats in general. However, future studies assessing bat activity at the lowest point of the rotor instead of at nacelle height are urgently needed, as well as studies that include additional variables such as proximity to bat roosts or the age of a forest.

Keywords Bat activity · Forests · Nacelle height · Wind farms · Germany · Collision risk · Acoustic surveys

H. Reers (✉) · S. Hartmann · J. Hurst · R. Brinkmann
Freiburg Institute for Applied Animal Ecology GmbH,
Dunantstraße 9, 79110 Freiburg, Germany
e-mail: reers@frinat.de

J. Köppel (ed.), *Wind Energy and Wildlife Interactions*,
DOI 10.1007/978-3-319-51272-3_5

Introduction

Renewable Energy and Increasing Wind Farms in Forests

The global trend in power being generated from renewable energy sources has led to increasing demands for wind power on a world-wide scale (Wang and Wang 2015; Valença and Bernard 2015; Bernard et al. 2014). In Germany, a new national energy concept was adopted in 2010. Therein, a target was defined: the renewable energy share of total electricity consumption should be increased to at least 35% by 2020 and 80% by 2050 (BMWI/BMU 2010). This target is well underway, in 2015 already 32.6% of total electricity consumption was produced from renewable energy sources, of which onshore wind farms provide 40.5% of the national renewable electricity production (Zentrum für Sonnenenergie- und Wasserstoff-Forschung Baden-Württemberg (ZSW) 2016). As a consequence, large numbers of wind power facilities are being installed all over Germany. While beneficial from an energy perspective, this can have severe consequences for wildlife, especially birds and bats (Voigt et al. 2015). Often, this leads to a typical "green-vs-green-dilemma" where sustainability and wildlife protection may not be compatible in all aspects (Köppel et al. 2014). The increasing demand for wind power requires decision-making despite the uncertainties that are involved.

So far, most wind farms have been constructed in northern Germany where wind yields are intrinsically higher due to geographical and climatic conditions. However, with an increasing demand for wind energy, more and more wind farms are being installed in middle and southern Germany. As the forested mountain ranges rank high among the sites due to a high wind resource, this leads to a larger proportion of wind farm developments in forested areas. However, information on the impacts of wind farms in forested areas is scarce compared to open landscapes. As forests play an important role for birds and bats, many experts claim that forested areas should be excluded completely from future wind farm planning (Richarz 2014) or claim that a minimum distance of 200 meters to forests should be preserved (Rodrigues et al. 2014). Such expert-opinions are so far mainly based on the precautionary principle and single case studies. For example, a study in the southern black forest found 35 dead bats of three species under 16 wind turbines in 2005 (Brinkmann et al. 2006), suggesting that collision risk with turbines might be high for bats in general.

The difficulty in predicting the impacts of wind farms in forests on bats is based mainly on two underlying uncertainties: the exact interaction of bats and wind farm infrastructure, and the role of forests for bat populations in general. Therefore, discussed below are both the knowledge and knowledge gaps with respect to these two topics.

Impacts of Wind Energy on Bats

The potential impacts of wind turbines on bats have been debated (Brinkmann et al. 1996; Rahmel et al. 1999) and studied when the first bat carcasses were found in Germany (Vierhaus 2000). However, it was not until wind farm construction increased significantly that awareness of the full potential impact of wind farms on bats arose, stimulating both initiatives in research and conservation (Brinkmann et al. 2011b; Arnett et al. 2008; Kunz et al. 2007; Behr et al. 2007). Many studies have provided evidence that large numbers of bats have died due to collisions with wind turbines world-wide (Barclay et al. 2007; Rydell et al. 2010a; Piorkowski and Timothy 2010; Lehnert et al. 2014; Baerwald et al. 2014; Barros et al. 2015).

A key issue that must be understood is the proximate and ultimate causes of death at wind turbines. Proximately, bats may die due to direct collision and subsequent fractures (Voigt et al. 2015; Brinkmann et al. 2006) or indirectly because of barotrauma (Baerwald et al. 2008). Ultimately, bats could be killed not only because wind farms are constructed within their natural foraging area or migration route, but also because wind farms could even attract bats from surrounding areas into the wind farms. Reasons for such an attraction could be that insects (a main source of food for bats) are attracted to the turbines (Rydell et al. 2010b), or that bats tend to swarm at tall structures (Cryan et al. 2014). This behavior seems to play a major role for migrating bat species (Jameson and Willis 2014). In summary, bats are probably the group of vertebrates which is most severely affected by wind farms (Voigt et al. 2015).

The manifold effects which lead to collisions between bats and turbines are increased by the bats' special biology and life-history: bats, in contrast to other taxonomic vertebrate groups such as similar sized passerine birds, are characterized by a long life-span and low reproductive rates (Kunz and Fenton 2003; Bernotat and Dierschke 2015). The death of a bat individual thus has a much larger effect on the local population dynamics than the death of a single passerine. Collisions at wind farms therefore impact population dynamics both through an increased mortality and a decreased reproductive rate. Despite their urgent necessity, studies on the impact of wind farms on a bat population level are still lacking (Voigt et al. 2015).

The occurrence of bats at a wind turbine can depend on a variety of factors, such as weather or location, but also on species characteristics (Schuster et al. 2015; Cryan and Brown 2007; Martin 2015; Leopold et al. 2014). Species hunting at greater height and in open air space are of highest concern (Zahn et al. 2014): In Germany, the largest proportion of bat fatalities at wind turbines stems from noctule *(Nyctalus noctula)*, both from local populations as well as migrating individuals (Lehnert et al. 2014; Dürr 2015; Voigt et al. 2012; Niermann et al. 2011a). The species with the second highest share is Nathusius's bat *(Pipistrellus nathusii)*, with most mortalities occurring during migration (Voigt et al. 2012; Dürr 2015; Niermann et al. 2011a). This underlines the special importance of Germany as a migration corridor due to its central geographic position. Germany therefore has an important responsibility for the preservation not only of local but also foreign bat populations (Lehnert et al. 2014; Voigt et al. 2015). National and international migration routes are of high importance and many of these cross wind farms (Cryan and Brown 2007; Baerwald et al. 2014). However, not only migrating species are concerned: the species with the third highest share in collision fatalities is the common pipistrelle (*P. pipistrellus)*, where individuals come from local populations (Dürr 2015; Voigt et al. 2012; Dietz et al. 2007; Niermann et al. 2011a). Other species that often collide with wind turbines are Leisler's bat (*N. leisleri)* and parti-coloured bat (*Vespertilio murinus*) (Dürr 2015).

Studies have demonstrated that the number of bat fatalities is strongly correlated with acoustic bat activity recorded at nacelle height, and that bat activity is lower when winds become stronger and temperatures become lower (Brinkmann et al. 2011b). Generally, species composition of bat fatalities can differ to some degree between different regions in Germany (Niermann et al. 2011b). If wind farms are increasingly built in forests, it is of crucial importance to understand whether and which of the above phenomena and correlations could be mediated by a forested landscape.

Special Habitat Characteristics of Forests

Forests play an important role for most bat species, as a wide range of bat species depends on forests to forage and roost (Lacki et al. 2007; Barclay and Kurta 2007). Of all 25 bat species recorded in Germany, 22 species use tree roosts, and all species use forests during hunting activities (Hurst et al. 2015; Dietz and Kiefer 2014). Especially rare and sensitive species rely on undisturbed old-growth forest, such as Bechstein's bat (*Myotis bechsteinii*) (Kerth et al. 2002; Dietz and Pir 2011). Such forests can also lead to mass concentrations of more common species, for example 600 noctules were observed to gather for hunting in a Polish forest (Polakowski et al. 2014).

Due to the importance of forests for bats, it is often assumed that bat activity and therefore collision risk is higher at forest sites than in open landscapes. Furthermore, the presence of a wider diversity of bat species in forests could also lead to a different species composition above the forest canopy. In addition, bat phenology could differ from that in open landscapes, e.g. because some forest-specific insects could display different behaviour at wind turbines in forests compared to open landscape. However, studies assessing the complex interaction of forest, wind farms and bats are still scarce (Niermann et al. 2012; Segers and Broders 2014). Some studies indicate a correlation of bat activity at turbine height with proximity to forests (Niermann et al. 2011c; Rydell et al. 2010a), whereas other studies found the opposite (Johnson et al. 2004). A generalization is further complicated because activity patterns and species distribution vary according to forest stratification and recording height (Staton and Poulton 2012; Müller et al. 2013).

Constructing wind farms in forests, which are special habitats for many bat species, generates further conservation issues which are less important in open landscapes. It can result in a loss, habitat deterioration or fragmentation of forested habitats. The clearance of forest patches at turbine construction sites and access roads thus leads to a loss and fragmentation of bat habitat (Farneda et al. 2015). The diversity of bats in forests decreases with an increasing degree of fragmentation (Lesinski et al. 2007). Recent studies show that habitat changes in forests due to wind farm construction can have species-specific effects on bats (Segers and Broders 2014; Morris et al. 2010). Studies that assess the interaction of bats, forests and wind farms on a larger scale and across a variety of bat species are therefore substantially required (BfN 2011; Wang and Wang 2015).

Implementation of European Law in Germany

The Birds and Habitats Directives are the cornerstones of the EU's biodiversity policy. According to the European Habitats directive, all 27 European member states have agreed to work together to conserve Europe's most valuable species and habitats across their entire natural range within the EU, irrespective of political or administrative boundaries. Article 12 and 13 of the Habitats Directive specify that it is prohibited for any member state to deliberately kill or disturb species in Annex IV. Since all European bats are protected by international and national legislation, any intentional killing is forbidden by law. Therefore, avoidance, or at least reduction to a minimum, of bat mortality by wind farms, is not only a priority for bat conservation, but also a legal obligation in Europe (Rodrigues et al. 2014). This implies an obligation to avoid or minimize bat collisions at wind turbines.

Researchers have provided some baseline information about bats and wind farms according to which avoidance and minimization measures have been developed. Guideline documents (European Commission 2010; Rodrigues et al. 2014) raise awareness amongst developers and planning agencies concerning the need to consider bats and bat conservation throughout the planning process of wind farms. These also prepare local and national authorities for the relevant steps in this process. Standardized impact assessments, such as post-construction monitoring, enable the design of a targeted avoidance and mitigation program. This program may include project abandonment, re-siting of the proposed turbines, site-specific use of blade feathering, higher turbine cut-in wind speeds and shutting down turbines temporarily to avoid or reduce bat mortality (Rodrigues et al. 2014). Key measures within these assessments are surveys of bat activity prior to and after wind farm construction (Hurst et al. 2015; Brinkmann et al. 2011b; Arnett et al. 2009; Willmott et al. 2015; Rodrigues et al. 2014).

Bat activity is heavily influenced by weather conditions, it decreases strongly with wind speeds above 6 m/s and temperatures below 10 °C (Brinkmann et al. 2011a). A model derived from a joint analysis of carcass searches, acoustic data and weather conditions now allows ecological consultants to predict the collision risk at specific wind farm sites. This can be used to develop turbine-specific curtailment algorithms that reduce bat mortality rates with a minimal loss of energy production

(Korner-Nievergelt et al. 2013). Following a national research project, site and weather-specific feathering (curtailment) strategies are now standard practice in Germany (Behr et al. 2015; Brinkmann et al. 2011a). Since 2008, the monitoring of bats at nacelle height has increased. By 2014, approximately 24,000 on-shore wind turbines have been installed in Germany (Berkhout et al. 2014), and the accompanying investigations have created an impressive dataset over the years. However, doubts remain whether these mitigation strategies derived from Brinkmann et al. (2011b) are effective in forests, as the underlying data used to develop the curtailment strategies stems mainly from wind farms in open landscapes (Behr et al. 2011b).

As part of this study, an extensive nation-wide dataset from nacelle surveys was gathered and analyzed in order to address the following questions:

1. Do bat species' compositions differ between forests and open landscapes?
2. Does bat activity and phenology differ between forests and open landscapes?
3. Does bat activity and composition differ between different regions in Germany?

Methods

Data Set

In total, data from 193 seasonal surveys (turbine-years) of 130 individual turbines were obtained (see Table 1).

Acoustic detectors, installed at the nacelle of the wind turbines, collected data during bat surveys to develop bat mitigation measures. The surveys were mostly run from April to October, with some exceptions whenever federal guidelines required different time periods or the detectors failed due to technical issues. Recording period spans across the years 2008 to 2014, with most survey data collected during the years 2012 to 2014. From 193 surveys, 106 surveys were conducted at wind turbines located in open landscapes and 87 facilities located in forests. Altogether, around 193,000 recordings were included in the analysis, of which around 93,000 have been recorded over open landscapes and around 100,000 over forests.

Most surveys were conducted with batcorders (n = 115) (ecoObs GmbH), followed by Anabat (n = 65) (Titley scientific) and Avisoft (n = 13) (Avisoft Bioacoustics). To analyse regional differences, data from these surveys was grouped into four regions. Those regions were derived from different Federal states as follows:

Table 1 Number of nacelle surveys per region, year and habitat type

	2008	2010	2011	2012	2013	2014	Open landscape	Forest	Total
North	0	0	0	7	14	13	34	0	34
East	0	0	15	7	12	12	27	19	46
South	0	0	0	3	5	8	8	8	16
West	6	1	3	31	45	11	37	60	97
Total	6	1	18	48	76	44	106	87	193

North: Lower Saxony (32 turbine-years) and North Rhine-Westphalia (2),
East: Mecklenburg-Vorpommern (2), Saxony-Anhalt (2), Brandenburg (38) and Saxony (4),
South: Bavaria (12) and Baden-Württemberg (4), and
West: Hesse (2), Rhineland-Palatinate (94) and Saarland (1).

As rotor radiuses have increased over time (>50 m for recent facilities), which may exceed recording capacities of the acoustic instruments (Behr et al. 2011a), this data does not cover the whole space impacted by the rotor blades. Further, because different bat species call at different intensities (Barataud 2015) not all bat calls across all species would have been fully captured equally. Thus, the results are to be interpreted as relative, not absolute activity data.

Data Processing

Recordings were identified to different levels for different taxa. Recordings belonging to genera that are hard to distinguish acoustically were lumped into groups. The Nyctaloids group consists of recordings that belong to bats from the genera *Eptesicus, Nyctalus* or *Vespertilio*. Recordings from species of the genus *Myotis* were lumped together. Recordings from the species of the genus *Plecotus* were similarly lumped into a *Plecotus*-group. Recordings from the genus *Pipistrellus* that were not identified to a species level were lumped into a group called Pipistrelloids. 144 recordings from the soprano bat *(P. pygmaeus)* were also placed in this Pipistrelloid group. Recordings from the Nathusius's pipistrelle and the common pipistrelle were considered at species level.

To level out differences between detector types, detector settings and recording sensitivities, we used presence-absence data for 10 min intervals. This was necessary to allow for a comparison between different detector types and detector settings, taking into consideration that information on absolute activity levels within 10 min intervals was lost. However as mentioned above, because the data was acquired through automated acoustic surveys, the number of calls cannot be interpreted absolutely anyway. For example, it is not possible to differentiate whether 20 recordings within 10 min stem from one hunting individual or 20 different individuals passing by. The 193,000 recordings translate into around 25,000 and 19,000 10 min intervals with activity over open landscape and forest, respectively.

Linear Mixed Effects Model

First, to investigate the impact of forest cover on bat activity using location as a gradual variable, we used a mixed effects model including forest cover in a 500 m radius in percent, latitude in degrees and longitude in degrees as covariates (fixed effects). Second, to investigate the impact of forest cover on bat activity for each region separately, we used a mixed effects model for each region including forest cover in a 500 m radius in percent. Initially, closest distance to large water bodies (more than 20 m wide ponds or more than 10 m wide streams) was included into the model, but was omitted in the final models due to insignificance. To account for non-independent data we included turbine identity, wind farm identity, survey year and detector type as random effects in all models. Prior to the analysis, the bat activity (as a percentage of 10 min intervals of bat activity, ranging from 0 to 100) was increased by 1 and log-transformed to approach normality of the residuals. The analyses were done by using the R-function *lmer* from the package *lme4* (Bates et al. 2015) using R.3.1.1 (R Core Team 2014).

Results

Descriptive Statistics

Species Composition

Between 50 and 60% of all 10 min intervals with bat activity could be attributed to the group of Nyctaloids, followed by Nathusius's bat and the common pipistrelle (Table 2).

Pipistrelloids and recordings from unidentified bats contributed to up to 8% of the 10 min intervals with bat activity. Interestingly, species that are strongly associated with forest as their feeding and/or roosting habitat (i.e. *Myotis*-group and *Plecotus*-group) were recorded in only a very small number of 10 min intervals. Only 22 10 min intervals with bat activity for the *Myotis*-group and only 37 for the *Plecotus*-group were registered during the entire surveys.

Table 2 Number of 10 min-intervals with bat activity for the different species and habitat types open landscape (N = 106 turbine-years) and forest (N = 87 turbine-years)

	Open habitat number of 10 min-intervals	Percentage	**Forest** number of 10 min-intervals	Percentage
Myotis group	15	0.06	7	0.04
Nyctaloids	15,555	62.80	9,262	48.93
Nathusius's bat	3,341	13.49	2,220	11.73
Common pipistrelle	4,390	17.72	5,060	26.73
Pipistrelloids	856	3.46	1,548	8.18
Plecotus group	7	0.03	30	0.16
Spec.	605	2.44	801	4.23
Total	24,769		18,928	

While looking at all data lumped together firstly, the following statistics also focus on the Nyctaloids, Nathusius's bat and common pipistrelle on a species group or a species level.

Overall Bat Activity Levels

In both habitats Nyctaloids contributed most to overall bat activity, while the Nathusius's bat appears to be recorded more often in open landscapes than in forests (Fig. 1).

For the common pipistrelle there appear to be more 10 min intervals with activity over forests than over open landscape. When investigating the species composition across the different regions, Nyctaloids contributed most to overall bat activity across all regions, but their contribution was more pronounced in the east and least in the west (Fig. 1). Similar differences can be found for Nathusius's bat and common pipistrelles. Whereas Nathusius's bats showed a higher presence than common pipistrelles in the north and east, common pipistrelles had a higher presence than Nathusius's bats in the west and south. These results give first indications towards regional differences in overall bat activity and species composition, with a lower effect of habitat on bat activity and species composition. However, since the data set consists of more wind farms in open landscape for the regions north and east and more wind farms in forests in the region west, the effects of habitat and region are intermingled. Therefore, the following mixed effects models address the statistical tests of these intermingled effects.

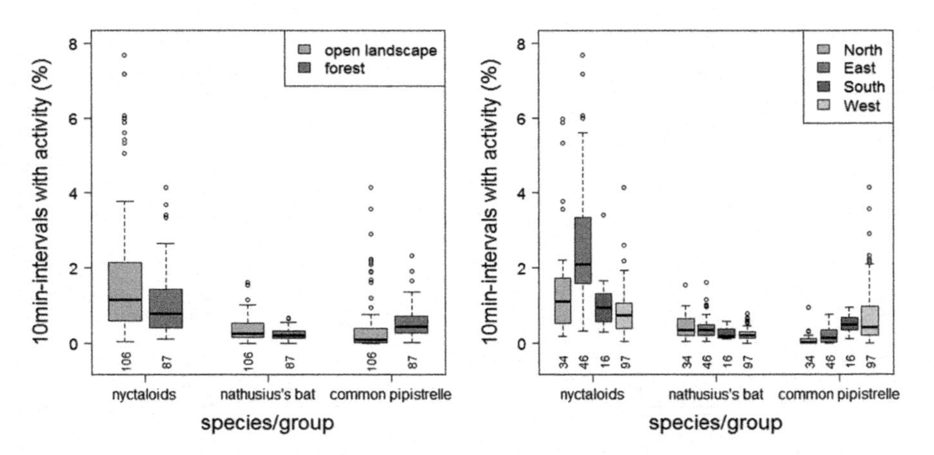

Fig. 1 Bat activity measured in percent of 10 min intervals for Nyctaloids, Nathusius's bat and common pipistrelle for the entire recording time period, depending on habitat and region. The *number* below the *boxplots* indicates the number of turbine-years

Annual Phenology of Bat Activity

Phenology patterns for all groups and species were very similar in open landscapes and forests. Activity in spring and autumn was very low with <1% activity in all recorded 10 min intervals. In both habitats and for all groups and species bat activity peaked in late summer and early autumn with up to 5% of all recorded 10 min intervals showing bat activity (Fig. 2).

In general, common pipistrelle activity was highest in July in both habitats, with a higher variance in open landscapes than in forests. Activity levels of Nathusius's bats were generally lower and later in the year than activity peaks of Nyctaloids and common pipistrelles. Both habitat types showed quite large variances between

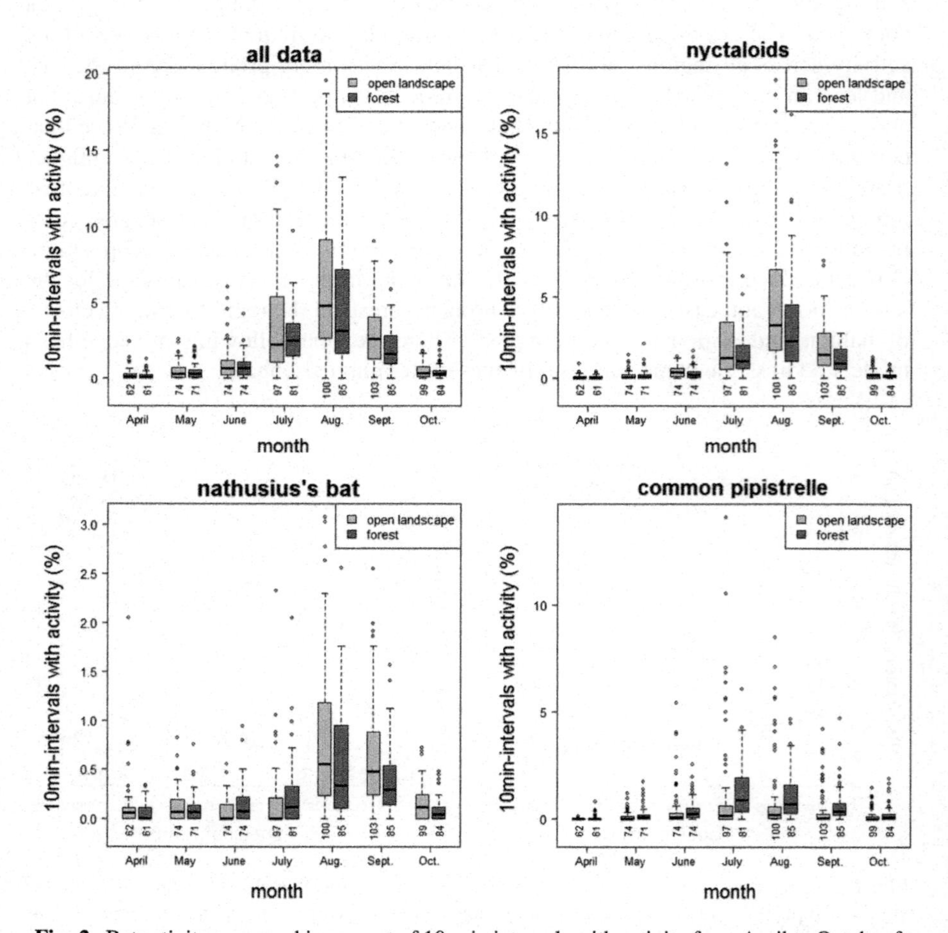

Fig. 2 Bat activity measured in percent of 10 min intervals with activity from April to October for all bat data, Nyctaloids, Nathusius's bat and common pipistrelle in open landscape and forests. The *number* below the *boxplots* indicates the number of turbine-years

different wind farms, but did not show different phenological patterns for any species in the two habitat types.

Annual phenology was also similar across regions, where highest activity was reached in all regions in late summer for all groups and species (Fig. 3).

However, the highest values for southern regions occurred in September, whereas for all other regions highest activity levels occurred in August. The activity levels for the Nyctaloids and the common pipistrelle also demonstrated the differences in species composition across regions. Highest overall Nyctaloid activity levels were found in the east, whereas the common pipistrelle showed highest activity levels in the south.

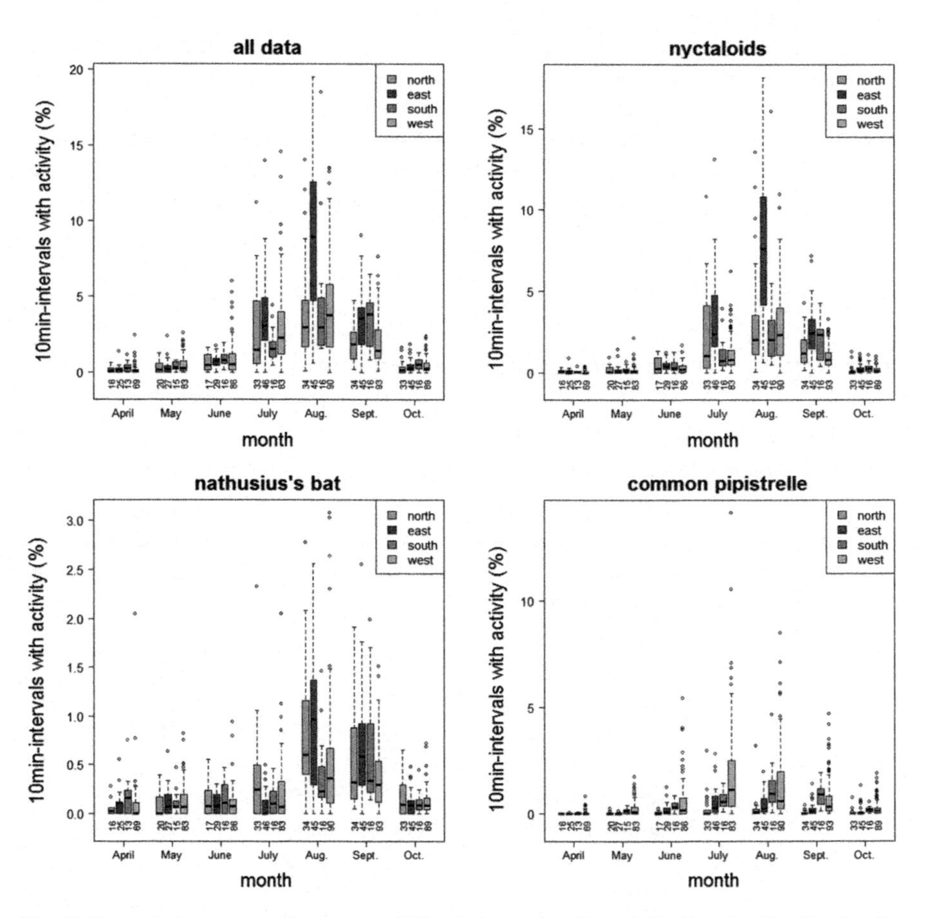

Fig. 3 Bat activity measured in percent of 10 min intervals with activity from April to October for all bat data, Nyctaloids, Nathusius' bat and common pipistrelle across all four regions. The *number* below the *boxplots* indicates the number of turbine-years

Linear Mixed Effects Model

The linear mixed effects model found no significant differences in overall bat activity between forests and open landscapes (Table 3).

Similarly, neither activity levels for Nyctaloids, Nathusius's bats nor common pipistrelle differed between forests and open landscapes (Table 3). The only factor that correlated with both overall bat activity as well as activity levels of specific species groups were longitude or latitude. Overall bat activity increased towards the east, which is mirrored by an increase in Nyctaloid activity towards the east (Table 3). In contrast, no east-west gradient was found for Nathusius's bats and common pipistrelles. In common pipistrelles, however, activity significantly increased towards the south. Activity levels of the Nathusius's bat were not correlated with geography at all.

When linear mixed effects models were used for each region separately, only assessing the effect of forest cover, there was a significant effect of higher common pipistrelle activity recorded over forests than over open landscapes within the west region, from which the largest data set was available (linear mixed effects model: coefficient: 0,00288; standard error: 0,00077; t-value = 3745; P-value = 0001).

Discussion

Bat Activity Over Open Landscape and Forests

The study revealed only small differences between bat activity at nacelle height over forest and open landscapes. Species composition was very similar between the

Table 3 Estimated effect sizes of forest cover in a 500 m radius, latitude and longitude on bat activity (logarithm of percentage 10 min intervals with activity increased by 1) for all bats, Nyctaloids, Nathusius's bat and common pipistrelle

	Variable	Estimate	Std. error	t-value	P-value
All data	Forest in 500 m (%)	−0.001	0.001	−1.362	0.177
	Latitude (°)	−0.040	0.033	−1.242	0.219
	Longitude (°)	**0.071**	**0.020**	**3.476**	**0.001**
Nyctaloids	Forest in 500 m (%)	−0.002	0.001	−1.921	0.058
	Latitude (°)	−0.002	0.030	−0.076	0.939
	Longitude (°)	**0.096**	**0.019**	**5.094**	**<0.0001**
Nathusius's bat	Forest in 500 m (%)	−0.001	0.000	−1.716	0.089
	Latitude (°)	0.015	0.013	1.158	0.251
	Longitude (°)	0.007	0.008	0.902	0.371
Common pipistrelle	Forest in 500 m (%)	0.000	0.001	−0.074	0.941
	Latitude (°)	**−0.009**	**0.024**	**−3.693**	**<0.001**
	Longitude (°)	−0.002	0.016	−1.420	0.162

In the model, we included turbine, wind park, year and detector type as random factors. Bold letters indicate significant effects as assessed by a t-test of the nullhypothesis that the effect is zero

two habitat types. Nyctaloids had a higher level of bat activity in both forest and open landscapes, followed by common pipistrelles and then Nathusius's bats. However, as both noctule as well as Leisler's bat rank among the bat species with the highest call intensity levels, they can be detected across a much larger distance (80–100 m) than Pipistrelloids (30 meters) (Barataud 2015; Rodrigues et al. 2014), and might thus be overrepresented in the dataset. Typical forest bat species of the genera *Myotis* and *Plecotus* have been registered extremely rarely despite their common occurrence at the forest floor, as revealed by acoustic surveys and mist-netting (own data). However, a medium to low call intensity (Barataud 2015) might have led to an underrepresentation in the data. This study shows that typical forest species are equally rare at nacelle height (mean height 140 m in this data set, thus a minimum of 100 m above the forest canopy) in forests as in open landscapes.

Given that the rotor radius and thus the lowest turbine outreach are increasing in modern wind farms and approaching the forest canopy, future acoustic surveys at the lowest turbine outreach will be furthermore adequate to model collision risk than nacelle height recordings. Studies which employ simultaneous acoustic surveys at forest floor, at the lowest turbine outreach and at nacelle height are therefore important to better assess potential differences in activity and subsequent collision risks over forests versus over open landscapes (Hurst et al. 2015).

As this dataset was based on existing acoustic surveys, data sampling was not equally distributed across regions (see Table 1). The inferences derived from this study thus warrant some caution. Despite pronounced similarities in bat activity between forests and open landscapes, overall variance within each habitat type was very high. This suggests that some variables, which could not be assessed in this study, could have a strong influence on bat activity at nacelle height. Probably, the proximity to bat roosts could explain differences in bat activity levels (Ferreira et al. 2015), as could proximity to important foraging sites, such as mass concentrations of insects (McCracken et al. 2008) or sites where major migration routes concentrate. Another important factor could be the age and type of forest around the wind farm (Niermann et al. 2011c). Future studies should therefore aim to include a variety of environmental factors, as they might influence bat composition and abundance (Humes et al. 1999; Meschede and Heller 2000; Bach et al. 2012; Müller et al. 2013).

Regional Differences Versus Differences in Forest Cover

A comparison of differences in bat activity between forests and open landscapes revealed that differences between regions were much more pronounced. The differences that emerge concern both bat activity as well as species composition. Overall activity levels are much higher in the east, which is probably due to the dominant group of Nyctaloids, as they have highest activity levels in the east. Higher Nyctaloid activity levels in the east are attributed to the noctules, because the nursery roosts of this species concentrate in the north-east of Germany

(Dietz et al. 2007). Regional differences are similarly pronounced for the common pipistrelle where activity levels decrease significantly towards the north. Interestingly the common pipistrelle was the only species or group to be affected by the forest cover surrounding the wind turbines and this only in the west region. The locations of the turbines from this region varied very strongly in their surrounding habitat as such that most turbines were placed either in agricultural areas with no forests for several kilometers or in densely forested hillsides. This suggests that forests can in fact have a direct or indirect effect on bat activity in nacelle height, however, this effect might be regional and limited to specific species.

The regional differences in species composition and activity levels confirm the findings of a past carcass search study (Niermann et al. 2011b). Remarkably, the Nathusius's bat ranks among the top bat fatality species across all regions, despite their small percentage in acoustic surveys. This observation is in line with Brinkmann et al. (2011b), where a disproportionally large amount of Nathusius's bats were found dead compared to their acoustic occurrence at wind farms.

Consequences for Future Wind Park Planning

The study found hardly any differences in bat species composition or activity at nacelle height between forests and open landscapes. Thus, forested areas must not per se be excluded from future wind farms development. However, suitability of sites might differ depending on forest type, age or other environmental factors, which were not assessed in this study. Strong regional differences in species composition, as revealed by this study, suggest that an interpretation of bat activity must occur within the frame of the specific region where acoustic surveys took place. Based on this study's findings, it is advocated that the curtailment algorithms developed (Behr et al. 2011c; Korner-Nievergelt et al. 2013) can be applied to wind farms in forests. This should lead to a similarly effective reduction of collision risk as in open landscapes. Further, the strong geographic differences in activity levels and species composition stress the importance of using regionally adapted algorithms, currently under development (Behr, personal communication).

Open Questions and Future Research Needs

The results strengthen the value of a scientific collection and analysis of large, standardized datasets. This study reviews the extensive dataset of 193 individual, standardized acoustic surveys stretching across a large spatial scale, thus allowing a simultaneous assessment of regional and environmental effects. Given that acoustic surveys are recommended or obliged for most wind farms in Germany, the size of the dataset will rapidly increase in the years to come. To fully tap the potential of such a dataset, it is recommended that the existing dataset continues to be maintained within the framework of a national-wide database, and to regularly evaluate this dataset scientifically. This study shows that already a relatively small proportion of the existing acoustic surveys can help to unravel some of the unknowns concerning wind farms and bats. Furthermore, the knowledge gained through such an investigation will increase tremendously with its sample size.

Future investigations for example could address the stratification of bat activity from forest floor to the uppermost turbine outreach. As rotor radiuses have increased over time, acoustic surveys from nacelle height capture less of the entire turbine radius, which severely limits their extrapolation concerning collision risk. Especially in wind farms of lower nacelle height combined with a large rotor radius, the lowest turbine reach can affect the forest canopy area, where bat activity is known to be higher than at nacelle height (Niermann et al. 2011c).

What further hinders the development of optimized mitigation measures is the lack of knowledge on the underlying reasons for the presence of bats at nacelle height. Basic research should therefore address the different hypotheses of why bats occur at different heights and how this phenomenon depends on weather conditions, time of the year or species-specific characteristics. A differentiation between the different hypotheses that are currently employed (social reasons, hunting behavior, etc.) would allow a better prediction and extrapolation across different sites. Furthermore, the large variance we find in this study points to the importance of some variables that so far remain unstudied. A larger dataset could allow for an increase in the number of investigated variables, for example: studying habitat quality in much finer detail than the forest cover in the surrounding area such as the type of forest or proximity to key structures such as roosts. The better we understand the complex interactions between meteorology, regional and site-specific characteristics, and the bat species itself, the more effective future mitigation strategies will become.

To further understand whether and to what extent habitat quality may decrease due to the proximity of wind farms, it is recommended that long-term studies be conducted across Germany. These studies should measure bat activity and behavior before and after wind farm construction. In this study, we have assessed activity at the nacelle as a proxy for collision risk only. However, it is known from other vertebrate groups such as birds that individuals can be displaced from their home-range in order to avoid wind farms (Gonzalez and Ena 2011). This suggests that disturbances or fragmentation effects due to wind farm construction might have negative influences on wildlife, including bats. Future studies could therefore assess whether bat activity and home-range use differ as a response to the construction of a wind farm and whether a subsequent decrease in habitat quality must be taken into account during the planning process.

Conclusions

Very small differences between forest and open landscape in terms of species composition, level of bat activity and phenology of bat activity were found in this study. Regional differences in bat activity and species composition outweigh the effect of differences in forest cover in the immediate surroundings of a wind turbine. This result therefore suggests that mitigation measures developed primarily on data from wind turbines in open landscape are also applicable to wind turbines in forests, as wind speed and temperature still seem the dominant factors determining bat activity. This study highlights the value of a compilation and analysis of a comprehensive central dataset, as the large sample size of the dataset provides an opportunity to address more complex questions than single case studies. The dataset generated through this study should thus be extended and be made available for future studies.

Acknowledgements We would like to thank all wind energy companies, especially juwi AG, and all colleagues from consulting offices for their support in contributing data to this project. The Federal Ministry for Economy and Energy commissioned this research project (FKZ 03MAP264).

References

Arnett E, Schirmacher M, Huso M, Hayes J (2009) Effectiveness of changing wind turbine cut-in speed to reduce bat fatalities at wind facilities. Bat and wind energy cooperative/Pennsylvania Game Commission

Arnett EB, Brown WK, Erickson WP, Fiedler JK, Hamilton BL, Henry TH, Jain A, Johnson GD, Kerns J, Koford RR, Nicholson CP, O'Connell TJ, Piorkowski MD, Tankersley RD (2008) Patterns of bat fatalities at wind energy facilities in North America. J Wildl Manag 72(1):61–78

Bach L, Bach P, Tillmann M, Zucchi H (2012) Fledermausaktivität in verschiedenen Straten eines Buchenwaldes in Nordwestdeutschland und Konsequenzen für Windenergieplanungen. Naturschutz und Biologische Vielfalt 128:147–158

Baerwald E, Patterson W, Barclay R (2014) Origins and migratory patterns of bats killed by wind turbines in southern Alberta: evidence from stable isotopes. Ecosphere 5(9):art 118

Baerwald EF, D'Amours GH, Klug BJ, Barclay RMR (2008) Barotrauma is a significant cause of bat fatalities at wind turbines. Current Biol 18(16)

Barataud M (2015) Acoustic ecology of european bats. Inventaires & biodiversité series, Paris

Barclay R, Kurta A (2007) Ecology and behavior of bats roosting in tree cavities and under bark. In: Lacki MJ, Hayes J, Kurta A (eds) Bats in forests. The John Hopkins University Press, Baltimore

Barclay RMR, Baerwald EF, Gruver JC (2007) Variation in bat and bird fatalities at wind energy facilities: assessing the effects of rotor size and tower height. Can J Zool 85:381–387

Barros MA, de Magalhães RG, Rui AM (2015) Species composition and mortality of bats at the Osório Wind Farm, southern Brazil. Stud Neotropical Fauna Environ 50(1):31–39

Bates D, Mächler M, Bolker B, Walker S (2015) Fitting linear mixed-effects models using lme4. J Stat Softw 67(1):1–48. doi:10.18637/jss.v067.i01

Behr O, Baumbauer L, Hochradel K, Hurst J, Mages J, Nagy M, Korner-Niervergelt F, Niermann I, Reers H, Simon R, Weber N, Brinkmann R (2015) "Bat-friendly" operation of wind turbines—the current status of knowledge and planning procedures in Germany. In: Conference on wind energy and wildlife impacts, 10–12 March 2015, Berlin

Behr O, Brinkmann R, Niermann I, Korner-Niervergelt F (2011a) Akustische Erfassung der Fledermausaktivität an Windenergieanlagen. In: Brinkmann R, Behr O, Niermann I, Reich M (eds) Entwicklung von Methoden zur Untersuchung und Reduktion des Kollisionsrisikos von Fledermäusen an Onshore-Windenergieanlagen. Umwelt und Recht, vol 4. Cuvillier Verlag, Göttingen, pp 177–286

Behr O, Brinkmann R, Niermann I, Korner-Niervergelt F (2011b) Fledermausfreundliche Betriebsalgorithmen für Windenergieanlagen. In: Brinkmann R, Behr O, Niermann I, Reich M (eds) Entwicklung von Methoden zur Untersuchung und Reduktion des Kollisionsrisikos von Fledermäusen an Onshore-Windenergieanlagen. Umwelt und Recht, vol 4. Cuvillier Verlag, Göttingen, pp 354–383

Behr O, Brinkmann R, Niermann I, Korner-Niervergelt F (2011c) Vorhersage der Fledermausaktivität an Windenergieanlagen. In: Brinkmann R, Behr O, Niermann I, Reich M (eds) Entwicklung von Methoden zur Untersuchung und Reduktion des Kollisionsrisikos von Fledermäusen an Onshore-Windenergieanlagen. Umwelt und Recht, vol 4. Cuvillier Verlag, Göttingen, pp 287–322

Behr O, Eder D, Marckmann U, Mette-Christ H, Reisinger N, Runkel V, Ov Helversen (2007) Akustisches monitoring im Rotorbereich von Windenergieanlagen und methodische Probleme beim Nachweis von Fledermaus-Schlagopfern- Ergebnisse aus Untersuchungen im mittleren und südlichen Schwarzwald. Nyctalus 12(2–3):115–117

Berkhout V, Faulstich S, Görg P, Hahn B, Linke K, Neuschäfer M, Pfaffel S, Rafik K, Rohrig K, Rothkegel R, Ziese M (2014) Wind Energie Report Deutschland 2013. Fraunhofer-Institut für Windenergie und Energiesystemtechnik -IWES-, Institutsteil Kassel, Stuttgart

Bernard E, Paese A, Machado RB, de Souza Aguiar LM (2014) Blown in the wind: bats and wind farms in Brazil. Natureza Conservação 12(2):106–111

Bernotat D, Dierschke V (2015) Übergeordnete Kriterien zur Bewertung der Mortalität wildlebender Tiere im Rahmen von Projekten und Eingriffen. 2. Fassung, Stand 25.11.2015, 463 Seiten

BfN (2011) Windkraft über Wald. Positionspapier des Bundesamtes für Naturschutz. Bonn

BMWI/BMU (2010) Energiekonzept 2050. Eine Vision für ein nachhaltiges Energiekonzept auf Basis von Energieeffizienz und 100% erneuerbaren Energien

Brinkmann R, Bach L, Dense C, Limpens H, Mäscher G, Rahmel U (1996) Fledermäuse in Naturschutz- und Eingriffsplanungen - Hinweise zur Erfassung, Bewertung und planerischen Integration. Naturschutz u Landschaftsplanung 28(8):229–236

Brinkmann R, Behr O, Korner-Niervergelt F, Mages J, Niermann I (2011a) Zusammenfassung der praxisrelevanten Ergebnisse und offene Fragen. In: Brinkmann R, Behr O, Niermann I, Reich M (eds) Entwicklung von Methoden zur Untersuchung und Reduktion des Kollisionsrisikos von Fledermäusen an Onshore-Windenergieanlagen. Umwelt und Recht, vol 4. Cuvillier Verlag, Göttingen, pp 425–457

Brinkmann R, Behr O, Niermann I, Reich M (2011b) Entwicklung von Methoden zur Untersuchung und Reduktion des Kollisionsrisikos von Fledermäusen an Onshore-Windenergieanlagen, vol 4. Cuvillier Verlag, Göttingen, Umwelt und Recht

Brinkmann R, Schauer-Weisshahn H, Bontadina F (2006) Untersuchungen zu möglichen betriebsbedingten Auswirkungen von Windkraftanlagen auf Fledermäuse im Regierungsbezirk Freiburg. Gutachten im Auftrag des Regierungspräsidiums Freiburg - Referat 56 Naturschutz und Landschaftspflege

Cryan PM, Brown AC (2007) Migration of bats past a remote island offers clues toward the problem of bat fatalities at wind turbines. Biol Conserv 139(1):1–11

Cryan PM, Gorresen PM, Hein CD, Schirmacher MR, Diehl RH, Huso MM, Hayman DT, Fricker PD, Bonaccorso FJ, Johnson DH (2014) Behavior of bats at wind turbines. Proc Natl Acad Sci 111(42):15126–15131

Dietz C, Helversen Ov, Nill D (2007) Handbuch der Fledermäuse Europas und Nordwestafrikas. Kosmos Naturführer

Dietz C, Kiefer A (2014) Die Fledermäuse Europas - kennen, bestimmen, schützen. Kosmos Verlag, Stuttgart

Dietz M, Pir J (2011) Distribution, ecology and habitat selection by Bechstein's bat (Myotis bechsteinii) in Luxembourg, vol 6. Ökologie der Säugetiere
Dürr T (2015) Fledermausverluste an Windenergieanlagen. Daten aus der zentra len Fundkartei der Staatlichen Vogelschutzwarte im Landesamt für Umwelt, Gesundheit und Verbraucherschutz Brandenburg. LUGV Brandenburg, Stand vom 1. Juni
European Commission (2010) Guidance document—wind energy developments and Natura 2000. European Commission
Farneda FZ, Rocha R, López-Baucells A, Groenenberg M, Silva I, Palmeirim JM, Bobrowiec PE, Meyer CF (2015) Trait-related responses to habitat fragmentation in Amazonian bats. J Appl Ecol 52(5):1381–1391
Ferreira D, Freixo C, Cabral JA, Santos R, Santos M (2015) Do habitat characteristics determine mortality risk for bats at wind farms? Modelling susceptible species activity patterns and anticipating possible mortality events. Ecol Inform 28:7–18
Gonzalez MA, Ena V (2011) Cantabrian Capercaillie signs disappeared after a wind farm construction. Chioglossa 3:63–74
Humes ML, Hayes JP, Collopy MW (1999) Bat activity in thinned, unthinned, and old-growth forests in western Oregon. J Wildl Manag 63(2):553–561
Hurst J, Balzer S, Biedermann M, Dietz C, Dietz M, Höhne E, Karst I, Petermann R, Schorcht W, Steck C, Brinkmann R (2015) Erfassungsstandards für Fledermäu se bei Windkraftprojekten in Wäldern - Diskussion aktueller Empfehlungen der Bundesländer. Natur und Landschaft 90 (4):157–169
Jameson JW, Willis CK (2014) Activity of tree bats at anthropogenic tall structures: implications for mortality of bats at wind turbines. Anim Behav 97:145–152
Johnson GDP, Matthew K, Erickson, Wallace P, Strickland MD (2004) Bat activity, composition, and collision mortality at a large wind plant in Minnesota. Wildlife Soc Bull 32 (4):1278–1288. doi:10.2193/0091-7648(2004)032[1278:BACACM]2.0.CO;2
Kerth G, Wagner M, Weissmann K, König B (2002) Habitat- und Quartiernutzung bei der Bechsteinfledermaus: Hinweise für den Artenschutz. Schriftenreihe für Landschaftspflege und Naturschutz 71:99–108
Köppel J, Dahmen M, Helfrich J, Schuster E, Bulling L (2014) Cautious but committed: moving toward adaptive planning and operation strategies for renewable energy's wildlife implications. Environ Manage 54(4):744–755
Korner-Nievergelt F, Brinkmann R, Niermann I, Behr O (2013) Estimating bat and bird mortality occurring at wind energy turbines from covariates and car cass searches using mixture models. PLoS ONE 8(7):e67997. doi:10.1371/journal.pone.0067997
Kunz TH, Arnett EB, Erickson WP, Hoar AR, Johnson GD, Larkin RP, Strickland MD, Thresher RW, Tuttle MD (2007) Ecological impacts of wind energy development on bats: questions, research needs, and hypotheses. Front Ecol Environ 5(6):315–324
Kunz TH, Fenton MB (2003) Bat ecology. University of Chicago Press, Chicago
Lacki MJ, Amelon SK, Baker MD (2007) Foraging ecology of bats in forests. In: Lacki MJ, Hayes J, Kurta A (eds) Bats in forests. The John Hopkins University Press, Baltimore
Lehnert LS, Kramer-Schadt S, Schönborn S, Lindecke O, Niermann I, Voigt CC (2014) Wind farm facilities in Germany kill noctule bats from near and far. PLoS ONE 9(8):e103106
Leopold M, Boonman M, Collier M, Davaasuren N, Jongbloed R, Lagerveld S, Wal vdJ, Scholl M (2014) A first approach to deal with cumulative effects on birds and bats of offshore wind farms and other human activities in the southern North Sea
Lesinski G, Kowalski M, Wojtowicz B, Gulatowska J, Lisowska A (2007) Bats on forest islands of different size in an agricultural landscape. Folia Zool 56(2):153
Martin C (2015) Effectiveness of operational mitigation in reducing bat mortality and an assessment of bat and bird fatalities at the Sheffield wind facility. Texas Tech University, Vermont
McCracken GF, Gillam EH, Westbrook JK, Lee Y-F, Jensen ML, Balsley BB (2008) Brazilian free-tailed bats (Tadarida brasiliensis: Molossidae, Chiroptera) at high altitude: links to migratory insect populations. Integr Comp Biol 48(1):107–118. doi:10.1093/icb/icn033

Meschede A, Heller K-G (2000) Ökologie und Schutz von Fledermäusen in Wäldern, vol 66. Schriftenreihe für Landschaftspflege und Naturschutz

Morris AD, Miller DA, Kalcounis-Rueppell MC (2010) Use of forested ges by bats in a managed pine forest landscape. J Wildl Manag 74(1):26–34

Müller J, Brandl R, Buchner J, Pretzsch H, Seifert S, Strätz C, Veith M, Fenton B (2013) From ground to above canopy—bat activity in mature forests is driven by vegetation density and height. For Ecol Manage 306:179–184

Niermann I, Brinkmann R, Hurst J (2012) Windenergieanlagen im Wald und mögliche Beeinträchtigungen von Fledermäusen - eine Literaturauswertung. Naturschutz und Biologische Vielfalt 128:159–184

Niermann I, Brinkmann R, Korner-Niervergelt F, Behr O (2011a) Systematische Schlagopfersuche - Methodische Rahmenbedingungen, statistische Analyseverfahren und Ergebnisse. In: Brinkmann R, Behr O, Niermann I, Reich M (eds) Entwicklung von Methoden zur Untersuchung und Reduktion des Kollisionsrisikos von Fledermäusen an Onshore-Windenergieanlagen. Umwelt und Recht, vol 4. Cuvillier Verlag, Göttingen, pp 40–115

Niermann I, Brinkmann R, Korner-Niervergelt F, Behr O (2011b) Windbedingte Verdriftung von Fledermausschlagopfern an Windenergieanlagen - ein Diskussi onsbeitrag zur Methodik der Schlagopfersuche. In: Brinkmann R, Behr O, Niermann I, Reich M (eds) Entwicklung von Methoden zur Untersuchung und Reduk tion des Kollisionsrisikos von Fledermäusen an Onshore-Windenergieanlagen. Umwelt und Recht, vol 4. Cuvillier Verlag, Göttingen, pp 116–129

Niermann I, Von Felten S, Korner-Niervergelt F, Brinkmann R, Behr O (2011c) Einfluss von Anlagen- und Landschaftsvariablen auf die Aktivität von Fledermäusen an Windenergieanlagen. In: Brinkmann R, Behr O, Niermann I, Reich M (eds) Entwicklung von Methoden zur Untersuchung und Reduktion des Kollisionsrisikos von Fledermäusen an Onshore-Windenergieanlagen. Umwelt und Recht, vol 4. Cuvillier Verlag, Göttingen, 384–405

Piorkowski MDOC, Timothy J (2010) Spatial pattern of summer bat mortality from collisions with wind turbines in mixed-grass Prairie. Am Midl Nat 164(2):260–269

Polakowski M, Broniszewska M, Ruczyński I (2014) Local concentration of foraging noctule bats (*Nyctalus noctula*) as a possible tool to assess the density of bats in large forest complexes. Turk J Zool 38(2):254–256

R Core Team (2014) R: A language and environment for statistical computing. R foundation for statistical computing, Vienna, Austria. URL http://www.R-project.org/

Rahmel U, Bach L, Brinkmann R, Dense C, Limpens H, Mäscher G, Reichenbach M, Rochen A (1999) Windkraftplanung und Fledermäuse - Konfliktfelder und Hinweise zur Erfassungsmethodik. Bremer Beiträge für Naturkunde und Naturschutz 4:155–160

Richarz K (2014) Energiewende und Naturschutz - Windenergie im Lebensraum Wald. Status report und Empfehlungen. Deutsche Wildtier Stiftung: 70 S

Rodrigues L, Bach L, Dubourg-Savage M-J, Karapandza B, Kovac D, Kervyn T, Dekker J, Kepel A, Bach P, Collins J, Harbusch C, Park K, Micevski J, Mindermann J (2014) Guidelines for consideration of bats in wind park projects-revision 2014. vol Eurobats Publication Series Nr. 6, Bonn, Germany

Rydell J, Bach L, Dubourg-Savage MJ, Green M, Rodrigues L, Hedenström A (2010a) Bat mortality at wind turbines in northwestern Europe. Acta Chiropterologica 12(2):261–274

Rydell J, Bach L, Dubourg-Savage M-J, Green M, Rodrigues L, Hedenström A (2010b) Mortality of bats at wind turbines links to nocturnal insect migration? Eur J Wildl Res 56(6):823–827

Schuster E, Bulling L, Köppel J (2015) Consolidating the state of knowledge: a synoptical review of wind energy's wildlife effects. Environ Manag 1–32

Segers J, Broders H (2014) Interspecific effects of forest fragmentation on bats. Can J Zool 92 (8):665–673

Staton T, Poulton S (2012) Seasonal variation in bat activity in relation to detector height: a case study. Acta Chiropterologica 14(2):401–408. doi:10.3161/150811012x661710

Valença RB, Bernard E (2015) Another blown in the wind: bats and the licensing of wind farms in Brazil. Natureza Conservação

Vierhaus H (2000) Neues von unseren Fledermäusen. ABU info 24(1):58–60
Voigt CC, Lehnert LS, Petersons G, Adorf F, Bach L (2015) Wildlife and renewable energy: German politics cross migratory bats. Eur J Wildl Res 61(2):213–219
Voigt CC, Popa-Lisseanu AG, Niermann I, Kramer-Schadt S (2012) The catchment area of wind farms for European bats: a plea for international regulations. Biol Conserv 153:80–86. doi:10.1016/j.biocon.2012.04.027
Wang S, Wang S (2015) Impacts of wind energy on environment: a review. Renew Sustain Energy Rev 49:437–443
Willmott JR, Forcey GM, Hooton LA (2015) Developing an automated risk management tool to minimize bird and bat mortality at wind facilities. Ambio 44(4):S557–S571
Zahn A, Lustig A, Hammer M (2014) Potenzielle Auswirkungen von Windenegieanlagen auf Fledermauspopulationen. Anliegen Natur 36(1):21–35
Zentrum für Sonnenenergie- und Wasserstoff-Forschung Baden-Württemberg (ZSW) (2016) Erneuerbare Energien in Deutschland. Daten zur Entwicklung im Jahr 2015. Berlin

Bird Mortality in Two Dutch Wind Farms: Effects of Location, Spatial Design and Interactions with Powerlines

Allix Brenninkmeijer and Erik Klop

Abstract Numerous field studies have assessed bird mortality rates in wind farms. However, results from different studies are often hard to compare due to differences in methodology. This makes it very difficult to draw conclusions and to use the results in the planning phase of new wind farms (e.g. how to mitigate impacts). In this study, it was attempted to assess how bird mortality rates are affected by (1) the location of the wind farm, (2) the spatial layout of the turbines, (3) the surrounding terrain and (4) the presence of other obstacles such as powerlines. This study involved the monitoring of 91 turbines in two contrasting wind farms in the Netherlands for five years. It used the same standardized search methodology, including experimental trials for carcass removal and search efficiency. The sites differ in location (coastal vs. inland), spatial layout, turbine dimensions, land use, bird community and flight intensity of birds. In addition, at one site powerlines were constructed halfway through the monitoring program. Any fatalities from these powerlines were also monitored in a separate monitoring program. This enabled a comparison of any differences in mortality rates or species composition between the turbine and powerline fatalities. The results show a major impact of turbine location on the number of bird fatalities, both within the same wind farm and between wind farms. Mortality rates at the coastal wind farm were three to five times higher than at the inland wind farm. By far the highest mortality rates were found at turbines close to high-tide roosts and at points where (during spring migration) migrating birds leave the coastline to cross the sea towards Germany or Scandinavia. At these turbines, mortality rates could rise up to several hundred of birds per turbine per year. When expressed in fatalities per ha, overall fatality rates of the powerlines were three times higher than of the turbines in the same area. This may be due to low visibility of the powerlines compared to wind turbines. Comparison of turbine versus powerline fatalities also showed major differences in species composition,

A. Brenninkmeijer (✉) · E. Klop
Altenburg & Wymenga Ecological Consultants, Suderwei 2,
9269 TZ Feanwâlden, The Netherlands
e-mail: a.brenninkmeijer@altwym.nl

E. Klop
e-mail: erik.klop@altwym.nl

J. Köppel (ed.), *Wind Energy and Wildlife Interactions*,
DOI 10.1007/978-3-319-51272-3_6

with powerline fatalities mostly consisting of passerines and waterfowl, and turbine fatalities being dominated by gulls. As several new wind farms are planned to be realized in the coming years, the results of this study can be used in spatial planning to both assess and mitigate potential impacts.

Keywords Bird mortality · Wind farms · Spatial design · Location · Roosts · Powerlines

Introduction

In the last two decades, a multitude of studies have been carried out to investigate bird mortality at wind farms. The results of these efforts show large variation in mortality rates, which raises the question of which driving forces are responsible for these differences. Although a number of meta-analyses have shed light on the impacts of relevant parameters such as turbine size, location, etc. (e.g. Hötker 2006; Rydell et al. 2012; Loss et al. 2013), major differences in the methodologies that were used in the monitoring programs make it difficult to draw conclusions on the causal factors behind mortality rates.

In this study, a comparison is made between the patterns in bird mortality at two wind farms in the Netherlands. They have been monitored using an identical methodology, but the windfarms differ in several key characteristics such as the surrounding terrain, spatial layout, location relative to the coast, etc. (Brenninkmeijer and Van der Weyde 2011; Klop and Brenninkmeijer 2014a). In contrast to many studies, in both wind farms fatality monitoring was carried out year-round (i.e. also outside the migration periods) and for a substantial period of time (five years). In addition, at one of these wind farms powerlines were constructed halfway through the monitoring program, and any fatalities from the powerlines were also monitored in a separate monitoring program (Klop and Brenninkmeijer 2014b). These conditions make it possible to make a sound assessment of the factors that govern bird mortality rates as well as the impact of the powerlines compared to wind turbines. Aside from mortality rates, differences with regard to species composition and species richness were studied.

Study Area and Methods

Study Area

The study areas, Eemshaven (wind farm and powerlines) and Delfzijl (wind farm), are located in the northern part of the Netherlands (Fig. 1).

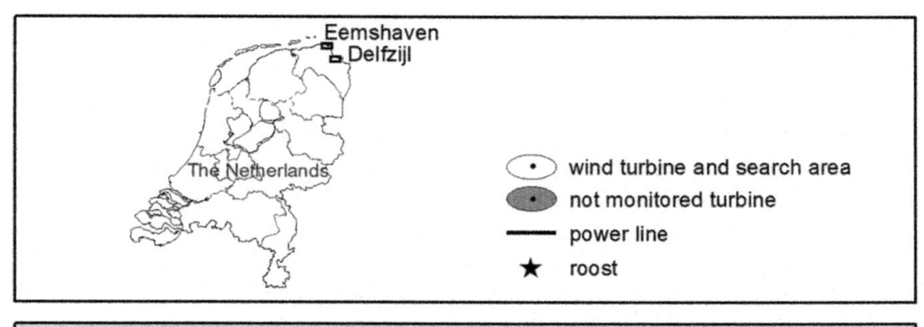

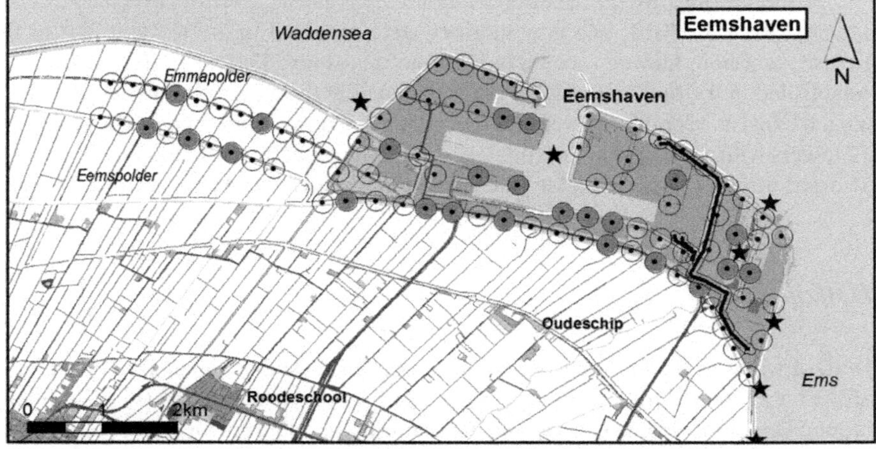

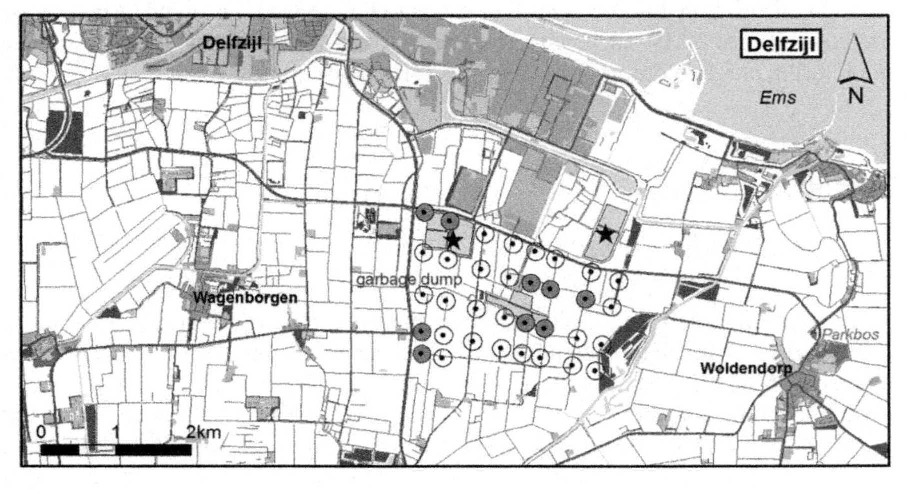

Fig. 1 Location of the study areas, Eemshaven (wind farm and powerline) and Delfzijl (wind farm), in the northeastern part of the Netherlands

Eemshaven

Eemshaven wind farm is one of the largest wind farms in the Netherlands and is situated in the north-east of Groningen province, adjacent to the Wadden Sea. The number of flight movements in this area is high, since it is located on a major migration route and very close to two major high-tide roosts (Fig. 1). In 2011 roughly 4 km of powerlines were constructed in the wind farm. The wind farm, situated in both industrial and agricultural area, consists of 88 turbines of 3 MW each, with a hub height of 100 m and tip height of 140 m.

The monitoring program of turbine fatalities was carried out from February 2009 to January 2014. In 2012, two new turbines were mounted in the western part of the Eemshaven; both turbines are excluded from this study. The powerline section was established in the first half of 2011 and is situated in the eastern industrial part of the area. It consists of approximately 4.3 km of 380 kV powerlines (height 10–50 m) and 12 towers with a width of approximately 30 m. The powerlines are located adjacent to 10 wind turbines. Powerline fatalities were monitored from August 2011 onwards.

Delfzijl

Delfzijl wind farm is situated 14 km southeast of Eemshaven in an agricultural area around a garbage dump, roughly 2.5 km south of the Wadden Sea (Fig. 1). In and around Delfzijl wind farm, there is no major migration route nor any major bird roosts, so flight intensity in this area is low compared to Eemshaven. The wind farm consists of 34 turbines of 2 MW each, with a hub height of 85 m and tip height of 120 m. The monitoring program of this wind farm was carried out from June 2006 to May 2011.

Methods

Wind Farm Monitoring

The monitoring programs in both wind farms were carried out during a period of five years using identical methodology. At Eemshaven, 68 of the 88 turbines and at Delfzijl, 25 of the 34 turbines were monitored monthly for fatalities (and twice a month in spring and autumn). Thus, 16 annual fatality searches were carried out (80 searches in 5 years). Similar to other studies (e.g. Winkelman 1992; Grünkorn et al. 2005) a circular area was searched with a radius equal to the turbine tip height (120 m in Delfzijl, 140 m in Eemshaven). For every turbine we quantified the area that could not be monitored (buildings, open water, roads) and excluded these parts from the search areas. When, during the growing season, vegetation height obstructed visibility of potential fatalities, the searchable area of the turbine was temporarily adjusted. The exact position of every fatality found within the search area was registered using a range finder and/or GPS.

Cause of Death

Three categories were used for the cause of death of the victims found. 'Certain turbine fatalities' includes birds with clear turbine-inflicted injuries. 'Possible turbine fatalities' includes birds with no visible injuries, scavenged birds, which are often no more than a bunch of feathers, as these birds could have collided first and been scavenged afterwards. Finally, 'Non-turbine fatalities', includes birds found outside the search area or birds with other causes of death, such as road traffic, predation by a non-scavenger such as Peregrine or Goshawk, oil, disease, etc. Birds in the last category were excluded from further analysis. To avoid double counts, all carcasses were marked with permanent, environmentally friendly paint.

Powerlines Monitoring

Monitoring of the powerlines in Eemshaven started in 2011 (also for a period of five years), about halfway through the turbines monitoring program. This provided an excellent opportunity to compare patterns in bird mortality between turbines and powerlines in the same area. Powerline and wind farm monitoring used—as far as possible—the same methodology, with monthly searches for fatalities and twice a month in spring and autumn. The area of 30 m directly under the powerlines as well as 50 m on both sides of the lines were searched for any fatalities (total search area 130 m). Previous studies used a comparable search area of 80–150 m (i.e. Janss 2000; Hartmann et al. 2010).

Some fatalities were found in areas where the search areas of powerlines and wind turbines overlapped. These birds were as far as possible assigned to either category based on location, injuries etc. For birds that could not be assigned with confidence to either turbines or powerlines, the cause of death was assigned at random to either of the two categories. In the comparison of turbine fatalities with powerline fatalities, the birds that were found inside the overlapping search areas were excluded.

Calculation of Collision Rate

Data from carcass searches needs to be corrected for search efficiency, scavenger removal and incomplete coverage of the search areas or monitoring period (Winkelman 1992). Non-searchable parts of the search area (buildings, open water, etc.) were quantified in GIS to calculate the correction factor for search area. Since all turbines were monitored for 12 months per year, correction for incomplete temporal coverage was not relevant. The correction factors for predation rates and search efficiency were determined experimentally using search and predation trials in the field (one in Delfzijl and three in Eemshaven).

Delfzijl

During the period from September to October 2006, a trial for search efficiency and predation rates was carried out in Delfzijl. On day 0 (12 September), 28 small (<100 g) and 30 large (>100 g) carcasses of both 'wild' birds and poultry were randomly placed within the search areas of 12 turbines (1.1 carcasses/ha). Carcass removal was monitored on 11 days between day 1 and day 29. To increase the sample size, a second search efficiency trial was established on 24 October 2006 with 24 small and 36 large carcasses. To increase the sample size and reduce confidence intervals, data from both trials was combined into one dataset afterwards.

Eemshaven

In the Eemshaven, three experimental trials with a total of 87 large and 98 small carcasses were carried out. On day 0 of every trial, starting in March–April 2009, October–November 2010 and May–June 2012, 22–41 small and 18–39 large carcasses were randomly placed within the search areas of 15 turbines (2009, 2010; 0.6 carcasses/ha) and approximately 4 km of powerlines (2012; 0.6 carcasses/ha). After placement, carcass removal was monitored 7–11 days between day 1 and day 31. All data of the three trials were combined afterwards.

Predation Rate

Calculation of carcass removal rates and the corresponding 95% confidence intervals was done using survival analysis in the software package R (R Core Team 2012, package 'survival'). The correction factor for predation was calculated separately for both size classes as the reciprocal of the Kaplan-Meier survival rate at a given day after placement. In both wind farms, the predation rate of small (<100 g) carcasses was much higher than of large (>100 g) carcasses (Fig. 2). In the first three days, roughly 50–60% of the small carcasses were removed compared to only 10–15% of the large carcasses. Predation rates were highest in the first week, but roughly stabilized thereafter. This might be due to a higher detection probability (by scent) of fresh carcasses, or reduced 'attractiveness' of older carcasses for predators

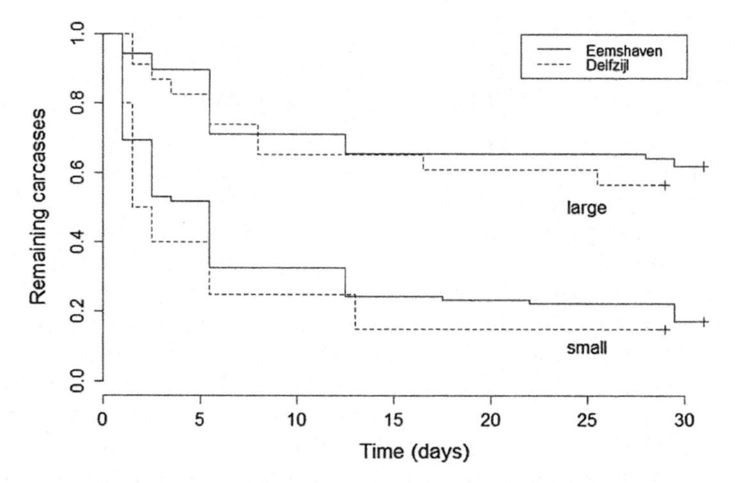

Fig. 2 Carcass removal rates of large (>100 g) and small (<100 g) birds in Eemshaven and Delfzijl. For simplicity, the 95% confidence intervals are not shown

Table 1 Search efficiency (P_F) of large birds (>100 g) and small birds (<100 g) in Eemshaven (three trials) and Delfzijl (two trials); present = number of carcasses present at day 1; found = number of carcasses retrieved at day 1

Search efficiency	Present	Found	P_F
Delfzijl			
Large birds	57	50	1.1
Small birds	47	29	1.6
Eemshaven			
Large birds	82	47	1.7
Small birds	68	12	5.7

(Arnett et al. 2005; Smallwood 2007; Ponce et al. 2010; Bispo et al. 2013). Predation rates in Delfzijl and Eemshaven were very similar, although removal rates of small birds were slightly higher in Delfzijl.

When monitoring turbine fatalities, birds may be hit by the turbines at any time between two consecutive monitoring rounds. In the analysis of corrected collision rates, the predation rate based on the the midpoint between two monitoring rounds was used (i.e. half the search interval; see Jain et al. 2009).

Search Efficiency

In addition to predation rates, the number of fatalities found during monitoring needed to be corrected for search efficiency (i.e. the ability to find dead birds on the ground). The search efficiency is affected by the size and color of the bird, the vegetation in which a dead bird is situated, the experience of the observer, etc. In both areas, the search efficiency for large birds was much higher than for small birds; there was also a major difference between both wind farms (Table 1).

Results

Spatial Distribution Patterns of Mortality Rates

Comparison of the collision rates (both the uncorrected numbers and those corrected for predation, search efficiency and searchable area) shows major differences between Eemshaven and Delfzijl wind farms (Table 2). The corrected fatality rates per turbine per year at Eemshaven are roughly six to ten times the number at Delfzijl. When looking at the uncorrected numbers (i.e. the number of birds found dead on the ground) the contrast is smaller but still a factor of two to three times the number at Delfzijl. This is probably due to the fact that at Delfzijl relatively few small birds were found that would need a large correction factor for predation and search efficiency.

Figure 3 shows a major impact of turbine location on the number of bird fatalities, both within the same wind farm and between wind farms. Eemshaven wind farm shows substantial spatial variation in (corrected) fatality rates, ranging from one to 156 fatalities per turbine per year. Of the 66 turbines however, only eight had fatality rates higher than 60 birds per year. The highest fatality rates were found at turbines bordering the Wadden Sea, especially those close to (high-tide) roosts. By far the highest fatality rates were found at the two turbines in the northeastern corner of the wind farm. Both turbines are positioned near a cooling water outlet, where high numbers of (subadult) gulls and other fish-eating birds roost and forage on fish that are periodically abundant after warm water discharge. After adapting the cooling water outlet in 2010, resulting in decreased fish availability, fatality rates at these two turbines dropped from 225 (c. 69% gulls) to 102 (24% gulls) per year.

Table 2 Corrected number of fatalities per turbine per day and per year in Eemshaven wind farm (88 turbines, mean 2009–2014) and Delfzijl wind farm (34 turbines, mean 2006–2011)

	Minimal estimate (certain victims)		Maximal estimate (certain + possible victims)	
	Mean	95% CI	Mean	95% CI
Eemshaven wind farm				
Uncorrected per turbine per year	1.1	0.1–2.1	4.2	1.9–6.4
Corrected per turbine per year	6.5	5.4–7.8	32.0	25.8–40.1
Corrected per year entire wind farm	583	484–706	2884	2324–3610
Delfzijl wind farm				
Uncorrected per turbine per year	0.5	0.0–1.1	1.6	0.5–2.7
Corrected per turbine per year	1.1	0.8–1.9	3.3	2.3–5.4
Corrected per year entire wind farm	36	24–61	112	75–182

The minimal estimate concerns only certain turbine victims, the maximal estimate concerns both certain and possible turbine victims

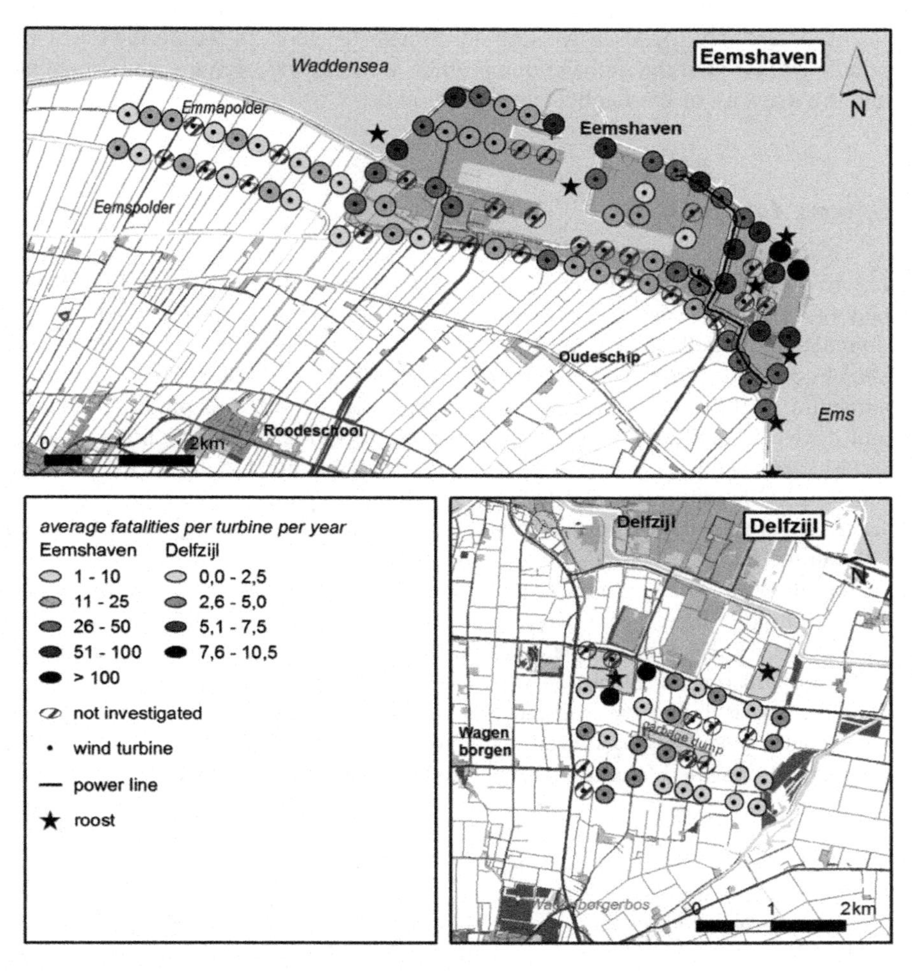

Fig. 3 Corrected number of average fatalities per turbine per year in Eemshaven and Delfzijl wind farms. Note the difference in fatality categories

Another reason for the high fatality rates at these turbines is their location at the corner position of the mainland. Around both turbines, being the most northeastern ones in the wind farm, many birds heading north and east during spring migration are 'hesitating' whether to cross the large water mass of the Eems estuary or to continue following the coast southwards. Apparently, these hesitating migrating birds become frequent victims. Besides large numbers of gulls, the fatalities here were dominated by migrating species such as Dunlin, Starling and a variety of thrushes (Blackbird, Song Thrush, Fieldfare, Redwing).

In sharp contrast to Eemshaven, Delfzijl wind farm showed low spatial variation in fatality rates, ranging from one to ten fatalities per turbine per year. The highest fatality rates were found at two turbines around the roost in the northwest of the wind farm. Around the garbage dump, gulls are frequently foraging and roosting, and the majority of the fatalities here were gulls.

Species Composition

Figure 4 compares the corrected fatalities per species group between Eemshaven and Delfzijl. The species composition in the coastal Eemshaven wind farm is dominated by gulls and passerines, and to a lesser extent by waterfowl and shorebirds. The more inland position of Delfzijl wind farm is illustrated by very few shorebird fatalities but relatively higher proportions of birds of prey and pigeons and doves. In addition, fewer passerines were recorded (mostly Starling) than at Eemshaven where migrating thrushes have been recorded frequently as turbine fatalities. Although migrating thrushes are usually reported to fly high over wind farms, apparently many thrushes come down at Eemshaven during both spring and autumn migration.

Based on the corrected fatality rates, songbirds (passerines) are by far the largest species group in Eemshaven wind farm with over 40% of all fatalities, compared to 25% in Delfzijl. The uncorrected data provides a very different picture: the number of found passerines comprise only 13% of all fatalities in Eemshaven and 10% in Delfzijl. Due to high predation rates and low search efficiency, the songbirds that were found have been corrected with high correction factors, resulting in high numbers of corrected fatalities. Due to the high correction factor for songbirds, the

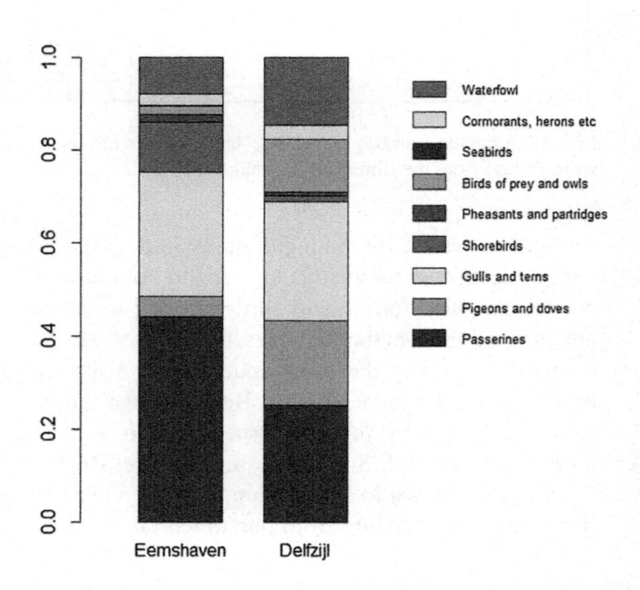

Fig. 4 Species composition of corrected fatalities in wind farms Eemshaven and Delfzijl

Table 3 Most frequently found species (with average found and corrected fatalities per turbine per year) in wind farms Eemshaven and Delfzijl

Eemshaven (88 turbines)	Fatalities/ turbine/year		Delfzijl (34 turbines)	Fatalities/ turbine/ year	
Top 5 species	Found	Corrected	Top 5 species	Found	Corrected
1. Herring Gull	0.92	4.1	1. Starling	0.04	0.45
2. Starling	0.08	3.1	2. Mallard	0.24	0.42
3. Black-headed Gull	0.57	2.7	3. Herring Gull	0.20	0.35
4. Song Thrush	0.07	2.0	4. Black-headed Gull	0.16	0.28
5. Dunlin	0.06	1.8	5. Domestic Pigeon	0.16	0.28

proportion of the other species groups (mostly large birds with low correction factors) also changed.

Table 3 provides a comparison of the corrected fatalities of the top five species of Eemshaven and Delfzijl. Three species have been found as fatalities frequently in both Eemshaven and Delfzijl: Herring Gull, Black-headed Gull and Starling. In Eemshaven, the Dunlin, a small shorebird species with a high correction factor, is also a frequent fatality. Dunlins roost in high numbers at the outer borders of the Eemshaven, such as the major high-tide roost immediately west of the wind farm. However, Dunlins in the area are also heavily predated by the resident pair of Peregrines, and part of the fatalities may have been predated rather than hit by a turbine. The number of Dunlin fatalities may have therefore been overestimated.

Surprisingly, Song Thrush is also among the top five of most frequent fatalities in Eemshaven, whereas no fatalities have been found in Delfzijl. As described above, the high number of thrush fatalities in Eemshaven may be mostly a result of the coastal position of this wind farm where large numbers of thrushes are funnelled through the area, whereas at Delfzijl migrating thrushes fly high over and are at no risk of being hit. In Delfzijl, Domestic Pigeon and Mallard are also top five fatalities; both species are abundant in the area.

Species Richness

As described in the previous sections, applying correction factors for predation and search efficiency have a major effect on calculated fatality rates. Using uncorrected numbers may severely underestimate fatality rates (Everaert 2008; Korner-Nievergelt et al. 2011, 2015; Péron et al. 2013). The fact that not all fatalities are found also has implications for the monitoring effort in terms of obtaining a reliable estimate of the number of species that are affected by the wind farm. In particular, small species are easily missed when searching for turbine fatalities (hence the large correction factors for search efficiency and predation), which means that species that are infrequently hit by a turbine are likely to go undetected, leading to biased estimates of the number of species affected by the wind farm.

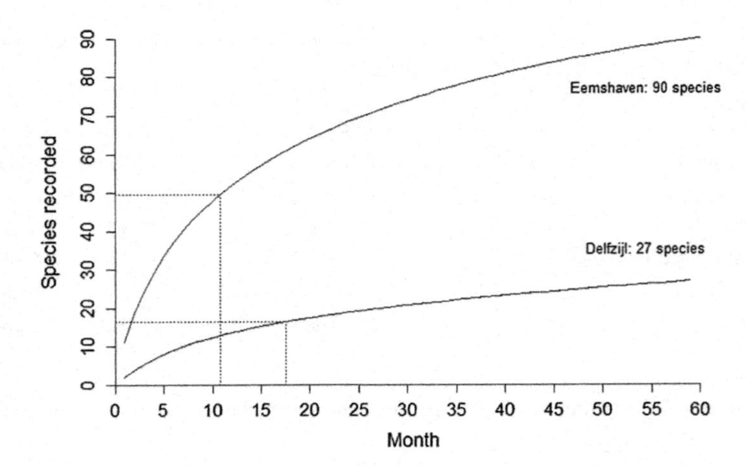

Fig. 5 Cumulative number of species found per month in Eemshaven and Delfzijl wind farms. The *dashed lines* indicate the time at which 50% of the estimated species richness was recorded. *Curves* are based on rarefaction analysis and were calculated using the program Estimates (Colwell 2013)

Species accumulation plots, in which the number of species is plotted against monitoring time, may give an indication of how many species collide with the turbines and how many species have been missed during fieldwork. In order to compare species richness of the fatalities between wind farms Eemshaven and Delfzijl, rarefaction analysis was used (Gotelli and Colwell 2001, 2010). This technique allows comparison of species richness between sites that have been sampled with unequal survey effort.

The species accumulation curves in Fig. 5 show that after five years of monitoring the curves have nearly flattened. However, there is still a slight increase, indicating that the actual number of species is likely marginally higher than what has been found in five years. The total number of species is estimated to be about 33 species at Delfzijl (95% confidence interval = 28–54) and 99 species at Eemshaven (95% confidence interval = 92–119). The dashed lines in Fig. 5 show that 50% of the estimated species richness was found in the first 11–18 months of monitoring. After five years of monitoring, 82% (Delfzijl) and 91% (Eemshaven) of the total species spectrum have been found effectively. These patterns illustrate the importance of having monitoring programs of sufficient length, in particular when fatalities of rare species or species of conservation concern need to be monitored.

Turbines Versus Powerlines

Halfway through the monitoring program in Eemshaven in 2011, over 4 km of powerlines were constructed in this wind farm. Similar to wind turbines, birds may

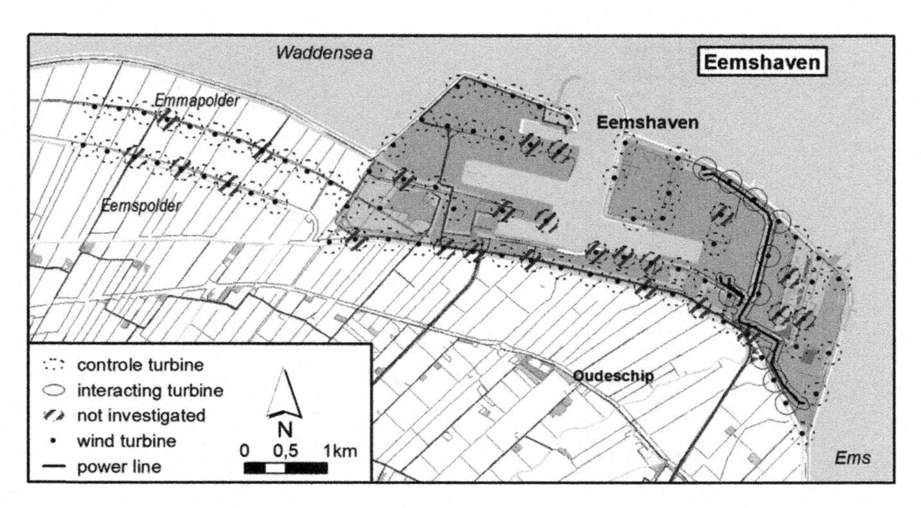

Fig. 6 The investigated powerline in the Eemshaven. Search areas of 10 turbines overlap partially with the search area of the powerlines

collide with powerlines and resulting fatality rates may be substantial (Koops 1979; Bevanger et al. 2011). Since 2011 any fatalities from the powerlines have been systematically monitored using similar methodology as for the wind turbines. The powerlines were constructed partially underneath ten turbines (Fig. 6). Here we compare patterns in powerline fatalities with turbine fatalities. In order to avoid bias due to incorrect classification of dead birds as either a turbine or powerline fatality, birds that were found in locations where the search areas of turbines and powerlines overlapped were excluded from the comparison. Hence, the comparison is based on the 56 turbines of which the search areas do not overlap with that of the powerlines. Likewise, only those sections of the powerline search area that do not overlap with the turbine search areas are included in the comparison. The analysis is based on found (uncorrected) numbers that were recalculated to mean numbers per hectare per year.

When expressed in uncorrected numbers per hectare, overall fatality rates of the powerlines were three times higher than of the turbines (turbines 1.15, powerlines 3.41 fatalities ha^{-1} yr^{-1}). The contrast between turbines and powerlines is strongly dependent on the species group (Fig. 7). In particular, waterfowl (including rallids, herons, etc.) and passerines seem to collide much more frequently with powerlines than with the turbines. Gulls and terns are the only group where turbine fatality rates were found to be higher than powerline fatality rates. This effect is somewhat inflated by the high numbers of gull fatalities near the cooling water outlet in the first two years of monitoring, although excluding these years from the analysis does not change the pattern.

The differences in species composition between turbines and powerlines are shown in Fig. 8. Powerline fatalities were dominated by various waterfowl and passerines; roughly half (51%) of all powerline fatalities consisted of only seven species, i.e. Mallard, Greylag Goose, Starling, Song Thrush, Redwing, Blackbird

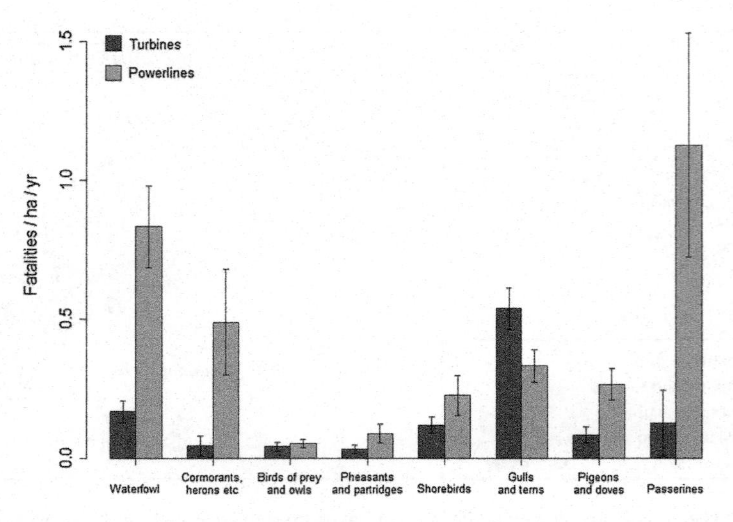

Fig. 7 The numbers of uncorrected fatalities per ha per year (±SE) for turbines and powerlines in Eemshaven. Fatalities found in overlapping search areas of turbines and powerlines were excluded from this analysis

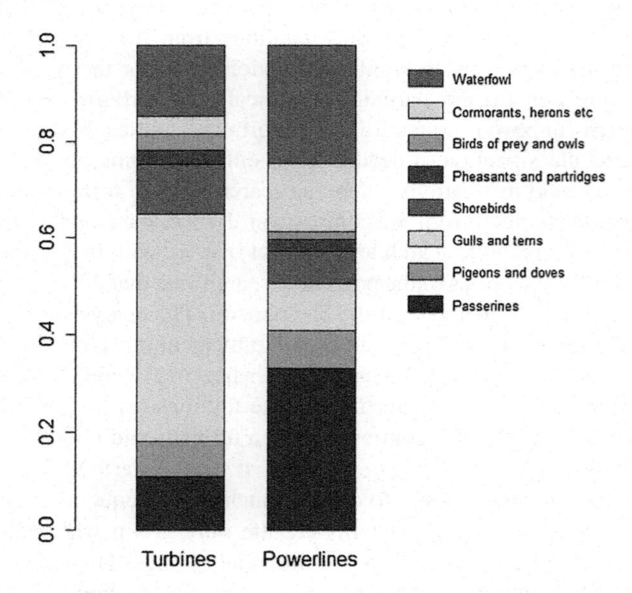

Fig. 8 Species composition of turbine and powerline fatalities, based on the uncorrected fatalities per ha per year in Eemshaven

and Moorhen. In contrast, these same species comprise no more than 13% of all turbine fatalities, which are dominated mainly by gulls and terns (47%). The majority of species among the powerline fatalities were found on average no more than once or twice per year.

It should be noted that Figs. 7 and 8 are based on the found (uncorrected) numbers, so the corrections for predation rate and search efficiency have not been taken into account. The corrected figures show proportionally more passerine fatalities because of the high correction factors for small birds (see also Fig. 4), but the differences between turbine and powerline fatalities remain regardless of whether the corrected or uncorrected figures are compared.

Discussion

Information on the factors that affect fatality rates can help in site selection and the spatial design of new wind farms, in order to minimize mortality among birds and bats. Many studies have identified 'high-risk' locations with high turbine fatality rates, such as mountain ridges, wetlands, shorelines and near breeding colonies of terns or gulls (Hötker 2006; Everaert and Stienen 2007; Drewitt and Langston 2008; Ledec et al. 2011). The results of this study show a major impact of turbine location on the number of bird fatalities, both within the same wind farm and between wind farms. Overall mortality rates at the coastal Eemshaven wind farm were three to five times higher than at the inland Delfzijl wind farm. However, the different turbines in Eemshaven show substantial variation in mortality rates, ranging from 1 to >100 fatalities per turbine per year. This variation illustrates the importance of the spatial layout of the wind turbines over the site, in particular when the site is close to major bird concentrations or areas with high flight intensity. By far the highest mortality rates were found at turbines close to high-tide roosts and at the corner position of the mainland, where (during spring migration) migrating birds leave the coastline to cross the sea towards Germany or Scandinavia. In addition, the presence of a cooling water outlet caused high mortality among (subadult) gulls and other fish-eating birds due to increased availability of fish. Similar patterns are visible at the wind farm in Zeebrugge at the Belgian coast, where some turbines close to a breeding colony of terns had fatality rates several times higher than the overall mean fatality rate for the entire wind farm (Everaert and Stienen 2007; Everaert 2014).

In contrast to Eemshaven, spatial variation in fatality rates at Delfzijl was low (range 1–10). This wind farm is located a few kilometers away from the coast on rather homogenous agricultural terrain, which is reflected by relatively low fatality rates that do not differ substantially between the turbines. In addition, the inland position of Delfzijl wind farm resulted in different species composition and lower species diversity of the fatalities, with proportionally fewer shorebirds and passerines but more pigeons and doves and birds of prey. However, some species were recorded frequently as fatalities in both wind farms, such as Starling, Mallard, Herring Gull and Black-headed Gull.

The construction of powerlines in Eemshaven wind farm provided the opportunity to compare any differences in mortality rates or species composition between the turbine and powerline fatalities. Overall, the uncorrected fatality rates (expressed as the mean number of found birds per hectare per year) of the powerlines were much higher than for turbines in the same area. This difference may be caused by low visibility of the powerlines compared to the turbines. In addition, the turbine rotors operate at greater heights (60–140 m) than the powerlines (10–50 m) which is of relevance to the height distribution and collision risk of flying birds in the area. The flight altitude of nocturnally migrating passerines is often too high to collide with turbines and especially powerlines, but birds may decrease flight altitude in headwind (e.g. Alerstam 1990; Pettersson 2011; Kahlert et al. 2012; Bowlin et al. 2015) or when hesitating to cross the Eems estuary northwards. Starling and various thrushes (Song Thrush, Blackbird, Fieldfare, Redwing) migrate in large numbers through the Eemshaven area, and all of these species regularly collide with both turbines and powerlines. Interestingly, powerline fatalities among thrushes are mostly concentrated during autumn migration (October) whereas turbine fatalities among thrushes are more evenly distributed over spring (April) and autumn migration peaks. These autumn fatalities are most likely due to birds using the Eemshaven as a stopover site to rest or feed at the bushes of Seaberry (*Hippophae rhamnoides*) in the area, thereby colliding with the powerlines.

Besides the difference in numbers, some notable differences were found in species composition between turbine fatalities and powerline fatalities: gulls were by far the most frequent turbine fatalities whereas waterfowl and passerines dominated the powerline fatalities.

As several new wind farms are planned to be realized in the coming years, the results of this study can be used in spatial planning to both assess and mitigate potential impacts.

Acknowledgements Jan van der Kamp and Klaas van Dijk carried out most fatality searches. Furthermore, Mark Koopmans, Kim Meijer, Mirte Greve, Marten Sikkema, Janne Ouwehand, Reinder Wissman, Daan Vreugdenhil, Olga Stoker, Ronald de Jong and Franske Hoekema contributed to collecting fatality data. Olga, Ronald and Franske also edited GIS data and Franske made the maps. The students Gijsbert Knol and Andrea Vos contributed to the search and predation trials. Christa van der Weyde, Leo Bruinzeel and Elske Tielens helped in analyses and reporting and Marcel Kersten provided statistical and methodological support. We are also grateful to two anonymous referees for their helpful comments on earlier versions of this chapter.

References

Alerstam T (1990) Bird migration. Cambridge University Press, Cambridge

Arnett EB, Erickson WP, Kerns J, Horn J (2005) Relationships between bats and wind turbines in Pennsylvania and West Virginia: an assessment of bat fatality search protocols, patterns of fatality, and behavioral interactions with wind turbines. A final report submitted to the bats and wind energy cooperative. Bat Conservation International. Austin, Texas, USA

Bevanger KM, Bartzke G, Brøseth H, Dahl EL, Gjershaug JO, Hanssen FO, Jacobsen KO, Kvaløy P, May RF, Meås R, Nygård T, Refsnæs S, Stokke S, Thomassen J (2011) Optimal design and routing of power lines; ecological, technical and economic perspectives (OPTIPOL). Progress report 2011, Norsk Institutt for Naturforskning

Bispo R, Bernardino J, Marques TA, Pestana D (2013) Modeling carcass removal time for avian mortality assessment in wind farms using survival analysis. Environ Ecol Stat 20:147–165

Bowlin MS, Enstrom DA, Murphy BJ, Plaza E, Jurich P, Cochran J (2015) Unexplained altitude changes in a migrating thrush: long-flight altitude data from radio-telemetry. Auk 132:808–816

Brenninkmeijer A, Van der Weyde C (2011) Monitoring aanvaringsslachtoffers Windpark Delfzijl-Zuid 2006–2011. Eindrapportage vijf jaar monitoring. A&W-rapport 1656. Altenburg & Wymenga ecologisch onderzoek bv, Feanwâlden

Colwell RK (2013). Estimates: statistical estimation of species richness and shared species from samples. Version 9, http://viceroy.eeb.uconn.edu/estimates/

Drewitt AL, Langston RHW (2008) Collision effects of wind-power generators and other obstacles on birds. Ann NY Acad Sci 1334:233–266

Everaert J (2008) Effecten van windturbines op de fauna in Vlaanderen. Onderzoeksresultaten, discussie en aanbevelingen. Rapportnr. INBO.R.2008.44. Instituut voor Natuur-en Bosonderzoek, Brussel

Everaert J (2014) Collision risk and micro-avoidance rates of birds with wind turbines in Flanders. Bird Study 61:220–230

Everaert J, Stienen EWM (2007) Impact of wind turbines on birds in Zeebrugge (Belgium). Significant effect on breeding tern colony due to collisions. Biodiv Cons 16:3345–3359

Gotelli N, Colwell RK (2001) Quantifying biodiversity: procedures and pitfalls in the measurement and comparison of species richness. Ecol Lett 4:379–391

Gotelli NJ, Colwell RK (2010) Estimating species richness. In: Magurran AE, McGill BJ (eds) Biological diversity: frontiers in measurement and assessment. Oxford University Press, Oxford, pp 39–54

Grünkorn T, Diederichs A, Stahl B, Poszig D, Nehls G (2005) Entwicklung einer Methode zur Abschätzung des Kollisionsrisikos von Vögeln an Windenergieanlagen. Unveröff. Gutachten Im Auftrag des Landesamtes für Natur und Umwelt Schleswig-Holstein

Hartman JC, Gyimesi A, Prinsen HAM (2010) Veldonderzoek naar draadslachtoffers en vliegbewegingen bij een gemarkeerde 150 kV hoogspanningslijn. Rapportnr. 10-082, Bureau Waardenburg bv, Culemborg

Hötker H (2006) Auswirkungen des 'Repowering' von Windkraftanlagen auf Vögel und Fledermäuse. Michael-Otto-Institut im NABU-Forschungs—und Bildungszentrum für Feuchtgebiete und Vogelschutz, Berghusen

Jain A, Kerlinger P, Curry R, Slobodnik L, Lehman M (2009) Annual report for the maple ridge wind power project. Post-construction bird and bat fatality study—2008. Seracuse, New York, USA

Janss GFE (2000) Avian mortality from power lines: a morphologic approach of a species-specific mortality. Biol Cons 95:353–359

Kahlert J, Leito A, Laubek B, Luigujõe L, Kuresoo A, Aaen K, Luud A (2012) Factors affecting the flight altitude of migrating waterbirds in Western Estonia. Ornis Fenn 89:241–253

Klop E, Brenninkmeijer A (2014a) Monitoring aanvaringsslachtoffers Windpark Eemshaven 2009–2014. Eindrapportage vijf jaar monitoring. A&W-rapport 1975. Altenburg & Wymenga ecologisch onderzoek bv, Feanwâlden

Klop E, Brenninkmeijer A (2014b) Vervolgmonitoring vogelslachtoffers hoogspanningslijnen Eemshaven. Jaarrapportage 2013–2014. A&W-rapport 2062. Altenburg & Wymenga ecologisch onderzoek bv, Feanwâlden

Koops FBJ (1979) Een miljoen draadslachtoffers, wat kunnen we er tegen doen? Lepelaar 63: 20–21

Korner-Nievergelt F, Korner-Nievergelt P, Behr O, Niermann I, Brinkmann R, Hellriegel B (2011) A new method to determine bird and bat fatality at wind energy turbines from carcass searches. Wildl Biol 17:350–363

Korner-Nievergelt F, Behr O, Brinkmann R, Etterson M, Huso MMP, Dalthorp D, Korner-Nievergelt P, Roth T, Niermann I (2015) Mortality estimation from carcass searches using the R-package carcass—a turorial. Wildl Biol 21:30–43

Ledec GC, Rapp KW, Aiello RG (2011) Greening the wind: environmental and social considerations for wind power development in Latin America and beyond (conference ed). The World Bank, Washington, DC. Report www.tinyurl.com/GreeningTheWind
Loss SR, Will T, Marra PP (2013) Estimates of bird collision mortality at wind facilities in the contiguous United States. Biol Cons 168:201–209
Péron G, Hines JE, Nichols JD, Kendall WL, Peters KA, Mizrahi DS (2013) Estimation of bird and bat mortality at wind-power farms with superpopulation models. J Appl Ecol 50:902–911
Pettersson J (2011) Night migration of songbirds and waterfowl at the Utgrunden off-shore wind farm—A radar-assisted study in southern Kalmar Sound. Report 6438, Swedish Environmental Protection Agency, Bromma, Sweden
Ponce C, Alonso JC, Argandoña G, García Fernández A, Carrasco M (2010) Carcass removal by scavengers and search accuracy affect bird mortality estimates at power lines. Anim Cons 13:603–612
R Core Team (2012) R: a language and environment for statistical computing. R Foundation for Statistical Computing, Vienna. http://www.R-project.org
Rydell J, Engström H, Hedenström A, Larsen JK, Pettersson J, Green M (2012) The effects of wind power on birds and bats: a synthesis. Report 6511, Swedish Environmental Protection Agency, Stockholm
Smallwood KS (2007) Estimating wind turbine-caused bird mortality. J Wildl Manage 71:2781–2791
Winkelman JE (1992) De invloed van de Sep-proefwindcentrale te Oosterbierum (Fr.) op vogels. 1 Aanvaringsslachtoffers. RIN-rapport 92, Rijksinstituut voor Natuurbeheer, Arnhem

Part IV
Mitigation, Compensation, Effectiveness of Measures

Radar Assisted Shutdown on Demand Ensures Zero Soaring Bird Mortality at a Wind Farm Located in a Migratory Flyway

Ricardo Tomé, Filipe Canário, Alexandre H. Leitão, Nadine Pires and Miguel Repas

Abstract Wind energy is considered a clean energy source, but produces negative impacts regarding avian mortality. The Barão de São João wind farm in Portugal's Sagres region is part of an important migratory flyway, crossed by 5000 individuals of 30 soaring bird species every autumn. The wind farm's licensing was conditioned to the implementation of rigorous mitigation procedures, namely a Radar Assisted Shutdown on Demand (RASOD) protocol to reduce the probability of bird casualties. A security perimeter with observers was aided by a radar system, detecting soaring birds approaching the wind farm. Turbines were to be turned-off when pre-defined criteria of intense migration or presence of threatened species were met. Turbine shutdown was operated by the wind farm staff after a request from the monitoring team (MT), or directly by the MT. Of the soaring birds crossing the wind farm, 55% were recorded at altitudes associated with high collision risk. However, due to RASOD, no soaring birds died from collisions during five consecutive autumns. The average annual shutdown period decreased continuously after the first year (105 h) reaching only 15 h when the MT was given direct access to shut down operations through SCADA (the remote system to monitor and control wind turbines). Shutdown period corresponded only to 0.2–1.2% of the equivalent hours in a year's wind farm activity. The use of radar, direct access to SCADA and cumulative experience by the MT improved the procedure's efficiency, allowing better judgments on the application of shutdown orders. Our results indicate that RASOD may be an essential tool in reconciling wind energy production with the conservation of soaring birds.

Keywords Shutdown · Wind farm · Turbine shutdown · Collision risk · Radar · Soaring birds

R. Tomé (✉) · F. Canário · A.H. Leitão · N. Pires · M. Repas
STRIX, Carcavelos, Portugal
e-mail: ricardo.tome@strix.pt

J. Köppel (ed.), *Wind Energy and Wildlife Interactions*,
DOI 10.1007/978-3-319-51272-3_7

Introduction

Growing concerns about climate change led nations to reduce greenhouse gas emissions by increasing the use of energy from renewable sources. This, along with decreasing costs associated with wind energy production led to a proliferation of wind farms around the world (Redlinger et al. 2002; Kaldellis and Zafirakis 2011).

However, wind farms may have adverse effects on birds (Orloff and Flannery 1992; de Lucas et al. 2008). The main potential hazards of wind farm construction and operation are loss of habitat, disturbance (Drewitt and Langston 2006; Dahl et al. 2012), and mortality resulting from collisions with turbines and associated power lines (Orloff and Flannery 1992; Drewitt and Langston 2006; Thelander and Smallwood 2007).

Although recorded casualties differ widely between wind farms, large soaring birds seem particularly vulnerable to collisions (Lekuona and Ursúa 2007; de Lucas et al. 2012; Ferrer et al. 2012). Although results regarding population level impacts are usually inconsistent (Kuvlesky et al. 2007), for species of conservation concern (especially those with low reproductive rates) even small increases in mortality may be significant (Dahl et al. 2012; Bellebaum et al. 2013). The problem increases if wind farms are located in areas with high concentrations of individuals such as migratory flyways and bottlenecks.

The Western African-Eurasian flyway has a major bottleneck site for soaring birds at the Strait of Gibraltar (Bernis 1980; Finlayson 1992; Zalles and Bildstein 2000). Some hundred thousand soaring birds cross this area each spring and autumn (Zalles and Bildstein 2000; de la Cruz et al. 2011). A portion of these birds (mainly juveniles and immatures) apparently fails to reach or cross the Strait on a first attempt, ending up in the Sagres region, the south-westernmost point of mainland Portugal and Europe (e.g. Mellone et al. 2011). Each autumn around 5000 soaring birds reach this region, which is also an important migration site for other species (Moreau and Monk 1957; Canário et al. 2012; Tomé et al. 1998).

Some birds of prey, which do not depend so much on thermals, may attempt to cross directly to Africa at this point (Strandberg et al. 2009; Vardanis et al. 2011). However, almost all soaring birds that are seen entering the Sagres area in their autumnal migration are observed later leaving the area (Tomé et al. 1998). Spring migration of soaring birds is negligible in this area.

This important bottleneck site is also the area of Portugal with the highest wind resource (Costa 2004). At present, nine wind farms with a total of 75 turbines have been installed in the region, with a total capacity of 138.5 MW. A few more are planned, thus creating a conflict between wind energy production and soaring bird conservation.

Although thorough pre-construction studies and strategic environmental impact assessment is the most effective way to reduce bird collisions with wind turbines (Drewitt and Langston 2006; Bright et al. 2006), mitigation measures aiming to reduce impacts on birds at wind farms have been applied occasionally. However, so far there is no conclusive evidence of their effectiveness (Drewitt and Langston 2006;

Johnson et al. 2008; Marques et al. 2014). Possibly, the main exception is the turbine shutdown on demand protocol, which has been applied in a few wind farms in southern Spain since 2008, producing encouraging results (de Lucas et al. 2012). This paper describes the application of a radar assisted shutdown on demand (RASOD) mitigation procedure implemented at the largest wind farm in the Sagres region, and its results on soaring bird mortality and wind energy production.

Methods

Study Area

The Barão de São João wind farm (hereafter designated as BSJ wind farm; Fig. 1) is located in a forest area in south-western Portugal (37° 08′N and 8° 48′W), in the Algarve region. The wind farm is composed of 25 turbines with an installed

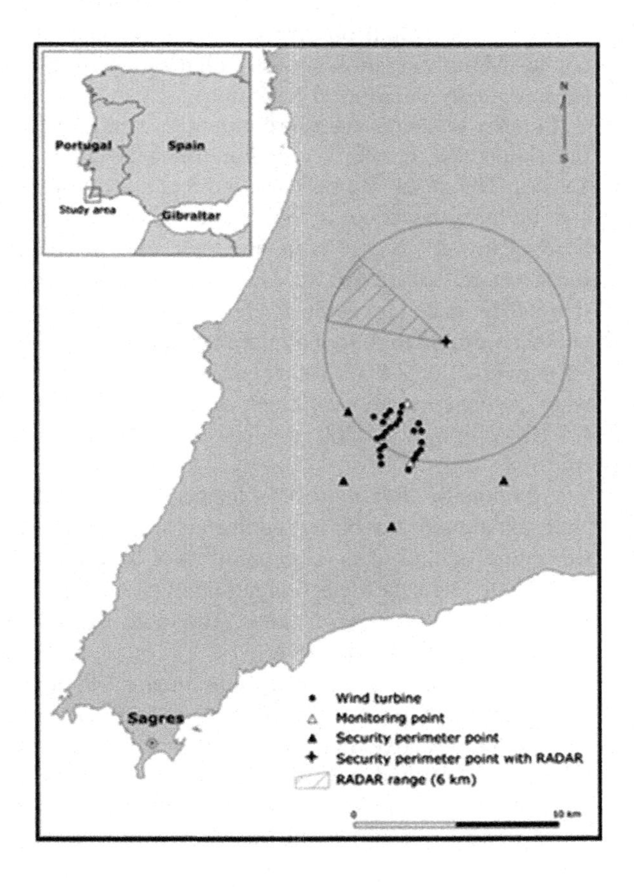

Fig. 1 Study area location and location of monitoring vantage points at the wind farm and at the security perimeter

capacity of 50 MW, being the largest wind farm in southern Portugal. The turbine blades swept area ranges from 35 to 125 m, and distances between turbines ranges from 273 to 539 m. This wind farm commenced operation in 2010. Because of its location in a sensitive area regarding soaring bird migration, the wind farm's licensing was conditioned to the development of extensive monitoring programs and to the implementation of mitigation procedures, namely a RASOD protocol.

Data Collection and Shutdown Protocol

Monitoring of soaring bird migration and mitigation programs have been conducted annually since 2010, from August 15 to November 30 (108 days). Results presented in this paper were collected from 2010 to 2014. Monitoring consisted of observations from two vantage points (VPs) located within the wind farm area (Fig. 1). In each of these points an observer equipped with binoculars and telescope recorded and accurately mapped every soaring bird detected (raptor, stork or raven *Corvus corax*). Both observers were in constant communication to avoid double counts. Monitoring took place simultaneously from approximately 9:00 to 6:00 PM (average daily duration: 6 h 52 min).

In order to characterize bird migration within the wind farm, the total number of individual soaring birds crossing the wind farm area was calculated for each species. This result did not have a direct correspondence with the actual number of soaring birds that occur in the region, since most birds crossing the wind farm were heading towards Sagres area, and then were detected once again leaving the area. Furthermore, some birds stay in the region for more than one day, being recorded more than once. In addition the number of movements recorded was used as another variable to describe migration patterns. A single movement could either be an individual or a flock (independent of the number of birds). The flight altitudes were estimated in five different classes: (1) <20 m, (2) 20–60 m, (3) 60–100 m, (4) 100–200 m, (5) >200 m. Classes (2)–(4) were considered to be of high collision risk.

Additionally, five more VPs located at 1.4–4.5 km from the wind farm central area constituted a security perimeter around it (Fig. 1). In one of these points (hereafter designated as radar point) an X-band marine radar was used (9410 MHz, 25 kW) to increase detection probability and the ability to follow approaching soaring birds at a larger distance. The radar was operated in horizontal mode with the range usually set at 6–8 km.

Observations from all security perimeter VPs took place from 15 September to 15 November, corresponding to the core migratory period. For the remaining period the number of active VPs varied. The radar point was active throughout the whole season in 2010 and only during the core periods onwards. Every observer was equipped with a walkie-talkie, ensuring communication between all VPs. Whenever soaring birds were at risk, the fieldwork coordinator decided whether or not some or

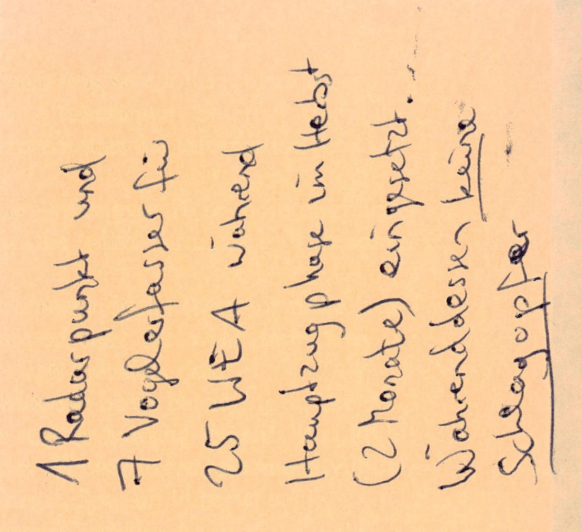

all the turbines must be shut down and also wh
This decision was based upon the follo
Environmental Impact Statement, taking also i
from the other observers:

A—Intense migratory flux of soaring bi
soaring birds were detected in one day ne
towards it;

B—Flocks of migrating soaring birds—
viduals) of migratory soaring birds were d
heading towards it at flight altitudes involvin

C—Threatened soaring bird species—
lowing species were detected in the wind far
altitudes involving high collision risk: black
Aegypius monachus, golden eagle *Aquila chr*
adalberti, Bonelli's eagle *Aquila fasciata*,
kestrel *Falco naumanni*.

D—Imminent collision risk—even when the previous criteria were not met, one or more turbines should be shutdown whenever there was an imminent risk of collision of a migratory soaring bird with one of the turbines.

From 2010 to 2012 turbine shutdown and restart orders were communicated via cell phone between the fieldwork coordinator and the wind farm technician responsible for the shutdown procedure. In 2013 and 2014 wind turbines were shut down directly by the field monitoring team through direct access to the SCADA system using an internet connection. The wind farms SCADA (Supervisory Control And Data Acquisition) is a system that provides real-time access and management of individual wind turbines and wind farms. This system allows users to monitor and control one or more wind turbines remotely, thus controlling e.g. shutdown and restart of wind turbines.

The duration of the shutdown periods was not defined a priori, depending on the verification of the criteria. During the shutdown period the team evaluated the prevalence of the conditions that triggered the application of the criteria and decided accordingly, determining the restart of some or all the turbines. Application of the criteria was not strict, since the fieldwork coordinator judged the risk level based on his/her experience. Often, shutdown orders were associated to the verification of more than one criterion.

Information regarding the relative importance of different vantage points in determining the verification of shutdown criteria started to be collected in 2011. Only the information collected between 2011 and 2014 and from 15 September to 15 November was used in the analyses.

Mortality searches were performed at each turbine every two weeks from August to February and monthly in the remaining months. Daily wind speed was also registered and correlated with RASOD operational results, namely shutdown periods. In order to standardise shutdown periods in each operation, the equivalent shutdown period was calculated, by multiplying each stoppage period by the proportion of turbines that were actually turned off.

Results

Bird Migration

The number of soaring bird species observed at the BSJ wind farm each year was close to 30 (Table 1). Several globally threatened species (IUCN 2015) were recorded, such as the Egyptian vulture *Neophron percnopterus*, Rueppell's vulture *Gyps rueppellii* and Spanish imperial eagle, as well as other that are near threatened at a global level, such as the cinereous vulture, the red kite *Milvus milvus* and the pallid harrier *Circus macrourus*. The annual number of movements ranged between 2658 and 3446, whereas the number of individuals counted ranged from a minimum of 8995, in 2014, to a maximum of 26,543 in 2010. The average number of birds counted annually was 18,300 (±6073) (Table 1). The species that accounted for the majority of movements was the short-toed-eagle (25.7% of the total number of movements). On the other hand, the most abundant species was the griffon vulture, which accounted for 77.8% of the total number of individuals counted between 2010 and 2014.

Collision Risk

Nearly three quarters (72%) of the observed movements were registered at height classes involving high collision risk. The proportion of movements observed in each of these three classes was very similar (Fig. 2).

Table 1 Number of soaring bird species, soaring bird movements and soaring bird individuals observed between 2010 and 2014

Year	Number of soaring bird species	Number of movements	Number of soaring birds
2010	28	2824	26,543
2011	30	2658	14,964
2012	28	3446	22,629
2013	28	3225	18,367
2014	27	2859	8995
Average	28	3002	18,300

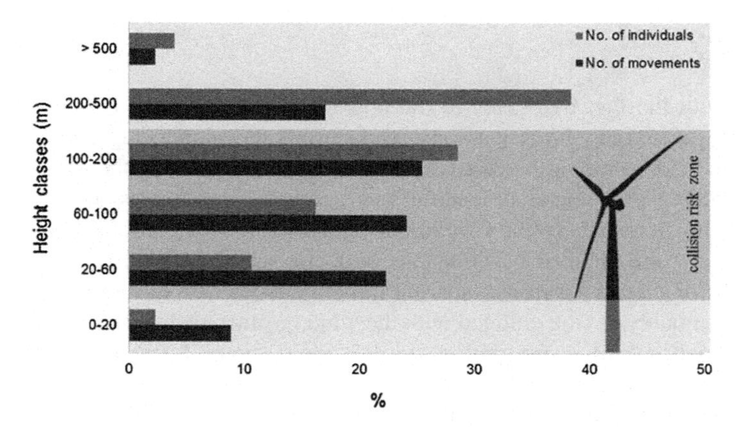

Fig. 2 Proportion of soaring bird movements (n = 14,892) and individuals (n = 89,458) observed at each flight height class during 2010–2014

Table 2 Proportion of the number of movements of soaring birds (n = 314) and of the number of soaring birds (n = 1958) that were detected flying at collision risk heights closer than 45 m from the turbines while the turbines were active or idle/stopped on account of wind absence or of the RASOD procedure

	% No. movements	% No. individuals
Turbines active	63.7	14.4
Turbines idle: no wind	10.5	2.2
Turbines stopped: RASOD	25.8	83.4

Regarding the number of individuals, the largest proportion was detected at the 200–500 m height class (38.4% of the total number of individuals; Fig. 2). Nevertheless, more than half (55.4%) of the total number of individuals flew over the wind farm area at high collision risk heights (Fig. 2).

Yearly, several dozens of soaring birds have been observed flying at height classes covered by the turbine's blades swept area and in close proximity to the wind turbines. In total, from 2010 to 2014, 314 of these movements occurred at 45 m (a blade's length) or less from the turbine's rotors (hereafter designated as 'high risk movements'). The number of individuals involved in those high risk movements reached 1958. The majority of the high risk movements occurred while the turbines were functioning, a quarter when they were stopped following a RASOD procedure, and a tenth when they were idle due to the absence of wind (Table 2).

Considering the number of individuals, the large majority of the soaring birds that crossed the area performing high risk movements did so when turbines had been shut down on the account of RASOD (Table 2). Of the 282 individuals (14.4%) involved in high risk movements when the turbines were active, only four belonged to endangered species which corresponds to a shutdown criteria (three Ospreys and one Black stork).

Mortality

Throughout the first five years of the operational phase of the BSJ wind farm no mortality of soaring birds was observed during the monitoring and surveillance activities or detected in the carcass searches conducted during the period in which the shutdown-on-demand procedure was applied. Nonetheless, non-soaring bird species, not targeted by the design and application of the shutdown on demand procedure, were affected by collisions with the wind turbines. In fact, eight carcasses belonging to seven non-soaring birds were detected during carcass searches after presumably having collided with the wind turbines. Additionally, the carcasses of four soaring birds—one griffon vulture, one common buzzard *Buteo buteo* and two Eurasian kestrels *Falco tinnunculus*—were found dead out of the autumn migration monitoring and surveillance period.

RASOD Application

Operational Results

The number of days in which at least one turbine shutdown operation was applied remained relatively unaltered throughout the five year period (Fig. 3a). Between 2010 and 2013 shutdown operations were ordered in 31–36 days, and only in 2014 this number decreased to 23. Therefore, and considering the overall duration of the annual period of RASOD application (108 days), shutdown operations were performed in 21% (2014) to 33% (2010, 2011 and 2013) of the days included in that period.

The annual number of shutdown operations (irrespectively of the number of turbines involved) decreased constantly from 2010 (74 shutdowns) to 2014 (30 shutdowns), with the single exception of 2013, when 64 shutdown operations were performed (Fig. 3b). Overall, on average at least one turbine was ordered to be shut down on 56 occasions per year (average 0.5 shutdown orders/day).

The average time it took since a shutdown order was given until the selected turbines became idle decreased markedly and successively from 4.5 min in 2010 to 24 s in 2014 (Fig. 3c). The annual total equivalent shutdown period decreased similarly, varying from 104 h 45 min in 2010 to 15 h 7 min in 2014 (Fig. 3d). Overall, the average annual equivalent shutdown period was 64 h 20 min, corresponding to 2.5% of the equivalent hours of the monitoring period and 0.7% of the equivalent hours in a year's wind farm activity.

The proportion of the overall shutdown period in each class of average daily wind speed followed approximately the distribution of the proportion of number of days in each class (Fig. 4). Yet, the highest proportion (30%) of shutdown hours was registered in the 5-6 m^{s-1} wind class, whilst the most available wind class was 4–5 m^{s-1}. Overall, 41% of the summed equivalent shutdown period during the five years happened in days with low (0–5 m^{s-1}) average wind speed.

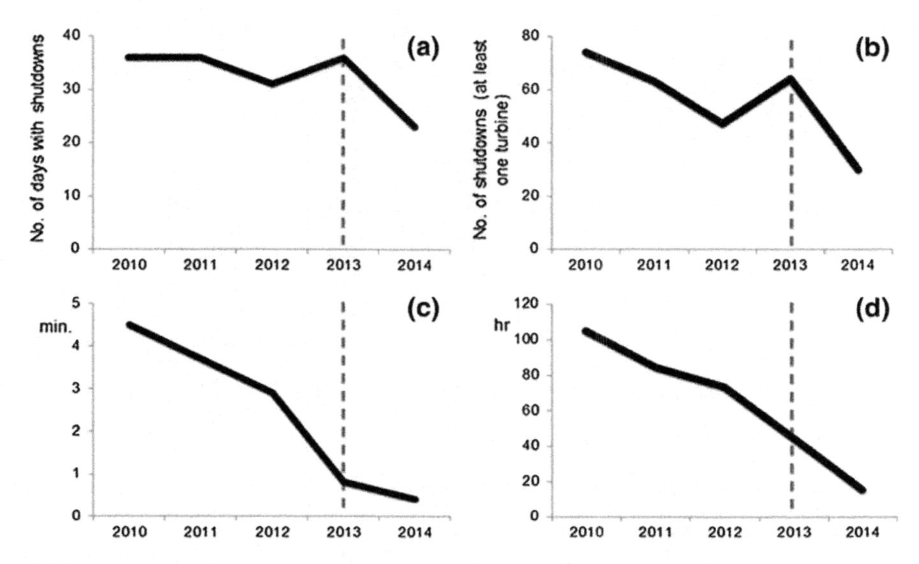

Fig. 3 **a** Number of days in which at least one turbine was shut down. **b** Number of shutdown periods (involving at least one turbine). **c** Average time until turbine shutdown after the order. **d** Total equivalent shutdown period; (*dashed line* represents the moment from which the field team accessed directly the SCADA system)

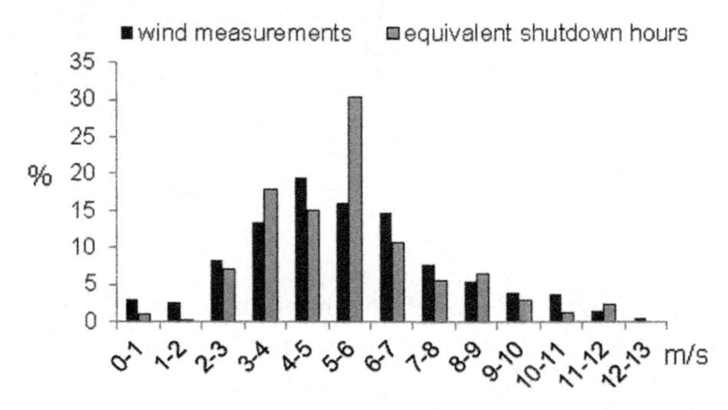

Fig. 4 Distribution of the proportions of the average annual number of days and of the equivalent shutdown hours, according to wind speed classes during 2010–2014

Shutdown Criteria

In most of the cases more than one shutdown criterion was verified simultaneously. The most frequently used criterion was "flocks of migratory soaring birds", followed by "intense migratory flux of soaring birds" and "threatened soaring bird species". The criterion of "imminent collision risk" was used on very few occasions

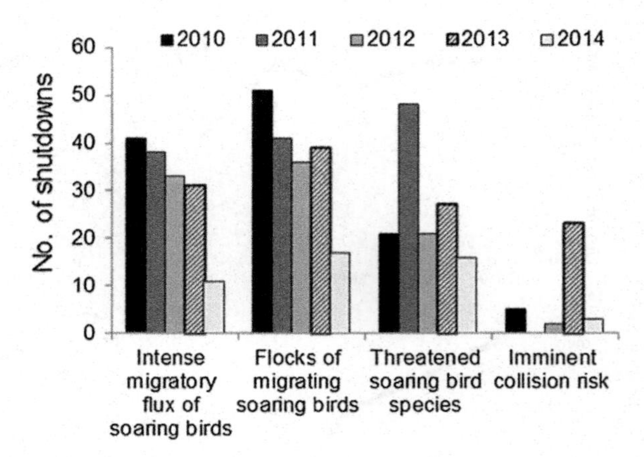

Fig. 5 Number of shutdown orders involving each of the criteria in different years (several shutdowns involved more than one criterion)

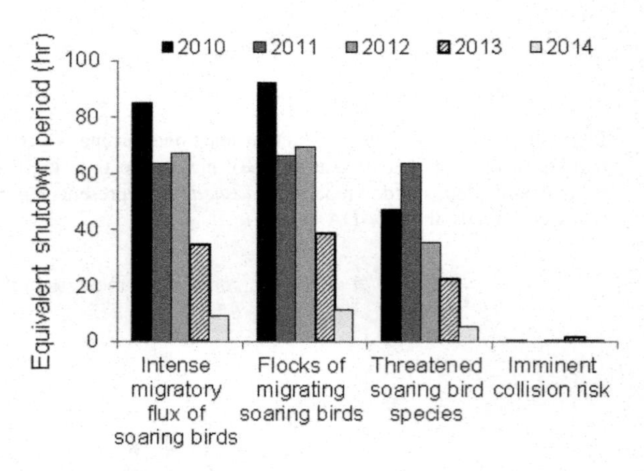

Fig. 6 Equivalent shutdown period resulting from the application of each criterion in different years (several shutdowns involved more than one criterion)

(Fig. 5). A similar pattern for the relative importance of the different criteria was obtained regarding the equivalent shutdown period. It should be noted that the application of "imminent collision risk" criterion was negligible in terms of shutdown duration (Fig. 6).

A great diversity of soaring bird species was included in the criteria that triggered turbine shutdown procedures. However, on average more than half (51.4%) of the yearly equivalent shutdown period involved exclusively or partially the occurrence of Griffon vultures (Fig. 7). The second species frequently associated to shutdown periods was the Black stork, whose occurrence was related to over 24% of the average yearly equivalent shutdown period (Fig. 7).

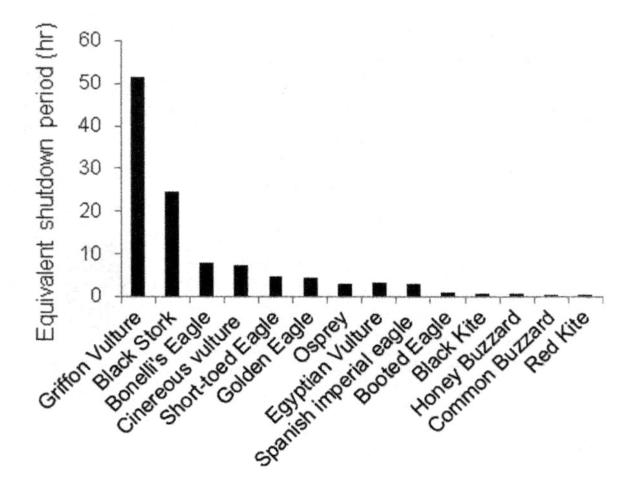

Fig. 7 Average annual equivalent shutdown period resulting from the presence of each species (some periods were associated to the presence of more than one species)

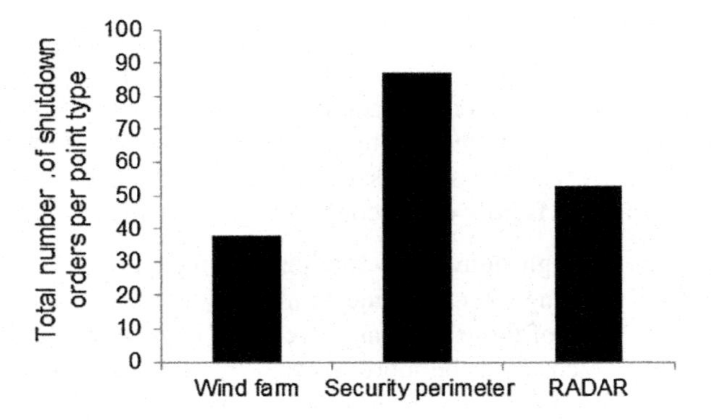

Fig. 8 Total number of times each vantage point type (wind farm monitoring points, security perimeter points without radar and radar point) first detected the soaring birds that triggered shutdown orders (n = 178)

Radar and Vantage Points

Nearly half of the shutdown operations were triggered after the detection of birds that constituted a shutdown criteria by VPs (without radar) located in the security perimeter (Fig. 1). In 30% of the operations the birds were first detected by the radar point, while the two VPs at the wind farm were the first to spot the birds in the remaining occasions. However, if the number of VPs in each category (wind farm, security perimeter and radar) is considered (two, four and one, respectively), their relative importance is changed (Fig. 8). In this case, the average number of shutdown operations triggered by detections by the radar point is higher, followed by the average values obtained from the remaining security perimeter points and from those at the wind farm (Fig. 9).

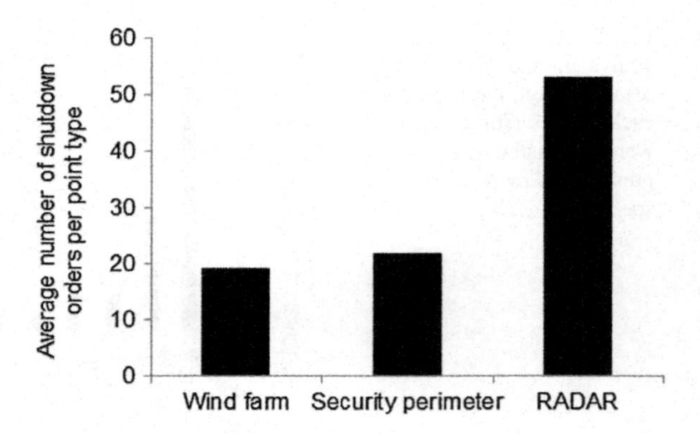

Fig. 9 Average number of times per point that each point type (wind farm monitoring points, security perimeter points without radar, security perimeter point with radar) first detected the soaring birds that triggered shutdown orders (n = 178)

Discussion

The application of the RASOD mitigation measure during the autumn migration period proved to be extremely effective in the BSJ wind farm, which is located at a major bottleneck site for migratory soaring birds.

The most relevant factor in evaluating the efficiency of the application of the RASOD procedure is the zero soaring bird mortality, resulting from the 35 fortnightly carcass surveys conducted in 2010–2014 during the migratory period. Different factors could contribute to mortality during this period:

1. the high number of soaring bird movements (3000) and individuals (18,300) crossing over the wind farm every autumn;
2. most of these birds are juveniles (Tomé et al. 1998, STRIX unpublished data), which might be more affected by collision mortality, as shown by the interactions with other man-made structures (Guil et al. 2011);
3. more than two-thirds and over a half of the soaring bird movements and individuals, respectively, occurred at high collision risk heights (between 20 and 200 m);
4. Griffon vultures (the most numerous migratory soaring bird in the area) are more likely to collide with turbine blades than most other avian species due to their large size, reduced manoeuvrability and visual field characteristics (Martin et al. 2012) and in fact show the highest raptor mortality rates from collisions in wind farms in Spain (e.g. Barrios and Rodríguez 2004; Lekuona and Ursúa 2007; de Lucas et al. 2008, 2012; Ferrer et al. 2012); and
5. since 2007 over 30 collision occurrences involving Griffon vultures have been recorded in wind farms in the vicinity of BSJ wind farm, mostly resulting in mortality.

A majority of the movements registered in the area at collision risk altitudes and closer than 45 m from the turbine's rotor occurred when the turbines were active. This results from the fact that most of the movements (85%) involved single individuals of non-threatened species that did not constitute per se a shutdown criterion. Most of these movements were recorded at the upper interval of the class height 100–200 m, hence a little above the blade swept area. Nonetheless, these movements were monitored carefully and turbine shutdown (following the criteria of imminent collision risk) was triggered whenever necessary. In contrast, 83.4% of the soaring birds involved in similar risk situations were registered while the turbines were shut down on account of the RASOD procedure, limiting dramatically the odds of mortality. Over the five year period only four individuals belonging to endangered species flew at theses distances from the turbines while these were active. As it would be expected from its abundance in the area, the Griffon vulture was responsible for most of the shutdown periods. Its gregariousness (aggregating in flocks that corresponded to a shutdown criteria) and flight behaviour, including long times to cross over the wind farm and frequent changes in course, also lead to an increase in the duration of shutdown periods.

The results of this study underline the importance of implementing a security perimeter around the wind farm and using a complementarily radar system to increase the efficiency of the procedure. Radar real-time data, in particular, provided an early detection and tracking of birds, improving the prediction of flight trajectories and anticipating their behaviour towards the wind farm, thus assisting the fieldwork coordinator with better quality information on which to base the shutdown decision. Furthermore, a pre-construction study carried out in the area showed that the radar detected at least twice the number of soaring birds that were detected by a visual observer (Pires 2008). Additionally, the cumulative experience by the monitoring team also affected positively the performance of RASOD application. This is shown by the continuous decrease on both the average time until shutdown after an order was given (a 91% decrease) and the average annual equivalent shutdown hours (86% decrease) over the period 2010–2014.

Moreover, the remote direct access to the wind farm SCADA by the monitoring team in the field, allowing for a more controlled application of RASOD, also resulted in significant decreases in the operational costs of the procedure on the wind farm productivity. In fact, without the need of intermediation by the wind farm company staff, and as turbine blades take only *ca.* 15 s to immobilize after the shutdown order is given in the SCADA, the operations coordinator in the field can wait until the last moment to decide upon the real need of shutting down. Likewise, restarting turbines can be ordered immediately after the birds that constituted a criterion abandon their close proximity. A new shutdown can be quickly implemented if a sudden change in the bird's direction is detected.

Increasing efficiency in RASOD application also lead to decreasing losses in wind farm productivity. While in 2010 the total equivalent shutdown period corresponded to 1.2% of the annual available equivalent time, in 2014 that value decreased to only 0.2% as a result of adaptive management. Moreover, more than 40% of the equivalent shutdown periods occurred when wind speed, and consequently the wind farm production, was low. Therefore, RASOD application resulted in negligible losses in energy production, especially if compared with nature conservation gains.

While the application of a temporary turbine shutdown in Spain has decreased the mortality rate of griffon vultures by 50% (de Lucas et al. 2012), the BSJ wind farm is the only known case where a similar procedure resulted in zero mortality of soaring birds. Although in this case RASOD made it possible to reconcile energy production with the conservation of migratory soaring birds in an important flyway, this mitigation measure should not be seen as a panacea to solve that potential conflict. Furthermore, where applicable, RASOD procedure should be fine-tuned to each case, through extensive pre-construction assessment. In particular, RASOD protocol, namely the number and layout of vantage points in the wind farm and in a security perimeter, and the number and location of radars, should be adapted according to the size of the windfarm, turbines distribution, targeted species and local movement patterns.

Acknowledgements We would like to thank the owner company of BSJ wind farm, E-ON, and all the collaborators that helped during fieldwork throughout this study.

References

Atienza JC, Fierro IM, Infante O et al (2011) Directrices para la evaluación del impacto de los parques eólicos en aves y murciélagos (guidelines for assessing the impact of wind farms on birds and bats). SEO/BirdLife, Madrid

Barrios L, Rodríguez A (2004) Behavioural and environmental correlates of soar ing-bird mortality at on-shore wind turbines. J Appl Ecol 41:72–81. doi:10.1111/j.1365-2664.2004.00876.x

Bellebaum J, Korner-Nievergelt F, Dürr T, Mammen U (2013) Wind turbine fatalities approach a level of concern in a raptor population. J Nat Conserv 21:394–400. doi:10.1016/j.jnc.2013.06.001

Bernis F (1980) La migración de aves en el Estrecho de Gibraltar. Volumen I: Aves planeadoras. Universidad Complutense, Madrid

Bright J, Langston R, Bullman R et al (2008) Map of bird sensitivities to wind farms in Scotland: a tool to aid planning and conservation. Biol Conserv 141:2342–2356. doi:10.1016/j.biocon.2008.06.029

Canário F, Leitão AH, Tomé R (2012) Predation attempts by short-eared and long-eared owls on migrating songbirds attracted to artificial lights. J Raptor Res 46:232–234

Costa PAS (2004) Atlas do potencial eólico para Portugal Continental. Dissertation, Universidade de Lisboa, Lisbon

Dahl EL, Bevanger K, Nygård T et al (2012) Reduced breeding success in white-tailed eagles at Smøla wind farm, western Norway, is caused by mortality and displacement. Biol Conserv 145:79–85. doi:10.1016/j.biocon.2011.10.012

de la Cruz A, Onrubia A, Perez B et al (2011) Seguimiento de la migración de las aves en el estrecho de Gibraltar: resultados del Programa Migres 2009. Migres 2:65–78

de Lucas M, Janss GFE, Whitfield DP, Ferrer M (2008) Collision fatality of raptors in wind farms does not depend on raptor abundance. J Appl Ecol 45:1695–1703. doi:10.1111/j.1365-2664.2008.01549.x

de Lucas M, Ferrer M, Bechard MJ, Muñoz AR (2012) Griffon vulture mortality at wind farms in southern Spain: distribution of fatalities and active mitigation measures. Biol Conserv. doi:10.1016/j.biocon.2011.12.029

Drewitt AL, Langston RHW (2006) Assessing the impacts of wind farms on birds. Ibis 148:29–42. doi:10.1111/j.1474-919x.2006.00516.x

Ferrer M, De Lucas M, Janss GFE et al (2012) Weak relationship between risk assessment studies and recorded mortality in wind farms. J Appl Ecol 49:38–46. doi:10.1111/j.1365-2664.2011.02054.x

Finlayson C (1992) Birds of the strait of Gibraltar. Academic Press Inc., San Die go, California

Guil F, Fernández-Olalla M, Moreno-Opo R et al (2011) Minimising mortality in endangered raptors due to power lines: the importance of spatial aggregation to optimize the application of mitigation measures. PLoS ONE 6:1–9. doi:10.1371/journal.pone.0028212

IUCN (2015) IUCN red list of threatened species. Version 2.1. Available via http://www.iucnredlist.org. Accessed on 01 April 2015

Johnson GD, Strickland MD, Erickson WP et al (2008) Use of data to develop mitigation measures for wind power development impacts to birds. In: de Lucas M, Janss GFE, Ferrer M (eds) Birds and wind farms: risk assessment and mitigation. Quercus, Madrid, Spain

Kaldellis JK, Zafirakis D (2011) The wind energy (r)evolution: a short review of a long history. Renew Energy 36:1887–1901. doi:10.1016/j.renene.2011.01.002

Kuvlesky WP, Brennan LA, Morrison ML et al (2007) Wind energy development and wildlife conservation: challenges and opportunities. J Wildl Manage 71:2487–2498. doi:10.2193/2007-248

Lekuona JM, Ursúa C (2007) Avian mortality in wind power plants of Navarra (northern Spain). In: Birds and wind farms: risk assessment and mitigation. pp 177–192

Marques AT, Batalha H, Rodrigues S et al (2014) Understanding bird collisions at wind farms: an updated review on the causes and possible mitigation strategies. Biol Conserv 179:40–52. doi:10.1016/j.biocon.2014.08.017

Martin GR, Portugal SJ, Murn CP (2012) Visual fields, foraging and collision vulnerability in Gyps vultures. Ibis (Lond 1859) 154:626–631. doi:10.1111/j.1474-919X.2012.01227.x

Mellone U, Limiñana R, Mallia E, Urios V (2011) Extremely detoured migration in an inexperienced bird: interplay of transport costs and social interactions. J Avian Biol 42: 468–472. doi:10.1111/j.1600-048X.2011.05454.x

Moreau JA, Monk J (1957) Autumn migration in south-west Portugal. Ibis 99:500–508

Newton I (2008) The migration ecology of birds. Academic Press, London

Orloff S, Flannery A (1992) Wind turbine effects on avian activity, habitat use, and mortality in Altamont Pass and Solano County Wind Resource Areas: 989–1991. California Energy Commission, Golden, CO, USA

Pires NM (2008) The use of radar as tool for studying bird migration and its role in environmental impact assessment—a pilot study in Portugal. Dissertation, Lisbon University, Lisbon

Redlinger RY, Dannemand P, Morthorst E (2002) Wind energy in the 21st century. Palagrave MacMillan, New York

Strandberg R, Klaassen RHG, Thorup K (2009) Spatio-temporal distribution of migrating raptors: a comparison of ringing and satellite tracking. J Avian Biol 40:500–510. doi:10.1111/j.1600-048X.2008.04571.x

Thelander CG, Smallwood KS (2007) The Altamont Pass wind resource area's effect on birds: a case history. In: de Lucas M, Janss GFE, Ferrer M (eds) Birds and wind farms: risk assessment and mitigation. Quercus, Madrid

Tomé R, Costa H, Leitão D (1998) A migração outonal de aves planadoras na região de Sagres—resultados da campanha de 1994. SPEA, Lisbon

Vardanis Y, Klaassen RHG, Strandberg R, Alerstam T (2011) Individuality in bird migration: routes and timing. Biol Lett 7:502–505. doi:10.1098/rsbl.2010.1180

Zalles JL, Bildstein K (2000) Raptor watch: a global directory of raptor migration sites. BirdLife Conserv, London

Mitigating Bat Mortality with Turbine-Specific Curtailment Algorithms: A Model Based Approach

Oliver Behr, Robert Brinkmann, Klaus Hochradel, Jürgen Mages, Fränzi Korner-Nievergelt, Ivo Niermann, Michael Reich, Ralph Simon, Natalie Weber and Martina Nagy

Abstract Alarmingly high numbers of bats are being killed at wind turbines worldwide, raising concerns about the cumulative effects of bat mortality on bat populations. Mitigation measures to effectively reduce bat mortality at wind turbines while maximising energy production are of paramount importance. Operational mitigation (i.e. feathering wind turbine rotors at times of high collision risk for bats) is currently the only strategy that has been shown to substantially reduce bat mortality. This study presents a model based approach for developing curtailment algorithms that account for differences in bat activity over the year and night-time and are specific to the activity level at a certain wind turbine. The results show that easily measurable variables (wind speed, month, time of night) can predict times of higher bat activity with a high temporal resolution. A recently published collision model that was developed based on an excessive carcass search study is then applied to predict bat collision rate based on the modelled bat activity. Using the ratio of wind energy revenue and collision rate, 10 min intervals were weighted, so that turbines are stopped when collision rate is high and loss in revenue is low. A threshold of two dead bats per year and turbine resulted in a mean loss in annual revenue of 1.4%. The presented approach of acoustic monitoring at the nacelle and turbine specific curtailment has become the standard method to mitigate collision risk of bats at wind turbines in Germany.

O. Behr (✉) · K. Hochradel · J. Mages · R. Simon · N. Weber · M. Nagy
Friedrich-Alexander-University Erlangen-Nürnberg, Erlangen, Germany
e-mail: oli.behr@fau.de

R. Brinkmann
Freiburg Institute of Applied Animal Ecology, Freiburg, Germany

F. Korner-Nievergelt
Oikostat GmbH, Ettiswil, Switzerland

F. Korner-Nievergelt
Swiss Ornithological Institute, Sempach, Switzerland

I. Niermann · M. Reich
Leibniz University Hannover, Institute of Environmental Planning, Hannover, Germany

J. Köppel (ed.), *Wind Energy and Wildlife Interactions*,
DOI 10.1007/978-3-319-51272-3_8

Keywords Acoustic activity · Bats · Collision risk · Central Europe · Operational mitigation · Wind turbine

Introduction

Bats are killed at wind turbines worldwide and frequently at high numbers (see overview in Arnett et al. 2016). Over a 12 year period (2000–2011) cumulative bat fatalities at wind turbines in the U.S. and Canada were estimated to range from 0.8 to 1.7 million bats (Arnett and Baerwald 2013) and similarly high fatality numbers have been suggested to occur in Germany (Voigt et al. 2015). Due to their long generation times and low reproductive rates, bats rely on high adult survival to maintain populations (Racey and Entwistle 2000; Barclay and Harder 2003). However, data on bat populations is scarce (O'Shea et al. 2003) and the known or suspected decline of some bat populations around the world due to numerous threats (e.g. loss of roosting and foraging habitat, climate change, white nose syndrome; e.g. Pierson 1998; Frick et al. 2010a, b; Winhold et al. 2008; O'Shea et al. 2016) raises warranted concerns about biologically significant additive mortality. Mitigation measures to reduce bat fatalities at wind turbines are thus critically important to maintain viable bat populations and their ecosystem services, and also for the environmental-friendly development and the public acceptance of wind energy.

Pre-construction estimation of bat collision risk at wind facilities is methodologically extremely difficult, involving considerable effort and expense and is associated with high prediction uncertainty. Hein et al. (2013) reviewed results from 12 study sites in the USA and failed to find a significant relationship between pre-construction acoustic activity (measured at ground level or up to 30 m above ground level) and post-construction fatalities of bats, suggesting that acoustic data gathered prior to the construction of wind facilities cannot reliably predict post-construction bat fatalities. A possible explanation for difficulties in predicting bat fatalities from pre-construction data could be that the presence of wind turbines alters the bats' habitat and behaviour. Bats may be attracted to wind turbines (Cryan et al. 2014) and/or change their habits of site use due to changes in habitat. These changes may include the creation of new hunting grounds (e.g. clearings and crane pads) and guidance structures (e.g. aisles, access routes) (Arnett et al. 2008; Cryan 2008; Cryan and Barclay 2009).

The evaluation of habitat characteristics (e.g. proximity to open water sources or known roosts) in the vicinity of planned wind energy facilities is commonly used in pre-construction fatality assessments to avoid building wind turbines close to bats' flight paths, roosts or foraging areas. However, most studies find weak or no correlations between bat fatalities or acoustic activity at the nacelle and habitat characteristics. These studies include land cover and distance to nearest wetland or woodlot/woodlands and forest (Johnson et al. 2004; Niermann et al. 2011) and the proximity of turbines to the coast or vegetation (Hull and Cawthen 2013). Thus, habitat characteristics do not seem to provide robust information on the risk of bat fatalities and their contribution to developing mitigating strategies seems to be limited.

Carcass searches have frequently been used to estimate the number of bats that die at existing wind turbines. However, the actual number of fatalities may be vastly underestimated if detection biases (i.e. scavenger removal, searcher efficiency and searchable area) are not accounted for (Kerns et al. 2005; Huso 2010; Korner-Nievergelt et al. 2011b). At many sites in Central Europe collision rate is difficult and expensive to assess with carcass searches, because in comparison to North America collision rates are often lower, removal rates by scavengers are high and searching conditions are often poor (Bispo et al. 2013; Korner-Nievergelt et al. 2013). Predicting collision rates from variables, like acoustic activity or wind speed, that are more easy to measure, seems more appropriate for many sites in Central Europe, including forest areas and off-shore sites.

Currently, only operational mitigation (i.e. stopping the rotors of wind turbines at times of high collision risk for bats) has been shown to substantially reduce the number of bats killed at wind energy facilities (Arnett et al. 2013, 2009). Peak numbers of bat fatalities have consistently been associated with low wind speeds and specific periods of the year, such as late summer to early fall in the temperate northern Hemisphere (Kerns et al. 2005; reviewed in Arnett et al. 2008; Schuster et al. 2015). Thus, a frequent and successful strategy to mitigate bat mortality consists in raising a turbine's cut-in wind speed (i.e. wind speed at which turbines start generating power to the utility system) above the manufacturer's cut-in wind speed (usually 2.5–4.0 ms^{-1}) and to allow only very slow movements of the rotor below the cut-in wind speed.

Increasing the cut-in speed should prevent the rotor from turning at a speed dangerous for bats, such as low wind speeds when bats are highly active. Usually this is done by feathering rotor blades at 90° until the cut-in speed is reached (Arnett et al. 2013). Most operational mitigation studies from North America and Europe demonstrated a substantial reduction in bat fatalities (frequently exceeding 50%) when raising the cut-in speed to 5.0−6.5 ms^{-1} (North America: e.g. Arnett et al. 2009; Baerwald et al. 2009; see also synthesis in Arnett et al. 2013; Europe: Behr and von Helversen 2006; Beucher et al. 2011). The few studies that estimated the costs of increasing cut-in speed reported that loss in power revenue due to curtailed operation of turbines during short periods (e.g. 75 days) of high collision risk would constitute 3–11% lost power output during this period. This is less than 1% of the total annual output (see synthesis in Arnett et al. 2013).

Substantial reductions in bat mortality at wind turbines can already be achieved with relatively unspecific operational curtailment based solely on wind speed (Arnett et al. 2013). However, further research is needed on more efficient operational mitigation that incorporates additional variables (e.g. time of night, bat activity) to define operation rules that are turbine specific and maximize energy production with the lowest possible collision risk for bats (Arnett et al. 2013; Weller and Baldwin 2011). Korner-Nievergelt et al. (2013) published a model-based approach to predict the collision rate of bats at wind turbines based on fatality search data, wind speed and acoustic activity measured at the nacelle. Once the model has been calibrated based on a sufficiently large data set, its predictors can be used to assess collision rate for new turbines with no need for carcass searches and to develop turbine-specific curtailment algorithms (Korner-Nievergelt et al. 2013).

This chapter outlines the development of turbine specific curtailment algorithms based on the same data set from German wind turbines used by Korner-Nievergelt et al. (2013). First, the data necessary to calculate the algorithms is outlined: acoustic activity data and predictive parameters like wind speed, time of year, and time of night. Then the modelling approach (GLM) is presented to predict bat activity based on parameters like wind speed, month, and time of night with a high time resolution (10 min intervals). Based on the predicted bat activity, collision risk and collision rate of bats will subsequently be estimated with the mixture model presented in Korner-Nievergelt et al. (2013). Finally, by calculating the ratio Q of power revenue (or wind speed to the third power that shows a linear correlation to the power produced) and estimated collision rate for each 10 min interval, the times with low power revenue and high collision rate for bats are identified (i.e. 10 min intervals with low values of Q). These are the times when turbines should preferably be stopped. Using the ratio Q and by setting a threshold for the number of accepted bat fatalities per turbine and year, the cut-in wind speeds that are specific for the bat activity level at a single turbine are calculated and then differentiated for different months and night-times.

Methods

Data-Set

In 2008, 70 wind turbines were sampled at 35 different sites (two turbines each) in four different natural regions in Germany (Fig. 1). All turbines sampled were Enercon turbines (ENERCON GmbH, Aurich, Germany) of the types E66, E70, and E82 with rotor diameters of 66, 70, and 82 m, respectively.

Batcorders (Model 1.0, ecoObs GmbH, Nürnberg, Germany) were installed by Enercon service teams in the nacelle of each turbine to continuously record acoustic bat activity in the rotor swept area. Batcorders were positioned inside and in the bottom of the nacelle between the rotor and the tower with the microphone pointing downwards, through the nacelle floor. Holes were drilled into the nacelle floor for this purpose (Fig. 2). Installation of detectors commenced in April, but by the end of April less than 10 detectors were successfully installed; and hence were excluded in April in this analysis. Data included into the analysis was recorded from 2008-05-01 until 2008-10-31 with a minimum of eight (beginning of May) and a maximum of 68 (in August) detectors sampling valid data during the same night.

Batcorders were operated with the following settings: Quality 20, threshold −36 dB, posttrigger 200 ms, and critical frequency 16 kHz. Batcorders ran continuously but produced valid data only during 71% of the nights sampled. Batcorder downtimes were mostly due to power, microphone or SD-card-failures, full

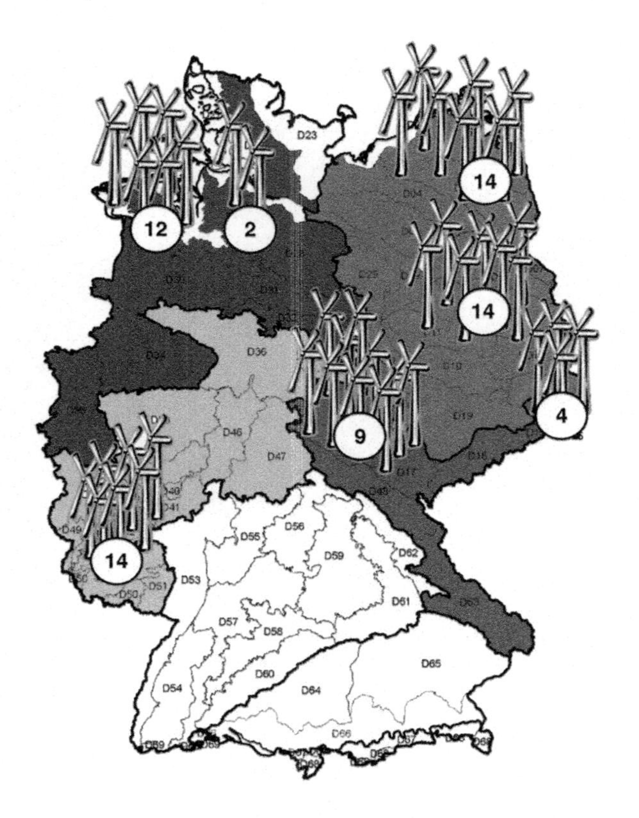

Fig. 1 Location of the wind turbines sampled in different natural regions of Germany. Numbers in *white circles* indicate the number of turbines sampled per area. Turbines were located in the Northwest German Plain (*blue*), the Northeast German Plain (*green*), the Western Central Uplands (*yellow*) and in the Eastern Central Uplands (*red*) (Color figure online)

SD-cards, or other technical problems with the detectors. Data from one turbine were excluded from the data set due to problems with microphone sensitivity. For the remaining 69 turbines, the mean number of nights with valid data was 126 per turbine (minimum 7, maximum 184) of a total of 184 nights sampled with a total sample time of 96,838 h. More than a million files were recorded (more than 400 Gbyte), of which 72,756 contained bat calls.

One temperature sensor was also installed (Sensor KTY81-110, Philips, Amsterdam, Netherlands), as well as one precipitation sensor (Sensor 5.4103.20.041, Adolf Thies GmbH, Göttingen, Germany) at most of the nacelles (68 temperature and 60 precipitation sensors). Temperature sensors were positioned in the nacelle floor at about half a meter distance from the detector microphone. Precipitation sensors were installed on top of the nacelle at the framework supporting turbine lighting and anemometer. The owners of the turbines provided access to the wind speed data measured with an anemometer recorded by the SCADA-System (Supervisory Control and Data Acquisition System) controlling the turbine. Wind speed data was available as mean values for 10 min intervals.

On two occasions at two different turbines, exceptionally high acoustic activity was recorded (1903 and 2071 recordings with bat calls, respectively) within a short

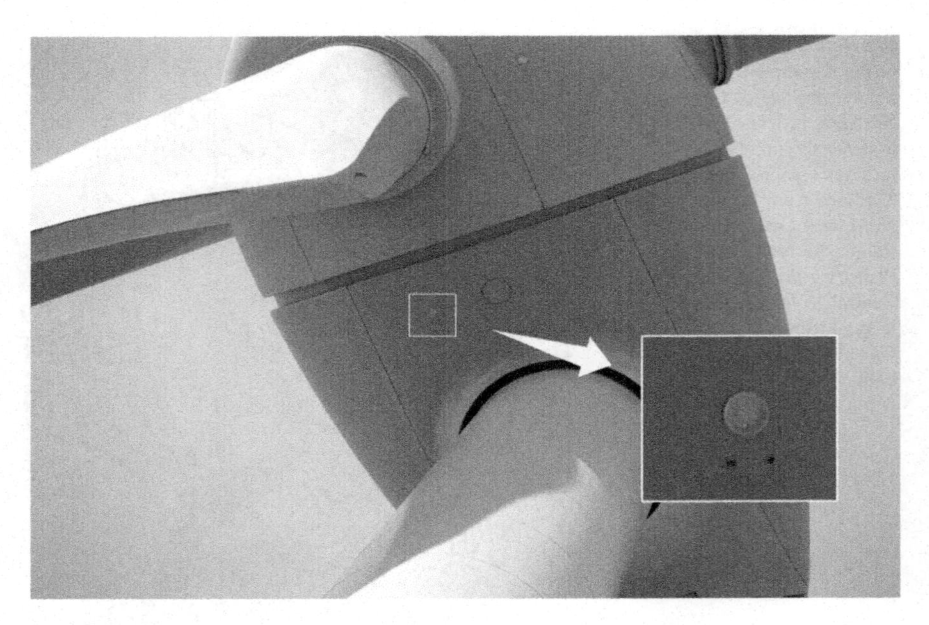

Fig. 2 Microphone mounted in a disc in the *bottom* of a nacelle (highlighted area on the *left*). The inlay on the *right* shows this area at a larger scale. The microphone disc has a *light grey* colour and the two *dark spots below* it are additional sensors (e.g. temperature) (Color figure online)

time period (1.3 and 2.8 h). This activity was mostly caused by calls of the Common Pipistrelle (*Pipistrellus pipistrellus*) and is most likely attributed to swarming behaviour. *P. pipistrellus* is well known for swarming behaviour, which can result in a short-term occurrence of a large number of bats, especially around existing or potential roosts (Simon et al. 2004). On both occasions wind speed was very low (max. 2.4 and 1.9 ms^{-1}) and so was the collision risk because rotors were only moving very slowly. Even a minor increase in wind speed of 0.5 ms^{-1} would have resulted in a large collision risk for the bats. Predictions of the frequency and occurrence of swarming behaviour in *P. pipistrellus* were not possible with such limited data (i.e. a sample size of 2 nights). Thus it was decided to exclude these swarming periods from the dataset, also because they appeared to be clear outliers in our analysis.

Identification of Species and Species Groups

To produce standardised results that can be compared to other studies, the software bcAdmin was used (Ver. 1.13, ecoObs; Call filter: Amplitude threshold 1.585, smoothness 2.00, Samples Hi 200, Min. call distance 15 ms, Min. call length

1.50 ms; Call Extraction: Min call interruption 1.10 ms, Forward MSE 0.060, Samples for regr. 8, Regression size 200 µs) and bcDiscriminator [Ver. 1.13, ecoObs; in combination with R 2.7.2 (R Core Team 2015) and the packages kernlab and RandomForest (Liaw and Wiener 2002)] to automatically identify bat calls in the batcorder recordings and also to identify species and species groups. All recordings that were automatically identified by the software as bat calls were manually checked and misclassified noise recordings were manually removed from this dataset.

Statistical Modelling of Bat Activity

Statistical modelling was used to predict bat activity (number of recordings; total activity of all bat species—Chiroptera) from predictive variables such as wind speed or temperature. Bat activity was modelled for 10 min intervals as wind speed data was available as mean values per 10 min intervals and because intervals of 10 min length were considered appropriate to map the temporal variability in bat activity and wind speed. Temperature and precipitation were also assigned to 10 min intervals as 10 min mean values. To attribute 10 min intervals to relative times of night, the center-time of that interval was used (5 min after its commencement).

All modelling was done in R (R Core Team 2015). Generalised linear models were used (function glm of the R-package stats) with Poisson error distribution and the logarithm link function. Different models were tested, including different combinations of predictive variables, including wind speed and wind speed^2 as continuous variables, temperature, precipitation, month, time of night, and turbine as categorical variables. The correlation of activity with temperature and precipitation did not follow a simple mathematical function but could be fitted when including them as categorical variables into the model (precipitation was categorised in pseudo-logarithmic categories with the margins 0.0001, 0.001, 0.01, 0.1, 1, 10, 100, 1000, 10,000 lx; temperature was categorised in steps of 5 °C from 0 to 40 °C). The turbine variable would usually be considered a random effect. Due to the high number of sampled nights per turbine the turbine variable could be included as a fixed effect (equal to the rest of the predictive variables). This increased the stability of the model fitting algorithm.

The full model contained wind-speed, the square of wind-speed, temperature, precipitation, month, time of night and turbine as predictor variables. Model selection was done based on AIC (Akaike information criterion, Burnham and Anderson 2002) and economic trade-offs. All models included the wind speed and turbine variables.

Studies quantifying the activity of bats with acoustic detectors make implicit assumptions about data structure (Hayes 2000; Sherwin et al. 2000; Gannon et al. 2003). In this study, it was assumed that recordings at a turbine within one night were correlated. Hence, autocorrelation was corrected by thinning to every 20th sample from the dataset. This selected dataset showed no relevant autocorrelation in the acf-plot. It was also assumed that recordings at different turbines and in different nights were independent.

Predicting the Collision Risk

The GLM model described in the previous chapter was used to differentiate times of low and high bat activity. Since access to wind speed data was obtained for all turbines for the entire year of 2008, the GLM model could be used to predict bat activity at all turbines, not only for times that had been sampled acoustically. As a result the month of April could be included into the dataset, applying the effect of this month measured at a turbine subset to all turbines. Time intervals without wind speed (Table 1) were extrapolated assuming that the effects of wind speed, month, time of night, and turbine did not differ between extrapolated and sample times.

The n-mixture-model published in Korner-Nievergelt et al. (2013) was used to calculate the collision rate from the acoustic activity predicted by the GLM (see also Table 4 on page 338, model type A in Korner-Nievergelt et al. 2011a). The n-mixture-model has been developed to predict the collision rate for entire nights from acoustic activity and from wind speed data. In order to use the model to predict the collision rate during 10 min intervals, subsequent calculations were undertaken and the following assumptions were made: (1) The activity predicted for a 10 min interval was extrapolated to the entire night using the length of the night (number of 10 min intervals) and weighted by the distribution of activity over night times shown in Fig. 4. (2) Then the collision rate for the entire night was calculated using the wind speed measured during the 10 min interval as predictor in the model. When applying the model to entire nights, the median of 10 min wind speed data per night was used as predictor. Therefore it was assumed that the effect of wind speed on fatality rates was the same during different times of night. (3) The predicted fatality rate during the entire night was then split into single 10 min intervals using, again, the length of the night (number of 10 min intervals) and the distribution of activity over night times shown in Fig. 4. Thus, it was assumed that the distribution of fatality risk over the night equals that of the acoustic activity.

Table 1 Sample size of the datasets for acoustic bat activity and meteorological parameters in 2008

Detector/sensor	Turbines	Nights	Hours
Batcorder	70	9.074	99.135
Wind	69	11.495	125.088
Temperature	68	9.878	109.644
Precipitation	60	7.238	84.588

Calculating the Loss in Power Production

The potential loss in revenue caused by curtailment algorithms was calculated from the turbine data provided by the owners. Data was available on the power production during 10 min intervals (the total time with valid data was the same as for wind speed, see Table 1) and for the entire year of 2008. Potential loss in power production was calculated by simply adding the power produced during 10 min intervals that were part of the curtailment periods defined by these algorithms. The loss in power production was calculated as a percentage of annual production, since these relative numbers are easier to compare for different sites and turbine types.

Results

Bat Species Recorded

Two species-groups together accounted for 86% of all recordings. The larger of these two groups (70.4% of all bat recordings) was the "Nyctaloid" group with larger species producing lower frequency echolocation calls. This group contained recordings mostly from the species *Nyctalus noctula* (Common Noctule, 15.0% of all bat recordings) and *Vespertilio murinus* (Parti-coloured Bat, 4.9%), and some *N. leisleri* (Lesser Noctule 0.2%), *Eptesicus nilssonii* (Northern Bat, 0.1%), and *E. serotinus* (Serotine Bat, 0.03%). The second was the "Pipistrelloid" group (16.1% of all recordings) with smaller species producing higher frequency calls. Recordings in this group contained calls mostly from *Pipistrellus pipistrellus* (Common Pipistrelle, 9.4%) und *P. nathusii* (Nathusius' Pipistrelle, 4.4%), and some *P. pygmaeus* (Soprano Pipistrelle, 0.1%) calls. Almost all remaining recordings (13%) were identified as bat calls by the software but could not be assigned to any species or species group.

Factors Affecting Bat Activity

Bat activity varied greatly depending on wind speed, temperature, precipitation, time of night, and month (Fig. 3).

Wind speed had a strong influence on bat activity (number of recordings) in all species and species groups and showed an approximately logarithmic trend. Only 15% of all bat activity (Chiroptera) was recorded at wind speeds ≥ 5 ms^{-1}, and only 6% at wind speeds ≥ 6 ms^{-1} (see Fig. 4). It is apparent from the log-plot that the decrease of activity with higher wind speeds was more pronounced in *P. pipistrellus* (6.4% of activity at wind speeds ≥ 5 ms^{-1}) than in the other species, while *P. nathusii* showed the greatest wind tolerance (18% of activity ≥ 5 ms^{-1}). The highest wind speed with activity recorded was 11.5 ms^{-1}. Bat activity

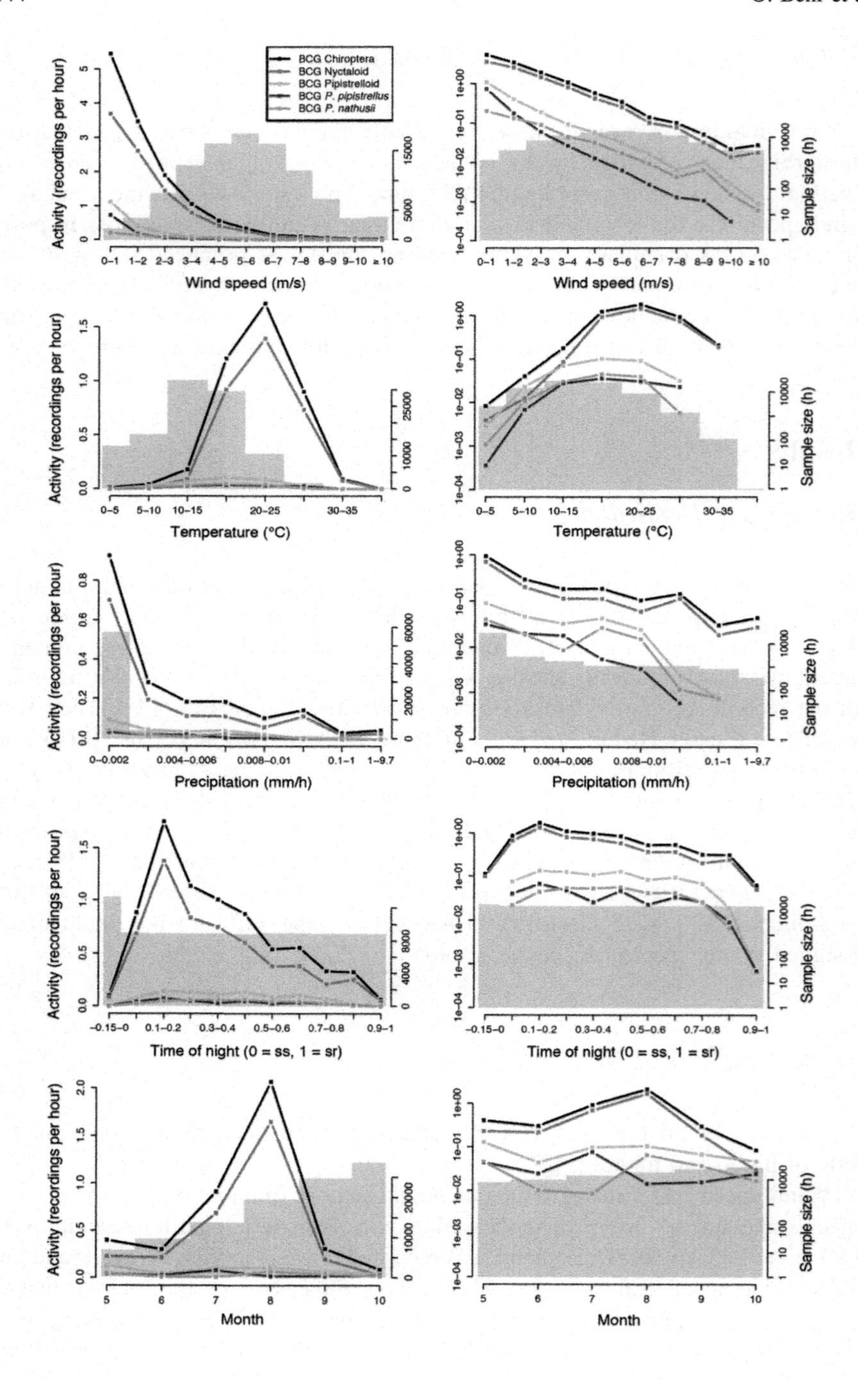
BCG Chiroptera
BCG Nyctaloid
BCG Pipistrelloid
BCG P. pipistrellus
BCG P. nathusii
Activity (recordings per hour)
Sample size (h)
Wind speed (m/s)
Temperature (°C)
Precipitation (mm/h)
Time of night (0 = ss, 1 = sr)
Month

◄**Fig. 3** Effect of different parameters (x-axes) on acoustic bat activity (number of recordings per hour) measured at the nacelle of 69 wind turbines at 35 sites in 5 different natural regions in Germany in 2008. The panels on the *left* have a linear y-axis and panels on the *right* show the same data with a logarithmic y-axis (zero values not shown). Lines of different colours show values for different species or species groups. Variables on the x-axis were factorised in intervals for the purpose of this plot: intervals for wind speed and temperature include the lower and exclude the upper margin shown. Precipitation was factorised in a pseudo-logarithmic scale (excluding the lower and including the upper margin). To compare nights of different lengths, relative times of night were used, from 0 (sunset, ss) to 1 (sunrise, sr). Then the night was split into 10 intervals of equal length (0–0.1, 0.1–0.2, etc.—lower margin included and upper margin excluded). In addition, an interval before sunset was included (−0.15 to 0) that had a 50% longer duration than the intervals during the night and included bat activity at dusk (no bat activity was recorded after sunrise). The months cover the time sampled from April to October. *Grey bars* in the background indicate the sample size in hours (right y-axes) for the respective intervals (Color figure online)

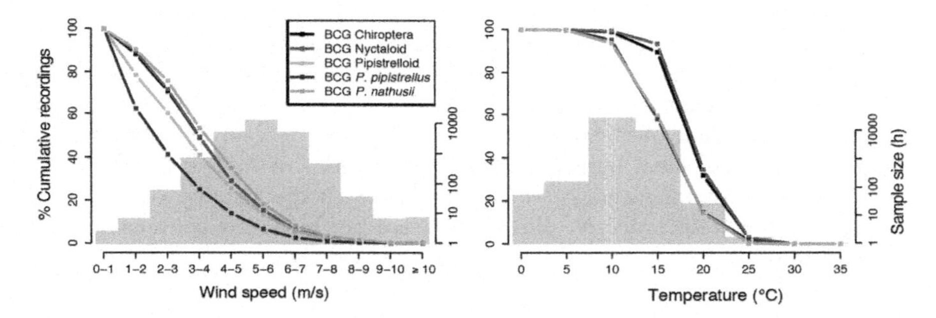

Fig. 4 Effect of wind speed and temperature on acoustic activity (recordings per hour) of bats at the nacelle of turbines. Activity is plotted cumulatively to show the remaining percentage of activity above a certain threshold of wind speed or temperature. The dataset was the same as in Fig. 3

increased markedly for **temperatures** between 10 and 25 °C. Low activity at temperatures above 25 °C is based on a small sample, was mostly recorded during the dusk interval before sunset (time of night −0.15 to 0), and is, hence, almost exclusively due to activity of the Common Noctule, a species frequently active before sunset.

Bat activity decreased with small rates of **precipitation** (fog or clouds with 0.002–0.004 mm min^{-1}). The sample size for higher rates of precipitation is low and rain periods amounted only to a small fraction of the entire time sampled. **Time of night**: bat activity peaked during the first half of the night. Activity of the Nyctaloid group (mostly Common Noctule) started before sunset and earlier than for the Pipistrelloid group. After the first quarter of the night, activity started to decrease continuously until sunrise, with a stronger decrease shortly before sunrise. At a few turbines a second, lower activity peak was recorded at the end of the night that was mostly due to activity of Nyctaloid species. Nathusius' Pipistrelle, again, showed a slightly different pattern than the other species: activity peaked during the middle of the night. Activity patterns during different **months** varied for the species recorded: all species and species groups showed a minor decrease of activity from May to June and a marked peak in late summer or autumn. For the Common Pipistrelle, this peak was recorded in July, while the activity of the Nyctaloid group peaked in August, and still later in August and September for Nathusius'Pipistrelle.

Predicting Bat Activity

Generalised linear modelling (GLM) was used to predict bat activity (number of recordings; total activity of all bat species) and, hence, times of high collision risk for bats at the wind turbines from the predictive variables wind speed, wind speed^2, temperature, precipitation, month, time of night, and turbine. The turbine effect described differences in the activity level between different turbines.

Interactions between variables either made no sense from a biological point of view (and were, therefore, not tested) or showed only small effects on the activity predicted. Interactions in each month were tested with each factor: wind speed, temperature, and time of night and of turbine with each of wind speed, month, and time of night. The final model used to predict times of high bat activity was calculated without interactions, which also increased the stability of the model fitting algorithm. Accordingly, differences were not modelled between different turbines in phenology or bat activity pattern over the night.

All predictive variables highly significantly improved the model fit (likelihood-ratio-test) and reduced AIC. Since the model is being used to calculate curtailment algorithms for wind turbines the decisions on whether to include predictive variables into the final model were also based on an economic trade-off. The cost of collecting the respective data during a site assessment (e.g. installation and maintenance of sensors, or data collection and analysis) was balanced against the higher energy yield when turbines operate with more sophisticated algorithms. This led to an exclusion of temperature and precipitation from the final model, since the increase in energy yield amounted to a mean of 13,970 kWh per turbine during 20 years of operation, which is not enough to justify the sampling of these variables. The coefficients of the final model are shown in Table 2. The final model used for the prediction of bat activity included wind-speed, the square of wind-speed, month, time of night and turbine as predictors.

Model effects reflected the univariate distributions shown in Fig. 3: Predicted activity sharply decreased with higher wind speeds, peaked in late summer and autumn, and in the first quarter of the night.

Predicting Collision Rate

The model presented in the previous section was used to predict the level of bat activity for all 10 min intervals during the entire period of 2008-04-01 to 2008-10-31

Table 2 Generalised linear model parameters for the prediction of bat activity (log of the number of recordings of all bat species per 10 min interval) from continuous variables (wind speed and wind speed^2) and categorical variables (temperature, precipitation, month, time of night, and turbine)

Effect	Total number of 10 min intervals	Number of 10 min intervals with bat activity	Number of recordings	Coefficient	p
(Intercept)	581,028	10,159	65,823	−5.03	<0.001
Wind-speed	581,028	10,159	65,823	−0.71	<0.001
Wind-speed^2	581,028	10,159	65,823	0.01	0.001
Month: May	38,338	475	2612	0.00	NA
Month: June	54,204	542	2800	−0.41	0.213
Month: July	76,380	1552	12,012	1.06	<0.001
Month: August	109,145	5279	38,996	1.74	<0.001
Month: September	139,770	1754	7094	−0.31	0.267
Month: October	163,191	557	2309	−1.36	<0.001
Time: −0.15 to 0	55,354	242	1043	0.00	NA
Time: 0–0.1	52,913	1494	7768	2.07	<0.001
Time: 0.1–0.2	52,670	2142	15,250	2.79	<0.001
Time: 0.2–0.3	52,716	1432	9839	2.40	<0.001
Time: 0.3–0.4	52,598	1176	8722	2.33	<0.001
Time: 0.4–0.5	51,952	1002	7506	2.16	<0.001
Time: 0.5–0.6	52,858	779	4751	1.73	<0.001
Time: 0.6–0.7	52,737	700	4829	1.73	<0.001
Time: 0.7–0.8	52,534	572	2830	1.21	<0.001
Time: 0.8–0.9	52,434	477	2747	1.16	<0.001
Time: 0.9–1	52,262	143	538	−0.48	0.188
Turbine: min	10,619	1	7	−1.44	NA
Turbine: 1st quartile	9636	41	213	1.90	0.989
Turbine: median	5712	106	454	2.78	0.296
Turbine: 3rd quartile	10,805	151	1316	3.36	<0.001
Turbine: max	9635	503	3571	4.69	<0.001

Reference levels of categorical variables have no level of significance assigned. From the effects of 69 turbines sampled, only the min, quartiles, and max are shown

(detailed in the methods section “Statistical Modelling of Bat Activity”). Then the n-mixture-model published in Korner-Nievergelt et al. (2013) was applied to calculate the collision risk from the predicted activity for each 10 min interval (for details and implicit assumptions see methods section “Predicting the Collision Risk”).

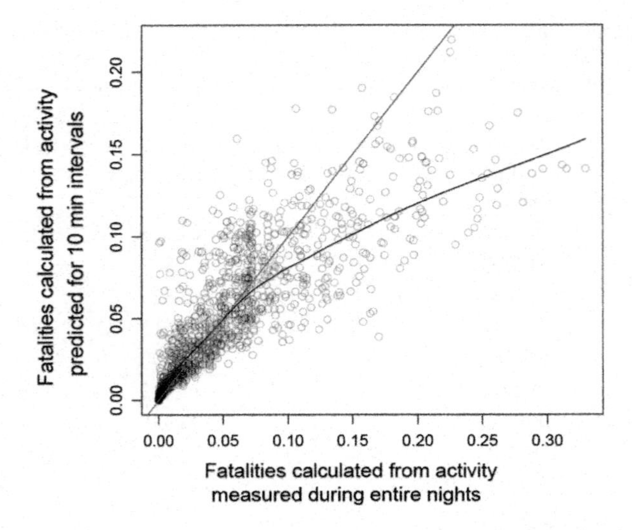

Fig. 5 The number of bat fatalities were calculated: (1) for 10 min intervals from the acoustic activity predicted by the GLM and with the method described here (fatality numbers then pooled for nights and turbines for comparison) on the y-axis, and (2) for entire nights from the acoustic activity measured on the x-axis. Calculations for the values on both axes are based on (Korner-Nievergelt et al. 2013). Transparent *circles* show figures for one night at a single turbine (*black* for three overlapping *circles*). The *red line* shows a perfect 1:1 correspondence, the *black line* a moving average (function loess in r-packet stats, span = 0.5) (Color figure online)

To test this approach and the underlying assumptions the number of fatalities predicted for 10 min intervals were pooled for each night and turbine and compared the result to the fatality numbers predicted when applying the model to entire nights and the activity measured (Fig. 5). The values generally showed a good correspondence, but very high activity during outlier nights was underestimated by the GLM and so was, in consequence, the fatality risk for these nights.

Curtailment Algorithms

To reduce the number of bat fatalities occurring at a wind turbine while minimising the curtailment cost, 10 min intervals were weighted. This was done by calculating the ratio Q of power revenue and estimating the collision rate per 10 min interval. When revenue data is not available, wind speed to the third power can be used that has a linear correlation to the power produced. Through this, times with low power revenue and high collision rates for bats (i.e. 10 min intervals with low values of Q) could then by identified. These are the times when turbines should preferably be stopped if necessary.

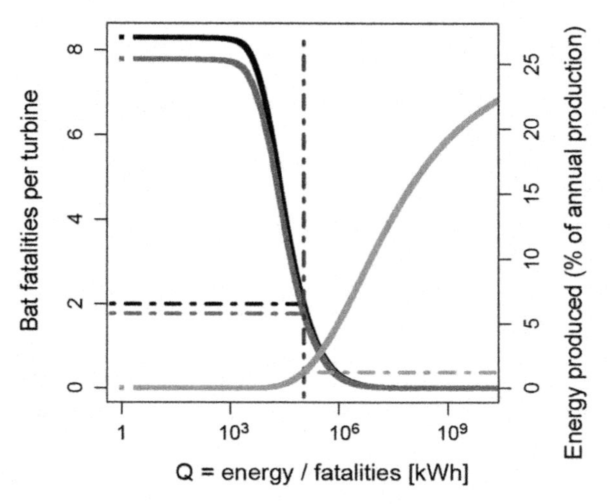

Fig. 6 Weighting of single 10 min intervals when running wind turbines with the curtailment algorithms described in this paper. Calculations are based on data from 2008-04-01 to 2008-10-31 (missing data have been intrapolated) and are mean values for the 69 turbines sampled. The **X-axis** in log-scale is the ratio of energy produced [kWh]/predicted number of bat fatalities (i.e. weighting factor Q for single 10 min intervals; plotting Q + 1 to avoid null-values on the logarithmic x-axis for intervals without energy produced. The upper limit of the x-axis was set to 10^{10} to render it easier to read. The **Y-axis on the left for the *black* and *red lines*** is the cumulative number of predicted fatalities (cumulated from high values of Q on the right to low values on the *left* of the x-axis). The *black line* shows the number of fatalities calculated from the activity numbers predicted by the GLM, the *red line* shows the slightly lower numbers calculated from the actual activity measured. The **Y-axis on the *right* for the *green line*** is the mean cumulative energy produced as a percentage of the annual revenue per turbine (cumulated from low values of Q on the left of the x-axis to high values of Q on the *right*). *Dashed lines* show the implementation of mitigation: a threshold of two dead bats per year and turbine on the left y-axis (*black line*) corresponds to a certain threshold for Q on the x-axis (*blue*) and a loss in revenue by mitigation und the right y-axis (*green*). The area *left* of the vertical *dashed blue line* contains 10 min intervals that have to be included into curtailment in order to keep the number of fatalities below the threshold (Color figure online)

To reduce the number of bat fatalities to a certain threshold (e.g. two dead bats per turbine per year—a threshold commonly used in Germany—dashed black line in Fig. 6), 10 min intervals have to be included into the curtailment periods starting with the intervals with the lowest values for Q (on the left hand side of Fig. 6). This has to be done until the cumulative fatality number predicted for the remaining intervals (Fig. 4, black line: based on activity predicted by the GLM; red line: based on measured activity), which are not included in curtailment periods, falls below the threshold. The value of Q at that point (dashed blue line in Fig. 6), then defines the criterion for single 10 min intervals to either slow the rotor to a speed not dangerous for bats (when Q for that 10 min interval is below the threshold) or to have the turbine run in normal mode (when Q is above the threshold).

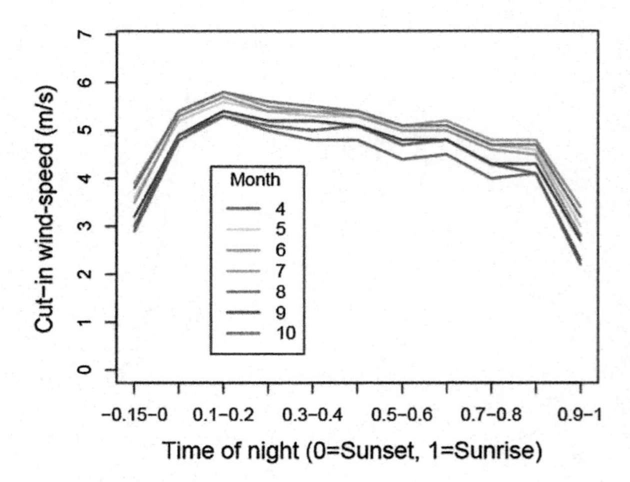

Fig. 7 Cut-in wind speeds defined by a curtailment algorithm for a sample turbine with a high level of bat activity to reduce the number of fatalities to two bats per year. In Germany turbines are frequently curtailed from April (4) to October (10). Cut-in wind speeds depend on the time of night (x-axis) and month (*colours*)

The value of Q for a specific 10 min interval depends on wind speed, on the fatality threshold, on the effects of month, time of night, and on the turbine effect on acoustic activity. Since the latter four are fixed and known, once the activity level of the turbine has been acoustically sampled, the models can be simplified to a cut-in wind speed for each combination of month and time of night (Fig. 7). These cut-in wind speeds are specific for the level of bat activity at a single turbine.

Curtailment Costs

The expected loss in power revenue can easily be calculated by adding up the power produced during all 10 min intervals included in curtailment periods (green line in Fig. 6). Losses in power production were calculated as a percentage of annual production (mean annual production was 2761 MWh for all turbines sampled) for different thresholds of bats killed per year and turbine (Fig. 8). With a threshold of two dead bats per year and turbine, the mean loss in power production was 1.4% (95% confidence interval: 1.0–1.8%, min 0.1%, max 4.2%) of annual revenue. Operational mitigation based solely on wind speed was more expensive (Fig. 8): for two dead bats per year and turbine the mean loss in power production was 1.8% (95% confidence interval: 1.7–2.3%, min 0.2%, max 5.8%) of annual revenue.

Discussion

The only method currently available to consistently reduce the number of bat fatalities at wind turbines is operational mitigation (Arnett et al. 2013), i.e. stopping the rotors of wind turbines at times of high collision risk for bats. Operational

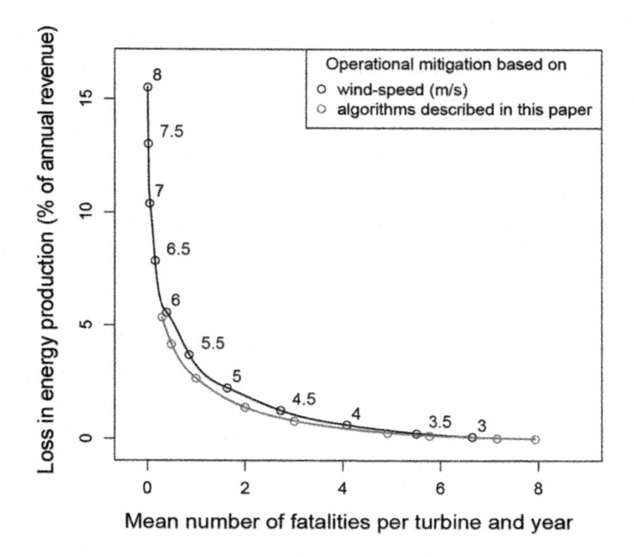

Fig. 8 Efficiency of operational mitigation based on wind speed only (*black line* and *black circles*—the numbers indicate the cut-in wind speed for each circle in ms^{-1}), or based on wind speed, time of night, and month (*red line* and *red circles*). *Circles* show the mean predicted loss in revenue (percentage of annual production) caused by curtailment algorithms for different thresholds of bats being killed per turbine and year with curtailment from April 1st to October 31st. Lines were drawn with R-function spline (Color figure online)

mitigation is usually achieved by feathering the rotor blades at low wind speeds. The exact effect on fatality numbers of curtailing up to a certain wind speed (e.g. 5.0–6.5 ms^{-1}) is, however, difficult to predict and, depending on the site specific fatality risk, may lead to very different results at different turbines.

Here, this study proposes a refinement of the operational mitigation that is hitherto based solely on wind speed, in two ways. First, times of bat activity are more precisely predicted and, hence, collision risk is more precisely predicted with a model that includes more predictive variables (time of night and month in addition to wind speed). Second, a turbine specific method is presented to reduce the fatality risk for bats at wind turbines in central Europe to a threshold set by the relevant authorities.

Applying Curtailment Algorithms to New Turbines

To apply this method to a new turbine, the level of acoustic activity at the new turbine has to be measured, which, in Germany, is done at the nacelle. These measurements usually last for at least two years, from April through October (although mandatory sample times differ in federal states). Acoustic data has to be comparable to the dataset presented here. Therefore, one of the three detector systems for which reference datasets were gathered (Anabat SD1 and SD2, Titley and Avisoft/BATmode system, Avisoft Bioacoustics/bat bioacoustictechnology GmbH in addition to the Batcorder described here) has to be used with a specific configuration of microphone sensitivity and settings (Behr et al. 2015). Also, the wind speed data measured at the nacelle of the turbine has to be provided by the turbine operator. The software-tool ProBat (available in German and English from http://windbat.techfak.fau.de) that is free of cost, estimates fatality rates, and calculates cut-in wind speeds based on acoustic activity data and wind speed and the methods described here.

During the first year of operation of a turbine (i.e. when no acoustic data is available yet), the turbine is often operated with a general curtailment algorithm (e.g. raising cut-in wind speed to 6.0 ms^{-1} at night). For the second year of operation turbine specific curtailment algorithms can be calculated (e.g. with ProBat) based on the activity measured in the first year. Subsequently, starting with the third year of turbine operation, the final set of cut-in wind speeds can be calculated based on two years of data on acoustic activity and wind speed.

Monitoring Acoustic Bat Activity at the Nacelle of Wind Turbines

Recording acoustic bat activity at the nacelle of wind turbines has become a standard method, at least in central Europe. Different acoustic detector systems have been used at a large number of different turbine types from all major manufacturers. The most common problems are microphone or detector failures and large numbers of noise files recorded due to the harsh conditions at the turbine nacelle. A daily status text containing information on microphone sensitivity, remaining memory, and a confirmation that the system is running properly is, therefore, best practice.

Generally, acoustic recording is suitable to detect the species most affected by collisions with wind turbines due to the properties of their echolocation calls – mostly a high level of sound pressure, allowing their detection over a large enough distance. There are, however, marked differences between species and species groups in peak frequency; for example, species of the Nyctaloid group have lower calls (mostly between 18 and 35 kHz) than those of the Pipistrelloid group (mostly 37–57 kHz; Skiba 2003). Since lower frequency calls are less attenuated in air, species of the Nyctaloid group can be detected over larger distances. At present differences in detectability cannot be accounted for, because in these models all species are pooled. This may cause bias in estimations of fatality numbers when these models are applied to sites with a different species composition (and, hence different call characteristics). Therefore, regional models are currently being developed that take into account the specific species composition in different areas in Central Europe.

Species Recorded

The most common species in the acoustic dataset (Common Noctule, *N. noctula*, Nathusius' Pipistrelle, *P. nathusii*, and Common Pipistrelle, *P. pipistrellus*) were also the species most commonly found in the fatality searches for this study. This is in parallel to the acoustic survey (Korner-Nievergelt et al. 2013) and to fatality searches in Germany in general (Dürr 2015). The Common and the Nathusius' Pipistrelle accounted for a lower percentage of the acoustic dataset as compared to their proportion in fatality numbers, which is probably due to their lower detectability caused by high call frequency. The number of recordings per fatality was still lower in the Nathusius' Pipistrelle than in the Common Pipistrelle.

Activity and Predictive Variables

A peak in bat activity was recorded from late July to mid-September, which corresponds to several reports of a peak in bat fatalities at wind turbines in late summer and autumn in Europe and also North America (see reviews in Arnett et al. 2008; Schuster et al. 2015). Time of night also had a strong influence on bat activity with a peak during the first quarter of the night and a continuous decrease until sunrise that is in accordance with findings from North America for bat recordings on towers (Arnett et al. 2006, 2007). Some turbines in the dataset showed a smaller second peak of activity during the morning hours that might reflect the morning activity at a nearby roost. Nathusius' Pipistrelle showed a different pattern with the main activity in the middle of the night and a decrease towards both, sunset and sunrise (a possible reason for this is discussed below). The Common Noctule was the only species to show substantial activity before sunset.

Bat activity steeply declined with higher wind speeds (only 15% of activity recorded at wind speeds $\geq 5\ \text{ms}^{-1}$). This was also the case in two acoustic studies in North America, where activity was measured at meteorological towers and declined by 4–13% (Redell et al. 2006) and 11–39% (Arnett et al. 2006) per 1 ms^{-1} increase in wind speed and in a thermal imaging study (Horn et al. 2008). For a review of the effect of wind and other metherological parameters, see Schuster et al. 2015. In California, Weller and Baldwin (2011) also reported higher activity at lower wind speeds for a dataset recorded at meteorological and portable towers. The authors also highlighted the improved fit of models that are based on several predictive variables, not just wind speed.

In this study's dataset, *P. pipistrellus* showed the steepest decline in activity with higher wind speeds. This small species has a relatively slow transfer-flight speed of 6 ms^{-1} (Kalko 1991; Simon et al. 2004) and might have difficulties in coping with wind. However, *P. nathusii*, a similarly small species, was the most wind-resistant species recorded. In contrary to the Common Pipistrelle, the Nathusius' Pipistrelle is a long-distance migrant and its activity at high wind speeds and deviating activity pattern during the night (see above) might indicate a high percentage of migration activity in the dataset (as opposed to foraging).

Bat activity increased with temperature up to approximately 25 °C but decreased with higher temperatures. Activity below 10 °C was very low. Similarly, Arnett et al. (2006) showed a peak of acoustic bat activity at around 20 °C measured at a tower 22 m above ground level and Reynolds (2006) also recorded very little activity below 10.5 °C in New York State. In this study, precipitation also had a strong influence on activity with almost no bat calls recorded at more than 0.002–0.004 mm/min (fog or clouds). This may, however, be partly due to a higher attenuation of sound in more humid air.

GLM

Generalised linear modelling (GLM) was used to predict bat activity and, hence, times of high collision risk from wind speed, month, time of night, and turbine (the latter described differences in the activity level among different turbines).

The results show that the tested predictive variables had a highly significant effect on the activity of bats at the turbines and that the model can be used to predict times of higher bat activity with a high temporal resolution of 10 min intervals. However, very high peaks of activity were underestimated. Concerning the calculations for mitigation algorithms presented here this is only a minor problem because the collision risk calculated for the few 10 min intervals in question is still very high resulting in an inclusion of those intervals in curtailment times. Generally, the GLM supplied a prediction of collision risk, not an exact prediction of the number of recordings for each 10 min interval.

Temperature and precipitation were excluded as predictive variables from the final model since the resulting increase in energy yield was not enough to justify the sampling of these variables. This may, however, be different for larger turbine types or for time periods with low temperatures very early or late in the year.

10 Minute Intervals as Temporal Units

Intervals of 10 min were used as temporal units to assess the collision risk and, if necessary, to stop the turbine rotor from moving at a speed that is dangerous for bats. Intervals of 10 min are commonly used by the SCADA systems running the turbines, which makes the implementation of the algorithms easier. Intervals of 10 min were considered short enough to react to changes in wind speed, the most variable of the factors influencing bat activity in this model. If shorter intervals were used, this would increase the number of cut-in and cut-out events and, hence, the wear on turbine components. The rotor will only be stopped after the value of Q (energy produced/collision risk) has fallen below the threshold during a 10 min interval. From the experience with this dataset it was considered that a delay of 10 min was short enough to react even to quick changes in bat activity and collision risk.

Implementation of Acoustic Surveys and Operational Mitigation in Germany

In Germany, acoustic monitoring at the nacelle and operational mitigation has become the standard method to assess and mitigate collision risk of bats at wind turbines. By the summer of 2015, seven of the sixteen German federal states (Bavaria, Brandenburg, Hesse, North Rhine-Westphalia, Rhineland-Palatinate, Saarland, Schleswig-Holstein) employed post-construction acoustic monitoring at the nacelle and gave very specific recommendations for conducting the acoustic survey or directly referred to the methods of the research project (final report in German): Brinkmann et al. (2011). The German federal states aforementioned also mandate operational mitigation (mostly the curtailment algorithms presented here) and often set thresholds for the number of accepted bat fatalities per year and turbine (e.g. one dead bat in Schleswig-Holstein, two dead bats in Bavaria, Rhineland-Palatinate and Saarland; Mayer et al. 2015).

Transferring These Methods to New Turbines

The GLM model presented here is based on a large dataset of 69 Enercon turbines with rotor diameters between 66 and 82 m. Datasets from site assessment studies are often small (one to two turbines sampled during one to two years) and, hence, strongly influenced by random effects (e.g. a rainy August, warm spring, etc.). For these cases it is recommended that the effects for the parameters influencing bat activity (wind speed, time of night, and month) should be taken from the GLM based on this large dataset. Only the level of activity at the respective turbine should be extracted from the dataset recorded at the turbine in question. For example this can be done by calculating an "offset" Model where all effects, except for "turbine", are pre-set. This is implemented in the software tool, ProBat.

If this method of acoustic sampling is adopted, the analyses and calculations shown here can usually be applied. There are, however some important restrictions: (1) Species at the new site should not differ to a great extend from the species recorded at this study's sites, (2) Rotor diameters should be similar to the mean in this dataset. A correction has been developed for differing rotor diameters that is implemented in ProBat (differences in tower height are less important since the activity is sampled at the nacelle), (3) Implementation at turbines of different manufacturers will depend on technical details of the respective turbine type.

The n-mixture model from Korner-Nievergelt et al. (2013) is based on simultaneous acoustic sampling and carcass searches at 30 turbines from July to September. In this data set, wind-speed had a higher predictive value for collision rate than acoustic activity. As a consequence, the current algorithm implemented in ProBat tends to overestimate collision rates during times with low bat activity, e.g. very early or late in the year. We are currently working on a new, hierarchical version of the collision rate model with a larger effect of bat activity on collision rate that is able to more precisely predict collision rates during times with low bat activity.

Effectiveness of Operational Mitigation

Although, due to differences in the conservation status of bats, operational mitigation is much more common in Europe (all bat species protected by European laws) than in North America (protection status is species specific), almost all publications (but see: Beucher et al. 2011; Behr and von Helversen 2006) on operational mitigation experiments are from the US and Canada (see review in Arnett et al. 2013). In all publications mentioned, curtailment was based on wind speed alone and the cut-in wind speed was chosen more or less arbitrarily between 5.0 and 6.5 ms^{-1} (the normal cut-in wind speed of the turbines was between 3.0 and 4.0 ms^{-1}).

At a site in southern France, Beucher et al. (2011) found 73 dead bats in 2008 and 98 bats in 2009 (all numbers at this site not corrected for biases). In the following year, a cut-in wind speed of 6.5 ms^{-1} was implemented and the lighting at the foot of the turbine towers was deactivated. This reduced the number of fatalities found to two in 2010 and three in 2011 (cut-in wind speed in 2011 was 5.5 ms^{-1}).

Most studies in North America (reviewed in Arnett et al. 2013) reported a reduction in fatality numbers of at least 50% when raising the cut-in wind speed at least 1.5 ms^{-1} above that of the manufacturer. Some studies did not show a comparable effect (this might, however, be explained by the wind speed conditions or the species present at this sites). Most studies that reported on the costs of mitigation estimated a loss of less than 1% of the yearly revenue for curtailment during the main times of collision risk.

The operational algorithms presented here also include the time of night and the month and have proven more efficient than the ones based on wind speed alone. A threshold of two dead bats per year results in a mean loss in power production of 1.4% of the annual revenue as opposed to 1.8% based on wind speed alone. Further costs arise from site assessment and implementation of the mitigation. Moreover, the collision risk for bats at a turbine was estimated from the acoustic activity recorded. Therefore, cut-in wind speeds are not set arbitrarily but tailored to the turbine specific collision risk. This is important because turbines differ vastly in collision risk. As a result unnecessary losses in energy production are avoided while meeting the reduction in collision risk required by the authorities.

Acknowledgements We thank the German Federal Ministry for the Environment, Nature Conservation and Nuclear Safety, particularly S. Hofmann and A. Radecke and the Project Management Jülich particularly G. Heider and T. Petrovic for funding and scientific feedback. We express our gratitude to the operators of wind turbines for granting us access to their turbines. We thank our project partner ENERCON GmbH (especially B. de Wolf, K. Einnolf, F. Kentler, U. Kleinoeder, M. Schellschmidt, R. Schulte, and several service teams) for technical support and for the installation of acoustic detectors and sensors at the wind turbines. This work profited greatly from manifold contributions in scientific discussions, meetings and workshops of numerous people represented here by T. Dürr, L. Bach, F. Bontadina, P. Korner, V. Runkel, and U. Marckmann.

References

Arnett E, Johnson G, Erickson W, Hein C (2013) A synthesis of operational mitigation studies to reduce bat fatalities at wind energy facilities in North America. A report submitted to the National Renewable Energy Laboratory. Bat Conservation International. Austin, Texas, USA. Austin, Texas, USA

Arnett EB, Baerwald EF (2013) Impacts of wind energy development on bats: implications for conservation. In: Bat evolution, ecology, and conservation. Springer, Berlin, pp 435–456

Arnett EB, Baerwald EF, Mathews F, Rodrigues L, Rodríguez-Durán A, Rydell J, Villegas-Patraca R, Voigt CC (2016) Impacts of wind energy development on bats: a global perspective. In: Bats in the anthropocene: conservation of bats in a changing world. Springer, Berlin, pp 295–323

Arnett EB, Brown WK, Erickson WP, Fiedler JK, Hamilton BL, Henry TH, Jain A, Johnson GD, Kerns J, Koford RR, Nicholson CP, O'Connell TJ, Piorkowski MD, Tankersley RD (2008) Patterns of bat fatalities at wind energy facilities in North America. J Wildlife Manage 72 (1):61–78. doi:10.2193/2007-221

Arnett EB, Hayes JP, Huso MP (2006) Patterns of pre-construction bat activity at a proposed wind facility in south-central Pennsylvania. An annual report submitted to the Bats and Wind Energy Cooperative. Bat Conservation International, Austin, Texas, USA

Arnett EB, Huso MP, Reynolds DS, Schirmacher M (2007) Patterns of pre-construction bat activity at a proposed wind facility in northwest Massachusetts. An annual report submitted to the Bats and Wind Energy Cooperative. Bat Conservation International, Austin, Texas, USA

Arnett EB, Schirmacher M, Huso M, Hayes J (2009) Effectiveness of changing wind turbine cut-in speed to reduce bat fatalities at wind facilities. An annual report submitted to the Bats and Wind Energy Cooperative Bat Conservation International Austin, Texas, USA

Baerwald EF, Edworthy J, Holder M, Barclay RMR (2009) A large-scale mitigation experiment to reduce bat fatalities at wind energy facilities. J Wildl Manage 73(7):1077–1081. doi:10.2193/2008-233

Barclay RM, Harder LD (2003) Life histories of bats: life in the slow lane. In: Kunz T, Fenton M (eds) Bat ecology, pp 209–253

Behr O, Simon R, Nagy M (2015) Leitfaden zur Durchführung einer akustischen Aktivitätserfassung an Windenergieanlagen und zur Berechnung fledermausfreundlicher Betriebsalgorithmen. In: Behr O, Adomeit U, Hochradel K et al. (eds) Reduktion des Kollisionsrisikos von Fledermäusen an Onshore-Windenergieanlagen. - Endbericht des Forschungsvorhabens gefördert durch das Bundesministerium für Umwelt, Naturschutz und Reaktorsicherheit (Förderkennzeichen 0327638C+D)

Behr O, von Helversen O (2006) Gutachten zur Beeinträchtigung im freien Luftraum jagender und ziehender Fledermäuse durch bestehende Windkraftanlagen - Wirkungskontrolle zum Windpark „Roßkopf" (Freiburg i. Br.) im Jahr 2005. unveröffentlichtes Gutachten im Auftrag der regiowind GmbH, Freiburg, Erlangen

Beucher Y, Kelm V, Albespy F, Geyelin M, Nazon L, Pick D (2011) Parc éolien de Castelnau-Pegayrols (12). Suivi d'impacts post-implantation sur les chauves souris. Bilan de campagne des 2ème et 3ème année d'exploitation (2009–2010)

Bispo R, Bernardino J, Marques TA, Pestana D (2013) Modeling carcass removal time for avian mortality assessment in wind farms using survival analysis. Environ Ecol Stat 20(1):147–165. doi:10.1007/s10651-012-0212-5

Brinkmann R, Behr O, Korner-Nievergelt F, Mages J, Niermann I, Reich M (2011) Entwicklung von Methoden zur Untersuchung und Reduktion des Kollisionsrisikos von Fledermäusen an Onshore-Windenergieanlagen, vol 4. Cuvillier Verlag, Göttingen

Burnham KP, Anderson DR (2002) Model selection and multimodel inference: a practical information-theoretic approach. Springer, Berlin

Cryan PM (2008) Mating behavior as a possible cause of bat fatalities at wind turbines. J Wildl Manage 72(3):845–849

Cryan PM, Barclay RMR (2009) Causes of bat fatalities at wind turbines: hypotheses and predictions. J Mamm 90(6):1330–1340

Cryan PM, Gorresen PM, Hein CD, Schirmacher MR, Diehl RH, Huso MM, Hayman DT, Fricker PD, Bonaccorso FJ, Johnson DH (2014) Behavior of bats at wind turbines. PNAS 111 (42):15126–15131

Dürr T (2015) Fledermausverluste an Windenergieanlagen – Daten aus der zentralen Fundkartei der Staatlichen Vogelschutzwarte im Landesamt für Umwelt, Gesundheit und Verbraucherschutz Brandenburg. (Accessed on 11/22/2015)

Frick WF, Pollock JF, Hicks AC, Langwig KE, Reynolds DS, Turner GG, Butchkoski CM, Kunz TH (2010a) An emerging disease causes regional population collapse of a common North American bat species. Science 329(5992):679–682. doi:10.1126/science.1188594

Frick WF, Reynolds DS, Kunz TH (2010b) Influence of climate and reproductive timing on demography of little brown myotis *Myotis lucifugus*. J Anim Ecol 79(1):128–136. doi:10.1111/j.1365-2656.2009.01615.x

Gannon WL, Sherwin RE, Haymond S (2003) On the importance of articulating assumptions when conducting acoustic studies of habitat use by bats. Wildl Soc Bull 31(1):45–61

Hayes JP (2000) Assumptions and practical considerations in the design and interpretation of echolocation-monitoring studies. Acta Chiropterologica 2(2):225–236

Hein C, Gruver J, Arnett E (2013) Relating pre-construction bat activity and post-construction bat fatality to predict risk at wind energy facilities: a synthesis. A report submitted to the National Renewable Energy Laboratory Bat Conservation International, Austin, TX, USA

Horn JW, Arnett EB, Kunz TH (2008) Behavioral responses of bats to operating wind turbines. J Wildl Manage 72(1):123–132

Hull CL, Cawthen L (2013) Bat fatalities at two wind farms in Tasmania, Australia: bat characteristics, and spatial and temporal patterns. New Zeal J Zool 40(1):5–15. doi:10.1080/03014223.2012.731006

Huso MMP (2010) An estimator of wildlife fatality from observed carcasses. Environmetrics. doi:10.1002/env.1052

Johnson GD, Perlik MK, Erickson WIP, Strickland JD (2004) Bat activity, composition, and collision mortality at a large wind plant in Minnesota. Wildl Soc Bull 32(4):1278–1288. doi:10.2193/0091-7648(2004)032[1278:Bacacm]2.0.Co;2

Kalko EVK (1991) Das Echoortungs- und Jagdverhalten der drei europäischen Zwergfledermausarten, *Pipistrellus pipistrellus* (Schreber 1774), *Pipistrellus nathusii* (Keyserling & Blasius 1939) und *Pipistrellus kuhli* (Kuhl 1819). Universität Tübingen, Tübingen

Kerns J, Erickson WP, Arnett EB (2005) Bat and bird fatality at wind energy facilities in Pennsylvania and West Virginia. Relationships between bats and wind turbines in

Pennsylvania and West Virginia: an assessment of fatality search protocols, patterns of fatality, and behavioral interactions with wind turbines Edited by EB Arnett The Bats and Wind Energy Cooperative, Bat Conservation International, Austin, Texas, pp 24–95
Korner-Nievergelt F, Behr O, Niermann I, Brinkmann R (2011a) Schätzung der Zahl verunglückter Fledermäuse an Windenergieanlagen mittels akustischer Aktivitätsmessungen und modifizierter N-mixture Modelle. In: Brinkmann R, Behr O, Niermann I, Reich M (eds) Entwicklung von Methoden zur Untersuchung und Reduktion des Kollisionsrisikos von Fledermäusen an Onshore-Windenergieanlagen., vol Umwelt und Raum, vol 4. Cuvillier Verlag, Göttingen, pp Umwelt und Raum Bd. 4:323–353
Korner-Nievergelt F, Brinkmann R, Niermann I, Behr O (2013) Estimating bat and bird mortality occurring at wind energy turbines from covariates and carcass searches using mixture models. Plos One 8(7). doi:ARTN e67997 10.1371/journal.pone.0067997
Korner-Nievergelt F, Korner-Nievergelt P, Behr O, Niermann I, Brinkmann R, Hellriegel B (2011b) A new method to determine bird and bat fatality at wind energy turbines from carcass searches. Wildl Biol 17:350–363
Liaw A, Wiener M (2002) Classification and regression by randomForest. R news 2(3):18–22
Mayer K, Hurst J, Niermann I, Reich M, Brinkmann R (2015) Recherche von behördlichen Planungsvorgaben und von Planungsempfehlungen nicht-staatlicher Verbände und Institutionen bezüglich der Beachtung des Fledermausschutzes bei Planung und Betrieb von Windenergieanlagen. In: Behr O, Adomeit U, Hochradel K et al (eds) Reduktion des Kollisionsrisikos von Fledermäusen an Onshore-Windenergieanlagen - Endbericht des Forschungsvorhabens gefördert durch das Bundesministerium für Umwelt, Naturschutz und Reaktorsicherheit (Förderkennzeichen 0327638C+D)
Niermann I, Velten Sv, Korner-Nievergelt F, Brinkmann R, Behr O (2011) Einfluss von Anlagen- und Landschaftsvariablen auf die Aktivität von Fledermäusen an Windenergieanlagen. In: Brinkmann R, Behr O, Niermann I, Reich M (eds) Entwicklung von Methoden zur Untersuchung und Reduktion des Kollisionsrisikos von Fledermäusen an Onshore-Windenergieanlagen. Umwelt und Raum, vol 4. Cuvillier Verlag, Göttingen, pp Umwelt und Raum Bd. 4:384–405
O'Shea TJ, Bogan MA, Ellison LE (2003) Monitoring trends in bat populations of the United States and Territories: status of the science and recommendations for the future. Wild Soc Bull 31:16–29
O'Shea TJ, Cryan PM, Hayman DT, Plowright RK, Streicker DG (2016) Multiple mortality events in bats: a global review. Mammal Rev. doi:10.1111/mam.12064
Pierson E (1998) Tall trees, deep holes, and scarred landscapes: conservation biology of North American bats. Bat biology and conservation. Smithsonian Institution Press, Washington, DC, pp 309–325
R Core Team (2015) R: A language and environment for statistical computing. R Foundation for Statistical Computing. Vienna, Austria
Racey PA, Entwistle AC (2000) Life-history and reproductive strategies of bats. Reproductive biology of bats, pp 363–414
Redell D, Arnett EB, Hayes JP, Huso M (2006) Patterns of pre-construction bat activity at a proposed wind facility in south-central Wisconsin. A final report submitted to the Bats and Wind Energy Cooperative. Bat Conservation International, Austin, Texas, USA
Reynolds DS (2006) Monitoring the potential impact of a wind development site on bats in the northeast. J Wildl Manage 70(5):1219–1227
Schuster E, Bulling L, Köppel J (2015) Consolidating the state of knowledge: a synoptical review of wind energy's wildlife effects. Environ Manage 56(2):300–331
Sherwin RE, Gannon WL, Haymond S (2000) The efficacy of acoustic techniques to infer differential use of habitat by bats. Acta Chiropterologica 2(2):145–153
Simon M, Hüttenbügel S, Smit-Viergutz J (2004) Ökologie und Schutz von Fledermäusen in Dörfern und Städten. Schriftenreihe für Landschaftspflege und Naturschutz, vol 76. BfN, Bonn - Bad Godesberg
Skiba R (2003) Europäische Fledermäuse. Neue Brehm-Bücherei Bd 648

Voigt CC, Lehnert LS, Petersons G, Adorf F, Bach L (2015) Wildlife and renewable energy: German politics cross migratory bats. Eur J Wildl Res 61(2):213–219. doi:10.1007/s10344-015-0903-y

Weller TJ, Baldwin JA (2011) Using echolocation monitoring to model bat occupancy and inform mitigations at wind energy facilities. J Wildl Manage

Winhold L, Kurtai A, Foster R (2008) Long-term change in an assemblage of North American bats: are eastern red bats declining? Acta Chiropterologica 10(2):359–366. doi:10.3161/150811008x414935

Is There a State-of-the-Art to Reduce Pile-Driving Noise?

Michael A. Bellmann, Jan Schuckenbrock, Siegfried Gündert, Michael Müller, Hauke Holst and Patrick Remmers

Abstract Underwater noise caused by pile-driving during the installation of offshore foundations is potentially harmful to marine life. In Germany, the regulation authority BSH set the following protection values: Sound Exposure Level SEL = 160 dB and Peak Level L_{Peak} = 190 dB for Harbor Porpoises that must be complied with at a distance of 750 m to the construction site in order to avoid temporal threshold shifts. The experience over the last years shows that underwater sound produced during pile-driving, depending on many parameters and measurements, shows values of up to 180 dB_{SEL} and up to 205 dB_{LPeak} in a distance of 750 m. Therefore, Noise Mitigation Systems (NMS) are required to significantly minimize the underwater sound. Since 2011, NMS must be applied during all noisy offshore construction work in Germany. The Institute of Technical and Applied Physics GmbH (itap) was involved in many offshore wind farm (OWF) projects with pile-driving activities (>1000 pile installations without and with different NMS). Based on these underwater noise measurements, the tested NMS were evaluated. In this paper, a general overview of existing and tested NMS including tested system variations is provided and the measured data and influencing factors on the resulting noise reduction are discussed. Additionally, combinations of two or more NMS are measured during the construction phase, if monopiles with diameters of up to 8 m are installed. It is demonstrated what level of effect one or more NMS have on the emitted noise. However, it is shown that it is possible to install monopiles with a diameter of >6 m with noise levels below 160 dB_{SEL} at a distance of 750 m, if combinations of suitable NMS are used. Nevertheless, any kind of noise mitigation will have a significant influence on the 'disturbed' or treated area

M.A. Bellmann (✉) · J. Schuckenbrock · S. Gündert
M. Müller · H. Holst · P. Remmers
itap — Institute of Technical and Applied Physics GmbH,
Marie-Curie-Str. 8, 26129 Oldenburg, Germany
e-mail: bellmann@itap.de

J. Köppel (ed.), *Wind Energy and Wildlife Interactions*,
DOI 10.1007/978-3-319-51272-3_9

from pile-driving noise for marine mammals. The question of whether a state-of-the-art measure to reduce pile-driving noise exists will be explored, based on measured data and experiences with NMS under real offshore conditions.

Keywords Noise mitigation system · Pile-driving · Noise reduction · Offshore wind farm · Monopiles

Introduction

The installation of renewable energy sources offshore is growing fast in Europe, especially in Germany, forced by the energy turnaround after 2011. Currently, 12 offshore wind farms (OWF) in Germany are in operation and several OWF are under construction or in the pre-construction phase. The demand for renewable energies has to go hand in hand with the awareness of sustainability issues, especially the conservation of nature and environmental systems. Besides other ecological topics, the hydro sound emissions have moved into focus due to the fact that most foundations are installed using impulse pile-driving. This installation method leads to enormous acoustic emissions (hydro or underwater sound), which are potentially harmful to marine life, especially for the hearing system of marine mammals like harbor porpoises and seals. For the conservation of the marine fauna, it is necessary to keep the noise levels in the water as low as possible during activities such as pile-driving. Therefore, in 2011, the German regulation authority Federal Maritime and Hydrographic Agency of Germany (BSH), as the first country worldwide, set the following protection (limiting) values based on preliminary works of the Federal Environmental Agency (UBA):

$$\text{Sound Exposure Level (SEL or } L_E) = 160 \text{ dB (re } 1\ \mu\text{ Pa}^2\text{ s) and}$$

$$\text{Peak Level (}L_{Peak}\text{ or zero-to-peak)} = 190 \text{ dB (re } 1\ \mu\text{ Pa}^2),$$

which must be complied with in a distance of 750 m to the construction site.

Existing underwater noise measurements over the last years show that hydro sound during pile-driving mainly depends on the pile diameter (currently, installations with a pile diameter of up to 8 m are conducted) and the blow energy used. In Fig. 1, measured data from itap for the acoustic parameters Peak Level (L_{Peak}) and 50th percentile (mean value) of the Sound Exposure Level (SEL_{50}) of different OWF construction projects over the last ten years are shown as a function of the used pile diameter. Each cross or triangle corresponds to an OWF, test pile, converter platform, supply station or met mast installation in the North Sea or Baltic Sea in Europe. In instances where the secondary influencing parameter — 'blow energy used' parameter — is taken into account, the variances can be reduced to ±2 dB (comparable to the common uncertainty of measurement).

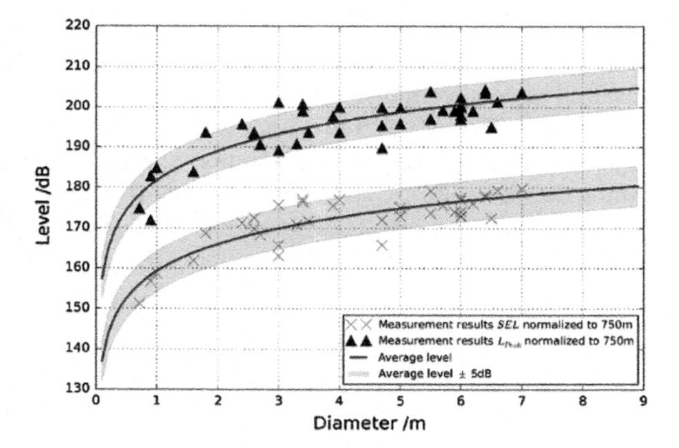

Fig. 1 Measured Peak Levels (L_{Peak}) and broadband Sound Exposure Levels (SEL_{50}) during pile-driving work without using noise mitigation systems (NMS) as a function of pile diameter measured by itap

The measurements show values of up to 180 dB for the SEL_5[1] and up to 205 dB for the L_{Peak}. Therefore, Noise Mitigation Systems (NMS) are necessary and requested in Germany to significantly minimize the hydro sound emissions. The requested noise reduction mainly depends on the used pile diameter and can achieve values of up to 20 dB. This corresponds to a reduction of the physical metric sound pressure by a factor 10.

Since 2011, NMS must be applied during noisy construction work at all OWF and converter platforms in Germany, especially during all pile-driving activities after the approval of the BSH.

One major problem was that many (theoretical) solutions or prototypes for noise mitigation measures existed in 2011, but most experiences were based on laboratory or near-shore studies (summary in e.g. Diederichs et al. 2014). Therefore, these measures did not correspond with first real offshore results (e.g. Wilke et al. 2012). Considerable research had to be conducted during the construction phase of OWF. Furthermore, many research projects dealing with noise reduction were performed in parallel with the construction phases, since no common state-of-the-art NMS were available at that time (e.g. founded research projects Diederichs et al. 2014; Nehls and Bellmann 2015).

The Marine Strategy Framework Directive (Directive 2008/56/EC) is currently the driving factor in Europe for evaluating the current status of the marine habitats. This Directive contains descriptor 11 'energy input', which includes underwater noise. Based on the Directive, several other EU countries developed and published their own protection values and underwater noise strategies, such as Denmark (2014) and the Netherlands (2015), which upcoming OWF projects must fulfill. This means, that not only in Germany, noise mitigation measures are applied during noise construction works. Both the approval authorities as well as the OWF project

---

[1] SEL_5: 95% percentile value of the SEL, which means 95% of all blows per installation have lower SEL values. The SEL_5 must be compared with the German limiting value 160 dB @750 m distance. Differences between SEL_{50} and SEL_{05} are usually 2–3 dB (empirical data).

companies are highly interested in setting up suitable and offshore-qualified noise mitigation concepts before starting pile-driving, in order to fulfill the requirements as well as to save money and time.

In this paper, a general overview of existing and tested NMS is provided. The main results regarding noise reduction and general findings regarding factors, that influence the noise reduction over the last five years, will be shown and discussed. Finally, it will be discussed, if a general state-of-the-art currently exists regarding noise reduction and NMS from the acoustic point of view.

Overview of Existing Noise Mitigation Systems (NMS)

Up until now (end of 2015), several NMS have been tested offshore (partly in serial) or prototypes are available. In any case, the use of a NMS will always have significant influence on the offshore logistics and produce costs (e.g. Schorcht 2015). The NMS that are currently applied most often and the NMS that are the most practicable are summarized in Table 1. Additionally, the range of tested system configurations for each NMS is shown (a brief description of all NMS is provided in e.g. Bellmann 2014). A general overview of existing noise mitigation measures, concepts and prototypes as well as alternative foundation installation procedures (other than pile-driving) is summarized in Koschinski and Lüdemann (2013).

In general, all NMS can be categorized into three different groups: (i) bubble curtains — air bubbles in the water like a Big Bubble Curtain; (ii) shell in shell systems — pile is driven inside a second tube like a noise mitigation screen or (iii) other techniques — such as Hydro Sound Damper (HSD) partly based on Helmholtz resonators or other effects.

Table 1 List of tested and evaluated noise mitigation systems (NMS) under offshore conditions during pile-driving, as well as, tested system configurations of each NMS

Noise mitigation system	Tested configurations	No. foundations (piles)
Single Big Bubble Curtain—BBC	Supplied air volume, diameter and length of the nozzle hose, hole configuration of the nozzle hose, distance to construction site, air feed-in (one- or double sided), ballast chain (inside/outside), pre-laying or post-laying, water depth DBBC: distance between nozzle hoses	>150 (300)
Double Big Bubble Curtain—DBBC		>150 (300)
Small Bubble Curtain—SBC	Supplied air volume, hole configuration of the nozzle hose	2 (2)
Hydro Sound Damper—HSD	Number and size of HSD elements	>10 (>10)
Noise Mitigation Screen—IHC-NMS	Space between inner and outer tube, additional BBC inside	>150 (150)

From 2013 onwards, combinations of the listed NMS were also tested to reduce noise, especially for projects with a large pile diameter.

Results

General Requirements on a Noise Mitigation System

For the protection of the marine fauna, it is essential to minimize the sound (energy) entry into the water, if possible, in order to avoid impairments to marine life. For this purpose, noise protection values for the Sound Exposure Level (SEL) and partly the Peak Level (L_{Peak}) were determined by different EU member states since 2011 (e.g. Germany, Denmark and the Netherlands). These noise protection values have particularly been developed for the protection of Harbor Porpoises and seals against auditory temporary or permanent threshold shifts (TTS or PTS).

Overall, more than 1000 piles were installed and measured by itap in the European North and Baltic Sea by using different NMS during normal construction phases of more than 12 OWF, several offshore supply stations and test piles. The water depth varied between 8 and 45 m. Certain parameters of the NMS were also investigated in three founded R&D projects (Wilke et al. 2012; Diederichs et al. 2014; Nehls and Bellmann 2015).

The following results are based on measured data at a distance of 750 m from the construction site (acoustical far-field and in line with the upcoming ISO Standard 18406). The measurements and the analysis of the data were conducted in line with German guidelines (BSH 2011, 2013a, b). A statistical approach for the evaluation of the noise reduction potential of a NMS was developed in the research project OWF BW II (Diederichs et al. 2014). This was due to the fact that some thousand blows per pile installation are usually needed and the resulting noise reduction can vary during the whole pile-driving activity. This means that the mean value as well as a kind of minimum and maximum of the noise reduction results for each NMS is calculated, based on several projects (Table 2).

Insertion loss or (sound) transmission loss is generally considered for the characterization of the effectiveness of a NMS. Transmission loss can be depicted broadband (as a single value) or frequency resolved (i.e. as 1/3 octave spectrum).

For broadband depiction, the broadband Sound Exposure Level (single SEL value) or the Peak Level (L_{Peak}) when using a NMS is subtracted from the broadband SEL of the reference condition (without a NMS), meaning the difference (ΔSEL or ΔL_{Peak}) is calculated (the greater the difference, the higher the insertion loss and the better the NMS). The advantage of this parameter is the possibility to record and describe the noise-reducing effect of a NMS by a single value. In case of changing the spectral shape of the produced pile-driving noise, the single value will increase the uncertainty for upcoming projects. Therefore, the frequency-resolved transmission loss for each frequency band would also be used for an evaluation of NMS (Fig. 3).

Table 2 Summary of the existing and most regularly used NMS including the (broadband) insertion loss of the best available system configuration tested in water depths ≤ 30 m

No.	Noise mitigation system	Optimized system configuration	ΔSEL (dB)	Piles
1	Single Big Bubble Curtain (BBC)	>0.3 m^3/(min m), ballast chain outside, optimized nozzle hose hole configuration for DBBC distance between nozzle hoses $\geq 1\times$ water depth	$10 \leq 13 \leq 15$	>300
2	Double Big Bubble Curtain (DBBC)		$14 \leq 17 \leq 18$	>300
3	Hydro-Sound-Damper (HSD)	Number and size of HSD elements	$8 \leq 10 \leq 13$	>10
4	Noise Mitigation Screen (IHC-NMS)	Space between inner and outer tube, additional BBC inside	$10 \leq 12 \leq 14$	>140
5	IHC-NMS + optimized BBC	See above	$17 \leq 19 \leq 23$	>90
6	HSD + optimized BBC	See above	$15 \leq 16 \leq 20$	>30
7	HSD + not optimized DBBC	Less air and ballast chain inside[a]	$14 \leq 16 \leq 22$	>20

Adapted from Bellmann (2014)

[a]The air-water-mixture is relevant for the resulting sound reduction of a Bubble Curtain System; the higher the air introduction into the water, the larger the sound reduction. Should the above-mentioned optimum air volume of 0.3 m^3/(m min) not be achieved by the Bubble Curtain System, a decrease of the above-mentioned average sound reduction cannot be excluded. In this case, the increase of the nozzle hose rings represents a possible and practice-proven alternative. Thus, effectively, less air volume per nozzle hose will be introduced into the water, however, due to the increased number of nozzle hoses, an elevated air introduction into the water is ensured

Overview of Measured (Single-Value) Insertion Loss

Table 2 summarizes the most used NMS including the (broadband single-value) measured insertion loss (ΔSEL) of the *best available system configuration* of each NMS. Usually, the resulting noise reduction for the L_{Peak} is a little higher than for the SEL (e.g. Bellmann 2014). In case of shortfalls or non-optimized system configurations used, this leads to significant lower sound reductions.

Results from two running projects and one closed project in 2015 at water depths up to 35 m are not included in Table 2, since the data are currently confidential.

If a combination of two NMS is used, the sound reductions of each single (separately) applied NMS do not add up (single-value), but sum up spectrally. For example two NMS with a 13 dB sound reduction at single application do not have a total of 26 dB in sound reduction at simultaneous application, but a lower overall sound reduction (e.g. combination of IHC-NMS and BBC $\sim$19 dB, Table 2).

The Big Bubble Curtain (BBC) is the most used NMS today, as it can be used for all common foundation structures like monopiles, jacket, triples and tripods. The air-water-mixture combination is relevant for the resulting sound reduction of a BBC. This relation between air volume and noise reduction is non-linear (e.g. Diederichs et al. 2014). A second important parameter is also the water depth, in which the bubble curtain is operating (refer to Table 3), as the static pressure of the

Table 3 Summary of the measured insertion loss of the best available system configuration tested for a Big Bubble Curtain in different water depths

No.	Noise mitigation system (system configuration)	Water depth (m)	ΔSEL (dB)	Piles
1	BBC (>0.3 m^3/(min m), ballast chain outside)	≤ 25	$10 \leq 13 \leq 15$	>300
2	DBBC (>0.3 m^3/(min m), ballast chain outside, distance between nozzle hoses ~ water depth)	≤ 25	$14 \leq 17 \leq 18$	>300
3	BBC (>0.3 m^3/(min m), ballast chain outside)	~30	$8 \leq 11 \leq 14$	>20
4	BBC (>0.3 m^3/(min m), ballast chain outside)	~40	$7 \leq 9 \leq 11$	>30
5	DBBC (>0.3 m^3/(min m), ballast chain outside, distance between nozzle hoses ~ water depth)	~40	$8 \leq 11 \leq 13$	8
6	DBBC (>0.4 m^3/(min m), ballast chain outside, distance between nozzle hoses ~ water depth)	~40	$12 \leq 14 \leq 17$	2

Adapted from Nehls and Bellmann (2015)

water on the air bubbles increases with depth (Nehls and Bellmann 2015). Only a few measured data sources with water depths around 40 m exist and some technical malfunctions have occurred; however, it is supposed that noise reduction decreases by more than 4 dB in water depths from 25 to 40 m (with the same system configuration).

By increasing the air volume from 0.3 m^3/(min m), this reduction can partly be compensated (Table 3). Measured data from presently running projects (data is currently not public; itap) confirm these findings (the maximum air volume for a BBC was ~0.55 m^3/(min m)). Because of the water depth and the experiences from current construction projects, it has to be verified, whether the above-mentioned air volume of >0.3 m^3/(min m) can possibly be raised to a maximum of 0.55 m^3/(min m). Thereby, the sound reduction will presumably only increase marginally (Table 3). However, it appears that in finalized construction projects, a significantly more constant sound reduction in all directions in space could be ensured. Furthermore, in relation to the fed compressed air, safety margins were available.

Overview of Measured Spectral Insertion Loss

In Fig. 2, an overview of the (spectral) insertion loss of different (publically available) NMS is given by presenting the different 1/3 octave spectra.

The general shape of all shown curves in Fig. 2 shows that with increasing frequency, the noise reduction (ΔSEL) increases up to a frequency of approximately 1–2 kHz. In this frequency range, the pile-driving noise can be reduced in several frequency bands by up to 40 dB with several NMS. At higher frequencies, the noise reduction decreases significantly, probably due to the low levels of piling noise in

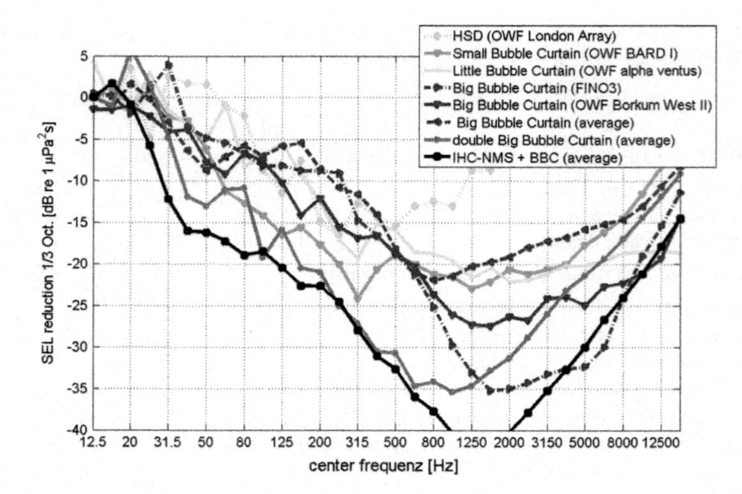

Fig. 2 Overview of the spectral 1/3 octave insertion loss (difference spectra) of different noise mitigation systems (including different system configurations of NMS)

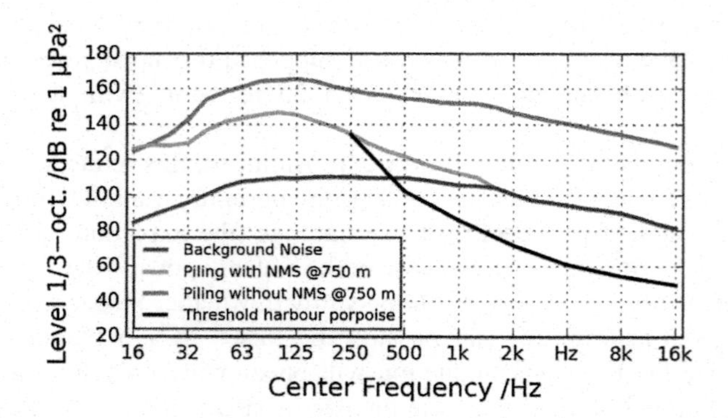

Fig. 3 Typical 1/3octave spectrum of an unmitigated (*red line*) pile-driving activity, a mitigated (*green line*, with a combination of close to and far from pile) pile-driving activity, the measured background noise before piling (*blue line*), and the absolute threshold of a Harbor Porpoise (*black line*), adapted from Kastelain et al. (2013) (Color figure online)

this frequency range (Fig. 3). However, the absolute noise reduction between different NMS or different system configurations of the same NMS (e.g. BBC) differs significantly over the whole frequency range. The frequency range between 80 and 400 Hz dominates the overall broadband (single) value for the insertion loss due to the acoustic energy distribution of the pile-driving activities (Fig. 3). Therefore, the overall noise reduction for a combination of NMS today is below 20 dB. The highest noise reduction was currently measured by using a combination of the IHC-NMS (type 6900) and an optimized BBC (Table 2).

Noise Mitigation by Reducing Pile-Driving Energy

A further opportunity for sound reductions is the reduction of the applied blow energy (maximum blow energy). Empirically, the acoustic parameters of pile-driving decrease in the North and Baltic Sea with water depths of less than 45 m by approximately 2.5 dB when the blow energy is halved (Gündert 2014; Bellmann 2014). By application of 'intelligent' pile-driving procedures, the lower blow energy can possibly be compensated by an increase in the blow frequency (blow rate). Here, the pile and the pile-driving hammer have to be checked for signs of material fatigue due to the impact of the higher number of blows. Moreover, the application of an intelligent pile-driving procedure depends on the soil resistance value, which highly depends on the penetration depth. In general, the higher the penetration depth, the higher the blow energy has to be.

The sound reduction potential of 'intelligent' pile-driving procedures is currently estimated, based on measured data of 2 dB maximum up to 4 dB, depending on the soil — if such a pile-driving procedure is practicable at all. This additional noise reduction can be added to the noise reduction results of the NMS or NMS combination.

Influence of Mitigated Pile-Driving Noise on Marine Mammals

In Fig. 3, a typical 1/3 octave spectrum of an unmitigated (*red line*) and a mitigated (*green line*) pile-driving activity measured at a distance of 750 m to the source is shown. The most acoustic energy of piling is emitted into the water between 50 and 400 Hz — depending on the pile diameter — without NMS. Additionally, the measured background noise before piling (*blue line*) and the measured absolute threshold of the harbor porpoise (*black line*; adapted from Kastelain et al. 2013) are shown. When using a combination of two independent NMS (one close to and one far from pile), the noise reduction increases from around 25 Hz to around 1.5 kHz onwards. For higher frequencies, the levels of mitigated piling noise are comparable to the measured background noise in this example, but can vary. This is typical for mitigated piling sequences and is the reason why the noise reduction of NMS in Fig. 2 decreases with increasing frequency from several kHz onwards, depending on the existing background noise. The measured absolute threshold curve for the harbor porpoises shows that the sensitivity of these marine mammals significantly decreases with decreasing frequency. In relation to the mitigated piling noise, most of the piling event is below the auditory threshold and well below the protection value preventing PTS. Nevertheless, higher frequencies of several kHz are almost overlapped by permanent background noise.

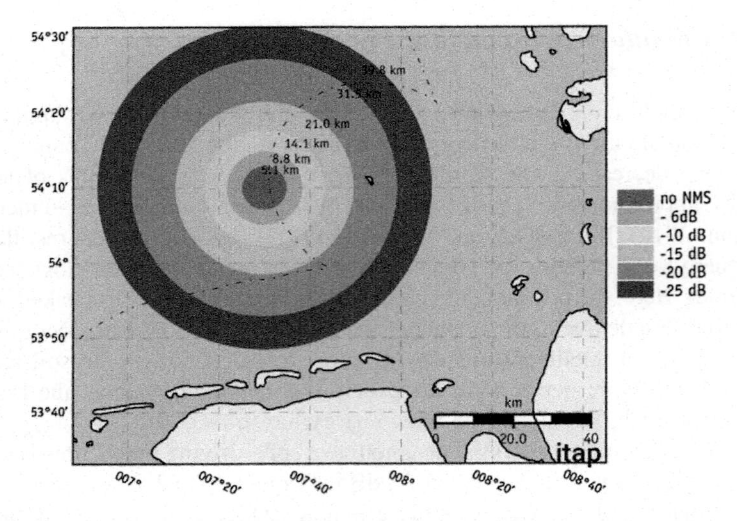

Fig. 4 The *circles* characterize the distances when the pile-driving noise reaches a level of 140 dB for the unmitigated and several mitigated (from −6 to −25 dB) cases, if an arbitrary pile-driving activity in the German North Sea with a Sound Exposure Level at 750 m (SEL) of SEL = 180 dB takes place

To demonstrate the effect of noise reduction by using NMS, the distances are demonstrated in Fig. 4, whereby an unmitigated piling sequence of SEL = 180 $dB_{@750m}$ and several mitigated scenarios decrease by absorption of the water (transmission loss) to a level of 140 dB (unweighted). This represents the level of disturbances for harbor porpoises according to the BMU strategy. For the transmission loss, the semi-empiric equation (by Thiele and Schellstede 1980) for the calm German North Sea was used, since the deviations between this semi-empiric transmission loss equation and real measured data as a function of distances is very low (e.g. shown in Nehls and Bellmann 2015). This equation shows that the absorption of underwater noise is non-linear over the distance. Influences of the bathymetry or other influencing factors on the transmission loss are not considered for this demonstration.

By using noise mitigation of 6–25 dB, the radiated area with levels of SEL $\geq$ 140 dB decrease significantly from (i) unmitigated case of 4976 km^2 (radius of 40 km) to (iv) minimum of 82 km^2 (r = 5.1 km). This means that the usage of any kind of NMS can help to avoid injury to marine mammals, but will also reduce the area where potential reaction or disturbances from marine mammals are likely or expected. In addition, the transmission loss will have a significant influence on the shown radiuses; therefore, the transmission loss must be selected for each individual area in case the German North Sea is not investigated.

Summary and Conclusions

- Several evaluated noise mitigation systems (NMS) to reduce the sound emission during pile-driving activities are available on the market. Noise Mitigation Screen (NMS), Hydro-Sound-Damper (HSD) and Big Bubble Curtains (BBC) are the most commonly used systems measured under real offshore conditions. Some other systems are currently under construction or are ready for an offshore test.
- The resulting noise reduction (insertion loss) of each evaluated NMS significantly depends on some system-relevant factors and parameters of the NMS, such as the air volume supplied for the BBC. For each application, the optimized system configuration, which has to be adapted to each single project, must be used to ensure a good noise reduction, as summarized in Table 2. In case of technical problems or non-optimal NMS configurations, the resultant noise reduction can decrease significantly. Resultant noise reduction can be positively increased by using a combination of noise mitigation measures, such as close to and far from pile and reduced blow energy.
- For water depths less than or equal to 25 m, a 'state-of-the-art' noise mitigation measure is available: (i) a single optimized noise mitigation measure can reduce the noise by minimum 10 dB (max. 15 dB); (ii) a combination of two systems leads to 15 dB (max. 20 dB), (iii) by using intelligent pile-driving methods (reducing the maximum blow energy), additional decibels (1–4 dB) can be produced.
- For water depths up to 40 m, no state-of-the-art noise mitigation is available, as for the time being only the BBC was partially used. Currently and in 2016, several other NMS will be tested at such water depths. The increasing water depth has a negative effect on the resulting noise reduction of air-based systems like the BBC due to the increasing static pressure on the air bubbles. This negative effect can partly be compensated by increasing the air volume. Nevertheless, with increasing water depths it is expected that the resulting noise reduction of NMS will slightly decrease.
- In any case, the usage of noise mitigation measures will significantly reduce the noise-induced area, especially for marine mammals, since all systems reduce the audible noise (>500 Hz) significantly. The audible noise is brought down close to the permanent background noise especially for frequencies above 1 kHz.

Acknowledgements Many thanks to all companies involved in the performed offshore tests within several funded research projects and normal construction phases, and especially to all OWF owners, who allowed us to conduct a lot of underwater noise measurements within their construction processes.

Great thanks go to all manufacturers of noise mitigation systems for their courage to develop and build different concepts to reduce pile-driving noise. Additionally, thank you very much for the constructive discussion during the improving and enhancing processes onshore and offshore; we enjoyed the experience.

References

Bellmann M (2014) Overview of existing noise mitigation systems for reducing pile-driving noise. Proceeding auf der Internoise 2014, Melbourne, Australia

BSH (2011) Offshore wind farms — measuring instruction for underwater noise measurement, technical report by Müller & Zerbst on behalf of BSH. Available via http://www.bsh.de/

BSH (2013a) Standard — investigation of the impacts of offshore wind turbines on marine environment (StUK4). Available via http://www.bsh.de/

BSH (2013b) Offshore wind farms — measuring specification for the quantitative determination of the effectiveness of noise control systems. Report M100004/05 by Müller & Zerbst on behalf of BSH. Available via http://www.bsh.de/

Diederichs A, Pehlke H, Nehls G, Bellmann MA, Gerke P, Oldeland J, Grunau C, Witte S (2014) Entwicklung und Erprobung des „Großen Blasenschleiers“ zur Minderung der Hydroschallemissionen bei Offshore-Rammarbeiten (HYDROSCHALL OFF BW II), technical final report, funded by BMWi and PTJ 0325309 A/B/C. Available via http://www.hydroschall.de

Gündert S (2014) Empirische Prognosemodelle für Hydroschallimmissionen zum Schutz des Gehörs und der Gesundheit von Meeressäugern. Masterarbeit an der Universität Oldenburg, Institut für Physik, AG Akustik

Kastelain RA, Gransier R, Hoek L, Rambags M (2013) Hearing frequency thresholds of a harbor porpoise (*Phocoena phocoena*) temporarily affected by a continuous 1.5 kHz tone. J Acoust Soc Am 134/3:2286. doi:10.1121/1.4816405. ISSN 00014966

Koschinski S, Lüdemann K (2013) Stand der Entwicklung schallminimierender Maßnahmen beim Bau von Offshore-Windenergieanlagen. Gutachten im Auftrag des Bundesamt für Naturschutz. Available via https://www.bfn.de/fileadmin/MDB/documents/themen/meeresundkuestenschutz/downloads/Berichte-und-Positionspapiere/Entwicklung-schallmindernder-Massnahmen-beim-Bau-von-Offshore-Windenergieanlagen-2013.pdf

Nehls G, Bellmann MA (2015) Weiterentwicklung und Erprobung des „Großen Blasenschleiers“ zur Minderung der Hydroschallemissionen bei Offshore-Rammarbeiten, technical final report, funded by BMWi and PTJ, FKZ 0325645A/B/C/D. Available via www.hydroschall.de

Schorcht S (2015) Technical noise mitigation during offshore-windfarm foundation examples from offshore windfarms: Meerwind Süd|Ost, Global Tech I, Nordsee Ost, DanTysk, EnBW Baltic 2, Borkum Riffgrund 1, Amrumbank West und Butendiek. Presented at the offshore wind R&D conference 2015 in Bremerhaven, 13/14 Oct 2015. Available via http://www.rave-conference.de/

Thiele & Schellstede (1980) Standardwerte zur Ausbreitungsdämpfung in der Nordsee. FWG-Bericht 1980-7, Forschungsanstalt der Bundeswehr für Wasserschall und Geophysik

Wilke F, Kloske K, Bellmann M (2012) Evaluation von Systemen zur Rammschallminderung an einem Offshore-Testpfahl, technical final report for the project ESRa, funded by BMWi and PTJ FKZ 0325307

Part V
Monitoring and Long-Term Effects

The Challenges of Addressing Wildlife Impacts When Repowering Wind Energy Projects

K. Shawn Smallwood

Abstract Industrial wind power expanded rapidly since the earliest projects, and with this rapid expansion came understanding of wind power's impacts on wildlife and how to measure and predict those impacts. Many of the earliest wind turbines began exceeding their operational lifespans >10 years ago, spawning plans for repowering with modern turbines. All wind turbines eventually wear out. Repowering can replace old turbines that have deteriorated to capacity factors as low as 4–12% with new wind turbines with capacity factors of 30–38%, and possibly sometimes better. At the same rated capacity, a repowered project can double and triple the energy generated from the project while reducing avian fatality rates by 60–90% when the new turbines are carefully sited. On the other hand, the grading needed for wider roads and larger pads can harm terrestrial biota, and can alter the ways that birds fly over the landscape. Larger turbines are usually mounted on taller towers, so the rotor-swept plane reaches higher into the sky and can kill species of birds and bats that were previously at lower risk. Slower cut-in speeds might increase bat fatalities, and faster cut-out speeds might increase bird fatalities. Repowering poses special problems to fatality monitoring and to estimating changes in collision rates. Differences in collision rate estimates before and after repowering can be due to climate or population cycles, changes in monitoring methods, and changes in wind turbine efficiency. Fatality monitoring could be more effective when it is (1) long-term, including when the older project was operating at peak efficiency, (2) executed experimentally, such as in a before-after, control-impact design, (3) largely consistent in methodology and otherwise adjusted for inconsistencies, and (4) sufficiently sampling the projects' installed capacity. Another challenge is overcoming public and regulator impatience over documented wildlife fatalities. Fatality monitoring before repowering necessarily reveals project impacts. Repowering can reduce those impacts, but this message needs to be delivered effectively to a public that might be skeptical after seeing the earlier impacts and will want to see trustworthy fatality predictions going forward.

K. Shawn Smallwood (✉)
Davis, CA, USA
e-mail: puma@dcn.org

J. Köppel (ed.), *Wind Energy and Wildlife Interactions*,
DOI 10.1007/978-3-319-51272-3_10

Accurately predicting impacts at repowered projects can be challenging because the often-used utilization survey has fared poorly at predicting impacts, and because flight patterns can shift in the face of larger wind turbines and an altered landscape.

Keywords Bats · Birds · Fatality monitoring · Repowering · Siting · Utilization survey · Wind energy · Wind turbine

Introduction

Industrial wind power expanded rapidly in the USA since the earliest projects in the 1980s (Fig. 1).

With this rapid expansion came greater understanding of wind power's impacts on wildlife and how to measure and predict those impacts (Orloff and Flannery 1992; Drewitt and Langston 2006; Smallwood and Thelander 2004; Smallwood 2007). Most research of wildlife impacts has been directed toward new projects, but research is beginning to shift toward the challenges posed by repowering (Hötker 2006) and how to minimize impacts going forward (Smallwood and Karas 2009; Smallwood et al. 2009a, 2016; Smallwood 2016).

Many of the earliest industrial-scale wind turbines began exceeding their operational lifespans by about 2004, spawning plans for repowering with modern wind turbines. Eventually, all modern wind turbines deteriorate to the point of needing to be decommissioned or replaced. Repowering can replace old-generation turbines that have deteriorated to capacity factors as low as 4–12% with new wind turbines

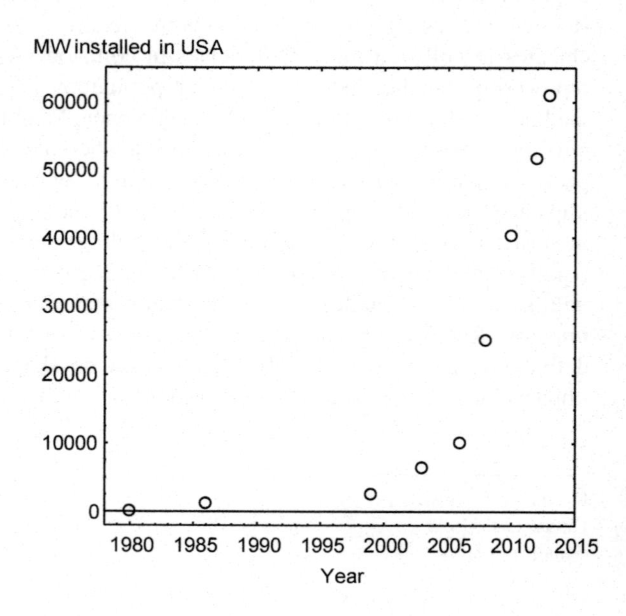

Fig. 1 Increase in installed capacity in the USA, 1980 through 2014 (*Source* American Wind Energy Association)

that can achieve capacity factors of 30–44%. Energy generated from a repowered wind project can, at the same rated capacity, double, triple or quadruple the energy generated from the old turbines. After four years of monitoring and research at one of the world's most notorious wind resource areas—the Altamont Pass Wind Resource Area—I concluded that repowering could potentially reduce avian fatality rates by 60–90% (Smallwood and Thelander 2004). After two years of monitoring at a repowered project in the Altamont Pass, avian fatalities were reduced substantially (Brown et al. 2014).

Repowering usually reduces the number of wind turbines in a project area, thereby opening up available airspace to flying birds (Figs. 2 and 3).

It also often moves electric distribution lines from above ground to below ground, thus reducing electrocutions and line collisions. Repowering also provides the opportunity to more carefully site new wind turbines to meet energy generation needs while minimizing collision risk with volant wildlife. This includes avoiding ridge saddles, breaks in slopes, relatively low-lying areas and ecological attractants such as wetlands.

On the other hand, repowering requires wider roads and larger pads. Grading for roads and pads can harm terrestrial plants and animals, and can alter the landscape and the ways that birds fly over the landscape. Grading to accommodate the pad of a new wind turbine can enhance the concave structure of a ridge saddle or can create a break in slope where a break did not exist before. Either of these changes to the landscape can channel avian and bat flights towards the turbines and increase the collision risk. In addition, larger wind turbines are usually mounted on taller towers, so the rotor-swept plane reaches higher into the sky and can kill species of birds and bats that were previously at lower risk.

Fig. 2 In this example in the Altamont Pass WRA, California, all of the small wind turbines can be replaced by the single large turbine at the hill peak to achieve the same rated capacity, although the capacity factor would be increased 3- to 4-fold and the available space available for careful siting increased by a factor of >20

Fig. 3 In this example in the Altamont Pass WRA, California, a 2.3 MW turbine was installed next to 7 120-KW Bonus turbines descending along a ridgeline into a ravine where raptor collision risk is generally higher. Nineteen of the Bonus turbines can be replaced by a single 2.3-MW turbine

The slower cut-in speeds of new turbines might increase bat fatalities, and the faster cut-out speeds might increase bird fatalities. The 'cut-in speed' is the speed at which the turbine first starts to rotate and generate power (typically between 3 and 4 m/s), and vice versa for 'cut-out speeds'. The larger wind turbines might also cause greater displacement impacts, although Hötker (2006) concluded that displacement effects are greater at the older, small turbines.

Measuring Effects of Repowering

Repowering poses special problems to monitoring and estimating impacts on birds and bats. Birds that were not displaced by the smaller old-generation turbines might be intimidated by the much larger sizes of the new wind turbines. Surveys that were performed to measure displacement caused by the old turbines, such as utilization surveys or pedestrian transects, might be located at sites that are less suitable for measuring displacement caused by the new wind turbines. For example, preserving a utilization survey station that overlooked old-generation wind turbines might be on the pad of the new wind turbine. The difference in size between the old and new turbines on the order of 100 KW–3 MW would force the observer to look upward into the airspace of the new turbine, whereas observers at the same station would have looked laterally to survey farther out along a row of old-generation turbines. Pedestrian transects that were sufficiently long enough to quantify displacement of grassland songbirds around old turbines could prove too short for doing so at the new, larger wind turbines. It could be physically impossible to keep the survey attributes between pre- and post-repowering, unless survey stations and transects are strategically located before repowering to accommodate comparisons of results before and after repowering.

In another example, evidence is growing of multi-annual cycles of avian fatalities where fatality monitoring has lasted long enough to test for long-term trends. If monitoring a species' fatalities corresponded with a peak in the fatality cycle before repowering and with the nadir of the cycle after repowering, then repowering can be over-credited for the measured change in fatality rates (Fig. 4).

If monitoring corresponded with the nadir of the cycle before repowering, then repowering might be under-credited because fatality rates might not appear to have

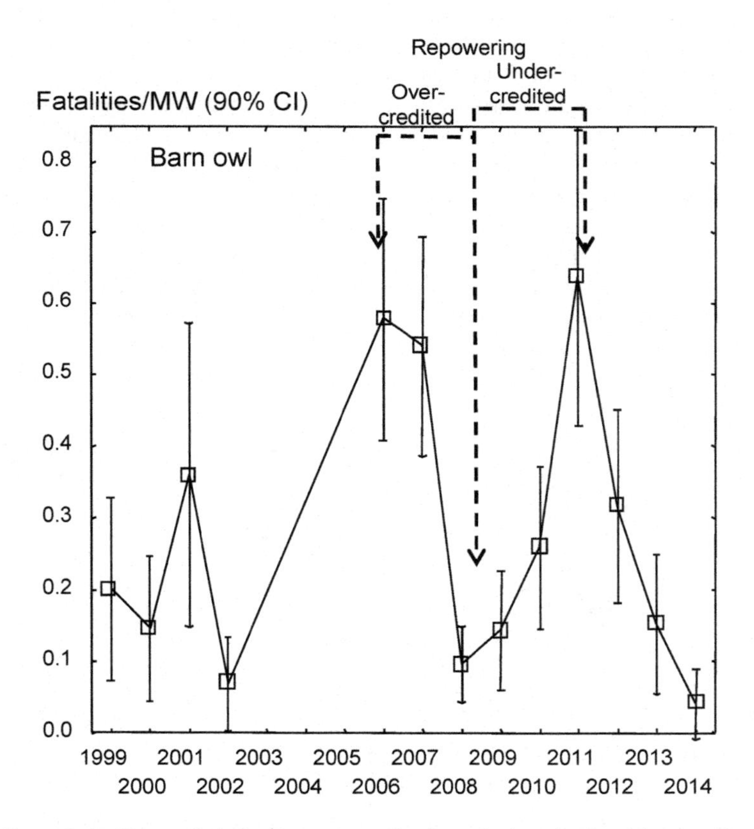

Fig. 4 Barn Owl (*Tytus alba*) fatality rates cycle through time in the Altamont Pass Wind Resource Area, so assuming pre- and post-construction monitoring periods of 1 or 2 years, too much credit would be given to repowering for a measured reduction in Barn Owl fatality rates had the fatality monitoring at old turbines ended in 2006 and the monitoring at new turbines commenced in 2008, and too little credit would be given to repowering had the fatality monitoring at old turbines ended in 2009 and commenced at new turbines in 2010

changed (Fig. 4). To overcome this potential source of confounding and bias, monitoring should last long enough to represent major portions of at least one cycle period. Optimally, fatality monitoring should last long enough to cover one full cycle period both before and after repowering.

Another strategy for preventing bias from monitoring before and after repowering at different portions of a fatality cycle is to concurrently monitor another nearby wind project before and after the focal project was repowered. Concurrent monitoring can provide the opportunity for comparing fatality rates in a before-after, control-impact (BACI) experimental design. Experimental designs, where planned or opportunistic, would benefit from replication and interspersion of groups of new and old turbines. They would also benefit from groups of new and old turbines being monitored at sufficient spatial and temporal scales to accumulate enough data to achieve more than minimal effects sizes.

Another confounding factor arises from fatality monitoring at the old wind turbines toward the end of their decline, followed by monitoring of the new turbines in peak operating condition (Fig. 5).

Comparing fatality rates from an old project operating at 9% capacity factor to the new project operating at 37–44% capacity factor can be misleading. For these reasons, fatality monitoring could be more effective when it is (1) long-term, including when the older project was operating at a higher capacity factor, and (2) executed in an experimental design, such as a BACI design. If an experimental design is implemented, then care must be taken to ensure that sufficient sample sizes of fatalities will be detected, and to minimize fatality contamination resulting from carcasses deposited by one turbine ending up within the search area of another turbine.

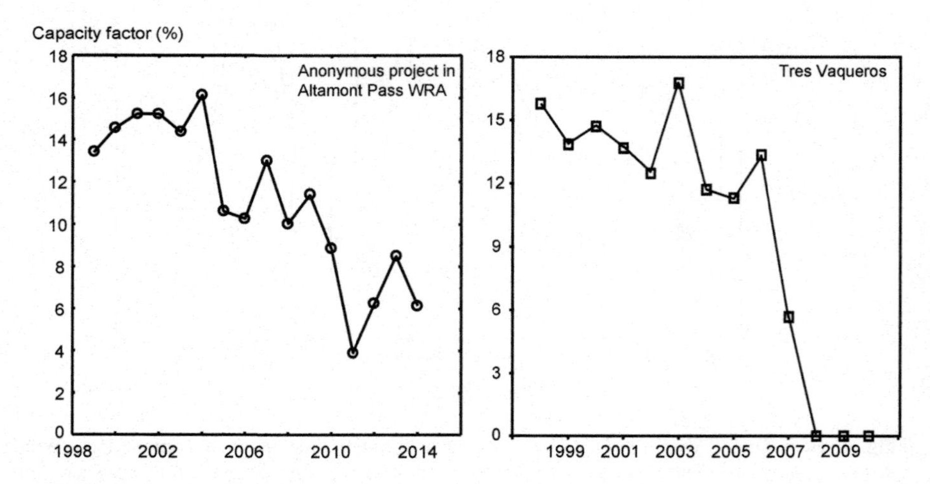

Fig. 5 The capacity factor of projects declined through time in the Altamont Pass Wind Resource Area, so the closer the fatality monitoring period to repowering, the less the operations of the old project will resemble its operations during peak performance

Compared to the older project, fatality monitoring at the repowered project usually will be performed to a much greater maximum search radius due to greater tower heights and larger rotor diameters. Monitoring at the new turbines often includes greater transect separation and shorter search intervals. Carcass detection probabilities change with these monitoring changes. For example, extending fatality searches from 50 m around old turbines to 120 m around new turbines can increase the proportion of the search area that is 'unsearchable' if it extends into open water, marshes, steep slopes, cliff faces, crops or thick vegetation. On the other hand, due to larger size, new wind turbines require wider access roads and larger pads to accommodate construction, which are portions of the search area where searchers most often find bat fatalities while the pads remain relatively clear of vegetation following construction. Repowering can increase the detection rates of bat and small bird carcasses within larger expanses of cleared ground, while also hiding more of them in larger expanses of thick vegetation farther away from the turbine.

Bat carcass detection rates can also increase with shorter search intervals, hence resulting in higher fatality rate estimates at repowered projects. This potential source of bias is more likely to express itself at repowered projects because search intervals have been lessening through time, and will likely be shorter at the repowered project compared to the same project prior to repowering.

Another challenge is overcoming impatience from the public and regulators over documented wildlife fatalities. Fatality monitoring before repowering necessarily reveals project impacts. Repowering can reduce some or most of those impacts, but this message needs to be delivered effectively to the public, who can become skeptical after seeing the earlier impacts. The public will want to see trustworthy fatality predictions going forward. Predicting impacts at repowered projects can be challenging because utilization surveys often used at wind projects have contributed poorly to predicting impacts (de Lucas et al. 2008; Ferrer et al. 2012). Also, flight patterns and behaviors can shift in the face of larger wind turbines and an altered landscape.

Careful Siting of Wind Turbines

Siting new wind turbines to minimize collision risk can benefit from learning the patterns of collisions among the old turbines, as well as from behavior observations. So long as the pre-repowering fatality monitoring was sufficient in duration and extent, investigators should be able to recognize topographic features associated with old wind turbines that caused disproportionate numbers of fatalities. Because the new turbines will likely be taller and larger than the old turbines, some differences might emerge in how fatalities at wind turbines relate to topography. These differences might be anticipated by observing flight behaviors of the local birds and bats in relation to wind patterns and terrain features.

Comparing fatality rates among old turbines to identify topographic associations can be challenging if the monitoring methods or effort levels vary among the turbines, or if the monitoring is of short duration. The magnitude and range of variation in fatality rate estimates diminishes as the duration of monitoring increases because the numerator (number of fatalities) of the fatality metric varies less than does the time period in the denominator (e.g. Fig. 6).

To compare fatality rates from various effort levels among wind turbines, a model can be fit to the data and used to predict fatality rates at each turbine after a time period when the fatality rates stabilize with monitoring duration. The variation in the predicted fatality rates would be more informative than the variation in the original fatality rates because the predicted rates will have accounted for the level-of-effort bias in the original fatality rates (Fig. 6).

Another useful metric that can be measured from the old wind project prior to repowering is the rate of wind turbine encounters (Fig. 7).

For a given species, observed near-misses with wind turbine blades or towers can inform the investigator about where and under what wind and topographic conditions collisions are more likely to occur. These near-misses can be especially predictive if they correlate with measured fatality rates among wind turbines. Near-misses can serve as a surrogate for fatalities, and sample sizes can be increased because near-misses can be counted more often than fatalities are found. Near-misses can be qualified as birds or bats flying within a certain distance of wind turbines, such as flying through the wind turbine's rotor, as evasive maneuvers

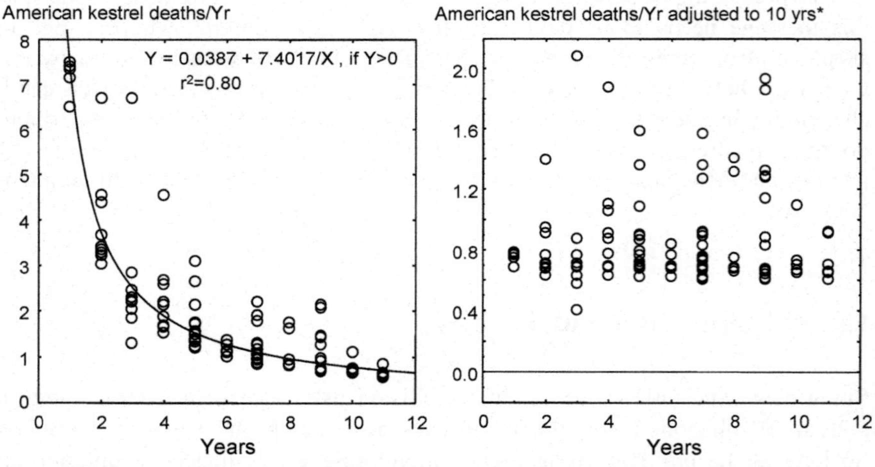

Fig. 6 As an example of an adjustment needed for monitoring duration, American kestrel (*Falco sparverius*) fatality rates among wind turbines in the Altamont Pass Wind Resource Area declined with increasing number of years the wind turbines were monitored over various periods from 1998 through 2014. The adjustment shown here was only for wind turbines where ≥ 1 fatality was found. For spatial comparisons of fatality rates among turbines, the analyst should compare actual and model-projected fatality rates at a common duration; in this example 10 years of monitoring was used as the standard for comparison

Fig. 7 A golden eagle (*Aquila chrysaetos*) narrowly misses collision with the moving blade of an operating wind turbine in the Altamont Pass Wind Resource Area, California

within the rotor plane to avoid collision, as jostling or tumbling caused by the blade's pressure wave, or as actual non-fatal contact. These near-misses can also be weighted to express their hazard levels. From 2012 through 2015, near-misses were tabulated for raptors in the Altamont Pass Wind Resource Area, resulting in hundreds of records that contributed to collision hazard models (Smallwood et al. 2016).

Using behavior patterns to guide the siting of new turbines in repowering requires some understanding of how the behaviors relate to collision risk. Whereas understanding has improved over how behaviors relate to wind turbine collisions, much remains to be learned. Every time a bird or bat crosses the plane of the turbine's combined rotor and tower, the likelihood of a collision rises from zero, so this threshold is fundamental toward predicting collision risk. Once the animal has entered the rotor plane of an operating turbine, then many other factors contribute to collision risk, including ground speed, the animal's size and profile, angle of entry and wind turbine attributes such as blade width, depth, rpm and tip speed (Tucker 1996a, b). Additionally, the animal's behavior affects collision risk, including whether evasive actions are taken to avoid a blade strike, and which actions are taken. Whether the animal enters the rotor plane during daytime or at night will affect the degree to which the blades can be seen. The animal's angle of entry with a terrain background or sky background affects its ability to view the blades, and this can also be affected by clouds, fog and time of day. Wind conditions can also affect collision risk, as bird species flying in stronger winds tend to be found more often as collision fatalities (Smallwood et al. 2009a), perhaps due to reduced flight control in strong winds. Behaviors involving distraction can also affect collision risk, such as predatory attacks, chasing competitors, fleeing from predators or competitors, playing or courtship displays.

The presence of a bird or bat in the general vicinity of a planned or existing wind turbine is thought to convey some collision risk, so utilization surveys have often been performed to quantify this risk. A typical survey for birds consists of an expanded point count or visual scan, where the observer records birds within a specified radius, often as far away as 800 m or even 1600 m. A typical utilization survey for bats consists of acoustical detectors mounted on turbine nacelles or met towers, picking up bat calls within the range of the detectors (usually within 30 m). However, evidence is building that utilization rates are poor predictors of fatality rates of birds (de Lucas et al. 2008; Ferrer et al. 2012; Smallwood 2016) and bats (Hein et al. 2012), probably because utilization rates of birds are measured with insufficient spatial resolution. For bats, utilization rates are measured too narrowly when detectors are mounted on wind turbines, and too coarsely when mounted on meteorological towers. The needed resolution is at the height and width of proposed new wind turbines, plus some buffer distance to account for error in estimating where a bird or bat is observed in the airspace. Data recorded at this resolution can consist of counts or flight paths per unit time. These passage rates, whether tabulated from daytime visual surveys, nocturnal surveys using thermal cameras or radar, or telemetry, include only those animals that are close to where the turbines exist or are planned, so they should relate more closely to fatality rates.

Measuring passage rates at every candidate wind turbine site could prove too costly, as sufficient survey effort would be needed to satisfactorily represent passage rates at all sites. A more efficient approach is to measure passage rates at a random selection of candidate turbines sites, construct predictive models from the data, and project the model predictions to candidate sites that were not selected for surveys. Another approach would be to tabulate passage rates over suites of terrain and wind conditions, construct predictive models, and then project the model predictions to the entire land surface of the wind project (e.g., Smallwood et al. 2009b). Passage rates can be even more predictive, however, if they represent specific behaviors that are believed responsible for increased collision risk.

The most predictive passage rates are those involving distractions. Many of the observed near-misses with wind turbines in the Altamont Pass Wind Resource Area involved birds fleeing from other birds or birds chasing or interacting with others. Passage rates including avian interactions composed one of the more predictive variables for repowering the Altamont Pass. Another useful passage rate was the number of flight paths intersecting segments of ridgelines classified by elevation, shape, orientation and downslope conditions affecting deflection updrafts.

Needed Changes to Monitoring Protocols

To more effectively site wind turbines in repowering efforts and to compare changes in impacts resulting from repowering, the following monitoring measures are recommended. Fatality monitoring could be more effective by including a time period when wind turbines of the older project operated at higher capacity factor than just prior to repowering. Longer-term monitoring not only represents the project when it operated more efficiently, but also captures inter-annual variation in fatality rates. Fatality monitoring would also be more effective when implemented experimentally, such as in a before-after, control-impact (BACI) design. A BACI design might be difficult to plan within a wind resource area due to various ownerships leading to project-by-project timing of repowering. Therefore, another approach would be to monitor as many of the wind projects in the wind resource area as possible to prepare for the emergence of opportunities to compare fatality rates in an unplanned BACI design.

When comparing fatality monitoring results before and after repowering, steps can be taken to prevent the effects of potentially large biases. For example, a greater maximum search radius can be used at the older project, even though the smaller turbines were typically searched with a much shorter search radius. If the new project will be monitored for fatalities within a radius of 110 m around the turbines, then the old project ought to be searched out to 110 m to be comparable. Furthermore, searching to the greater maximum distance would facilitate hypothesis-testing related to the effects of survey radius on fatality rate estimates.

The average transect separation should also be consistent pre- and post-repowering. Using a transect separation of about 6 m at the old turbines and a transect separation of 10 m at the new turbines could bias the comparison of fatality rates to an unknown degree. Similarly, if the average search interval is going to be weekly at the new wind turbines, then it ought to be weekly at the old turbines. To account for the improved ground visibility nearby repowered turbines due to wider access roads and larger, cleared pads, integrated detection trials should be implemented at both pre-and post-repowered projects (Smallwood 2013; Smallwood et al. 2013). Integrated detection trials are randomized placements of bird and bat carcasses throughout fatality monitoring for quantifying the overall proportions of placed birds and bats ultimately found by searchers. In other words, the integrated detection trials should replace the usual separate trials performed for searcher detection and carcass persistence, and every effort should be made to realistically simulate the detection probabilities of wind turbine fatalities. This approach will account for differences in ground cover pre- and post-repowering.

Utilization surveys should be replaced by behavior surveys performed by qualified behavioral ecologists, or by careful use of radar or telemetry. Visual surveys for behavior should be performed day and night because much of the collision risk for both birds and bats is at night. Before repowering, special attention should be directed toward interactions between volant wildlife and the old-generation wind turbines. All behavior observations should be related to wind conditions and terrain so that these relationships can be used for siting the wind turbines in repowering.

Concluding Remarks

The pace of repowering will soon equal or exceed new projects. The challenges introduced herein will soon be common. Repowering provides opportunities to lessen project impacts, but we also need to be prepared to measure the changes in project impacts. Waiting until the older project deteriorates to the point of very low capacity factor before implementing appropriate monitoring will make it more difficult to measure changes in project impacts. All fatality, relative abundance and behavior monitoring should be planned from now on to anticipate comparisons between extant and future wind projects before and after repowering.

The data that best informs siting to minimize the impacts of repowering include fatalities at the old turbines, as well as near-misses with wind turbines, passage rates and especially passage rates involving certain behaviors including interactions with other birds or bats. In the absence of fatality data, behavior data or passage rate data, new wind turbines should not be located in relatively lower portions of the landscape, on breaks in slope, on terrain features receiving strong deflection updrafts (such as from concave slopes oriented toward the prevailing wind direction), downwind of a berm or small ridge that will force birds and bats to fly at the height of the rotor, or along edges such as coastlines, or between two or more types of vegetation cover. Siting should also anticipate the effects of grading to accommodate access roads and tower pads, as well as nearness to meteorological towers, transmission towers, and trees. Electric distribution lines should be installed underground.

For comparing impacts before and after repowering, monitoring needs to be long-term, of sufficient spatial extent, and of sufficiently short intervals between searches to accumulate adequate sample sizes. Comparisons would also be improved by setting the stage for planned or opportune BACI experimental designs. Conventional carcass persistence and searcher detection trials should also be replaced with integrated detection trials to account for differences in ground cover and other conditions before and after repowering.

References

Brown K, Smallwood KS, Szewczak, J, Karas, B (2014) Final 2012–2013 annual report Avian and Bat monitoring project Vasco winds, LLC. Report to NextEra Energy Resources, Livermore, California

De Lucas M, Janss GFE, Whitfield DB, Ferrer M (2008) Collision fatality of raptors in wind farms does not depend on raptor abundance. J Appl Ecol 45:1695–1703

Drewitt AL, Langston RHW (2006) Assessing the impacts of wind farms on birds. Ibis 148:29–42

Ferrer M, de Lucas M, Janss GFE, Casado E, Munoz AR, Bechard MJ, Calabuig CP (2012) Weak relationship between risk assessment studies and recorded mortality in wind farms. J Appl Ecol 49:38–46

Hein C, Erickson W, Gruver J, Bay K, Arnett EB (2012) Relating pre-construction bat activity and post-construction fatality to predict risk at wind energy facilities. In: Schwartz SS (ed) Proceedings of the wind-wildlife research meeting IX. Broomfield, CO. November 28–30, 2012. Wildlife Workgroup of the National Wind Coordinating Collaborative by the American Wind Wildlife Institute, Washington, DC

Hötker H (2006) The impact of repowering of wind farms on birds and bats. Michael-Otto-Institute within NABU—Research and Education Centre for Wetlands and Bird Protection. Available via http://www.sofnet.org/1.0.1.0/1267/download_916.php. Accessed on 11 Dec 2015

Orloff S, Flannery A (1992) Wind turbine effects on avian activity, habitat use, and mortality in Altamont Pass and Solano County Wind Resource Areas: 1989–1991. Report to California Energy Commission, Sacramento, California

Smallwood KS (2007) Estimating wind turbine-caused bird mortality. J Wildl Manage 71: 2781–2791

Smallwood KS (2013) Comparing bird and bat fatality-rate estimates among North American wind-energy projects. Wildl Soc Bull 37:19–33. +Online Supplemental Material

Smallwood KS (2016) Monitoring birds. In Perrow M (ed) Wildlife and wind farms: conflicts and solutions. Pelagic Publishing (in press)

Smallwood KS, Thelander C (2004) Developing methods to reduce bird mortality in the Altamont Pass Wind Resource Area. Final Report to the California Energy Commission, Public Interest Energy Research—Environmental Area, Contract No. 500-01-019. Sacramento, California

Smallwood KS, Karas B (2009) Avian and bat fatality rates at old-generation and repowered wind turbines in California. J Wildl Manage 73:1062–1071

Smallwood KS, Neher L, Bell DA (2009a) Map-based repowering and reorganization of a wind resource area to minimize burrowing owl and other bird fatalities. Energies 2:915–943. Available via http://www.mdpi.com/1996-1073/2/4/915

Smallwood KS, Rugge L, Morrison ML (2009b) Influence of behavior on bird mortality in wind energy developments: the Altamont Pass wind resource area, California. J Wildl Manage 73:1082–1098

Smallwood KS, Bell DA, Karas B, Snyder SA (2013) Response to Huso and Erickson comments on novel Scavenger removal trials. J Wildl Manage 77:216–225

Smallwood KS, Neher L, Bell DA (2016) Siting to minimize raptor collisions: an example from the repowering Altamont pass wind resource area. In Perrow M (ed) Wildlife and wind farms: conflicts and solutions. Pelagic Publishing (in press)

Tucker VA (1996a) A mathematical model of bird collisions with wind turbine rotors. J Sol Energy Eng 118:253–262

Tucker VA (1996b) Using a collision model to design safer turbine rotors for birds. J Sol Energy Eng 118:263–269

Part VI
Planning and Siting

Wind Farms in Areas of High Ornithological Value—Conflicts, Solutions, Challenges: The Case of Thrace, Greece

Alkis Kafetzis, Elzbieta Kret, Dora Skartsi, Dimitris Vasilakis and Ioli Christopoulou

Abstract Thrace, in northeastern Greece, is a crucial habitat of European value for the survival of the continent's birds of prey, and it is also designated as a priority area for the development of wind farms. Consequently, it is a valuable case study for identifying the conditions under which common ground can be found. A proactive planning system grounded on sound research and analysis and robust environmental impact assessments will direct wind farm development to the most suitable locations. It will also reduce the negative impacts of wind farms and reduce the time and resources required for the approval of individual projects, while improving the reputation of the wind energy sector.

Keywords Wind turbine collision · Strategic approach · Spatial planning · Policy · Black vulture

Introduction

Mitigating climate change and halting biodiversity loss constitute global cornerstones in environmental protection. For each of these environmental priorities, separate international, regional and national level legal frameworks have been established in order to meet their objectives without prioritizing one against the other. However, additional efforts are needed in order to create inter-environmental linkages and synergies.

A case in point is the development of wind farms in areas of high ornithological value. As noted by the European Commission (EC 2010) wind farms can have a significant negative impact on birds, especially when wind farm site selection falls within areas of increased ornithological interest, such as the Special Protection Areas (SPA) of the European Union's Bird Directive (2009/147/EC) (WWF Greece 2014). According to Langston and Pullan (2003) the impact can be more critical

A. Kafetzis (✉) · E. Kret · D. Skartsi · D. Vasilakis · I. Christopoulou
Dadia, 68400 Soufli, Greece

J. Köppel (ed.), *Wind Energy and Wildlife Interactions*,
DOI 10.1007/978-3-319-51272-3_11

when these areas host small populations of long-lived, large species, with low annual productivity and slow maturity (Martínez-Abraín et al. 2012).

In such situations of potential conflict between actions targeted at climate change mitigation and biodiversity conservation a proper spatial planning strategy is the major way for the negative impacts of wind farms on rare bird species to be proactively addressed (Bright et al. 2008; Dimalexis et al. 2010; WWF Greece 2013; Vasilakis et al. 2016). Such a planning strategy must be based on the best available scientific data regarding bird interactions with wind farms and on spatially explicit zones that set specific requirements on wind farm development. This approach can provide the conditions that maximize the wind energy potential of an area while minimizing impacts on endangered bird species.

Thrace, a region in north Greece (Fig. 1), is both an area of great ecological value due to the rare avifauna found there and an area of significant value to wind farm development, due to its high wind potential (Dimalexis et al. 2010; Poirazidis et al. 2011). As a result, Thrace constitutes a valuable case study for identifying the conditions under which common ground can be found. This paper draws on the long-standing engagement of WWF Greece in the region and presents: (a) the data collection, analysis approach and mapping that was undertaken in order to understand the issues surrounding wind farm development in a region of high ornithological importance; and (b) the tools that were proposed in order to seek ecologically sound wind farm development in the region of Thrace. The paper

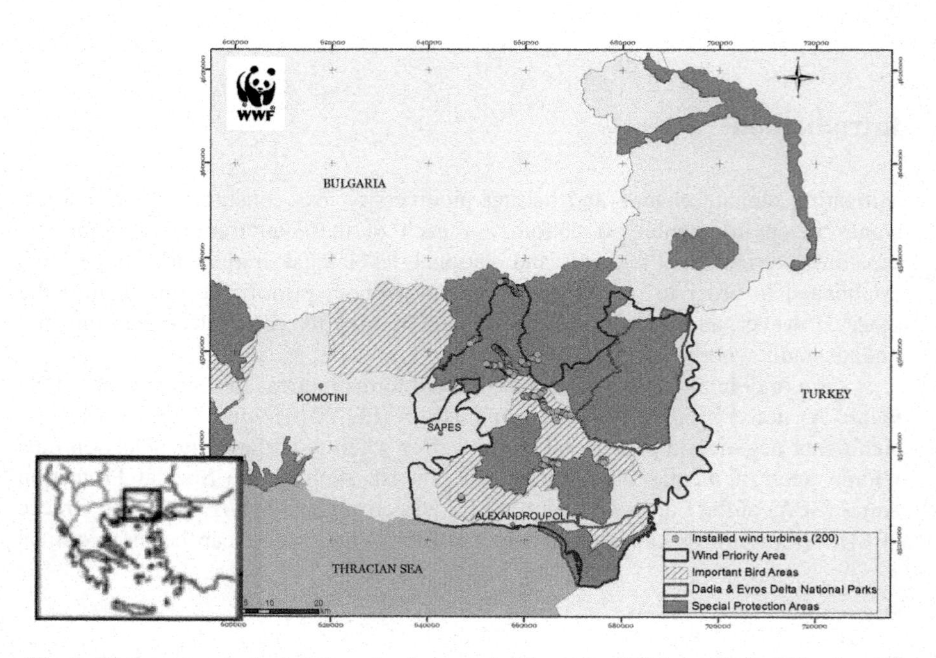

Fig. 1 Thrace: a case of wind farm development inside areas of high ornithological value. The region includes 2 National Parks, 5 SPAs and 5 Important Bird Areas

concludes with insights on the challenges such efforts face, as well as lessons learned that could aid similar endeavors.

The Region of Thrace

Ornithological Value

The region of Thrace is of exceptional ornithological importance, given that it hosts habitats that are of European-wide significance, mainly for large birds of prey and aquatic birds. In the area, five SPAs have been designated as well as two national parks: the Dadia-Lefkimi-Soufli Forest (hereafter Dadia National Park) and the Evros Delta National Parks. Dadia National Park, rightfully, has been characterized as "the land of the birds of prey", as 36 of the 38 diurnal birds of prey of Europe have been observed there (Poirazidis et al. 2011). It is one of the few places in eastern Europe where three of the four European vulture species breed: the Black Vulture (*Aegypius monachus*), the Egyptian Vulture (*Neophron percnopterus*), and the Griffon Vulture (*Gyps fulvus*), of which the first two are internationally "Near Threatened" and "Endangered" respectively.

The local Black Vulture population is the last breeding population that remains of the formerly large Balkan population (Skartsi et al. 2008). Currently its size is estimated at 120 individuals, forming each year around 28–35 breeding pairs. It has a breeding success that fluctuates between 50 and 60% (Zakkak et al. 2014; Zakkak and Babakas 2015). The Egyptian Vulture is considered nowadays as one of the most endangered species in the country. In 2015, only eight pairs were recorded in Greece, out of which five were in Thrace. The individuals nesting in Greece are part of the larger Balkan population, which thirty years ago formed a single, more or less spatially continuous population. Lastly, the Griffon Vulture that nests predominantly outside the Dadia National Park forms, inside one of the SPAs, the biggest colony on mainland Greece with 13 reproductive couples (WWF Greece 2013).

Moreover, the area is used for nesting, wintering or passage by rare birds of prey such as the Imperial Eagle (*Aquila heliaca*), the White-tailed Eagle (*Haliaeetus ablicilla*), the Greater Spotted Eagle (*Aquilla clanga*) and the Golden Eagle (*Aquila chrysaetus*) (Poirazidis et al. 2011; Schindler et al. 2014). These species are included in Annex I of the Birds Directive (2009/147/EC) and all of them are characterized as under threat in the Red Book of Endangered Animals in Greece (Legakis and Maragou 2009). It is equally worth noting that the vultures nesting in Dadia National Park seek out their food over a large area stretching almost all the way to the Evros Delta in the south and Rodopi in the west and often entering the neighbouring countries of Bulgaria and Turkey (Vasilakis et al. 2008; Noidou and Vasilakis 2011; Vasilakis et al. 2016). Thus it is of great importance to recognize the region (consisting of two National Parks, five SPAs and five Important Bird Areas) not as separate areas of distinct environmental value but as a closely-knit ecological network.

Wind Power Capacity

Aside from its importance to biodiversity conservation, the region has a significant role to play in the efforts to tackle climate change. A large part of the region has been selected as priority area for the development of wind energy sources as it is one of the areas with the highest wind capacity in mainland Greece. Specifically, the biggest part of the Regional Unit (RU) of Evros and a part of the RU of Rodopi have been delineated as Wind Priority Area 1 (WPA 1) under the National Renewable Energy Spatial Plan framework (Fig. 1). The WPA 1 covers about half of the region's Natura 2000 sites, including the two National Parks, and overlaps with the area used by the birds of prey in the region. Half of the WPA 1 (53%) falls within the core area of the Black Vulture population and also envelops the Strictly Protected Area of the Dadia National Park. Since 2000, 200 wind turbines, with a total capacity of roughly 250 MW, have been installed in 15 wind farms. The overall capacity of the WPA 1 is estimated to reach 960 MW.

Wind farm development in Thrace epitomizes the dilemma between policies that promote renewable energy resources and policies intended to protect rare and endangered species and their habitats. Even more so, as the emblematic Black Vulture population in the region is unique and significantly small and the designated area for wind farm development coincides to a large extent with crucial habitats for large birds of prey. Finding a solution to this dilemma is paramount for preserving Thrace's biodiversity. Such a solution must address the wind farms' cumulative impacts, must be grounded on scientifically robust information and tools that consider the ecological links between seemingly distinct ecosystems, and must lead to specific proposals according to the precautionary principle.

Understanding the Effects of Wind Farm Development: Data Collection, Analysis and Mapping

During the last decade WWF Greece has produced various data regarding the birds of prey of Thrace and their interactions with the wind farms operating in the region. More specifically the research conducted was on: (i) the extent of bird mortality incidents due to collision with wind turbines, (ii) bird behavior in the vicinity of wind farms, (iii) the spatial use of the Black Vulture population estimated through satellite and radio telemetry studies, and (iv) the nesting sites of territorial birds of prey, including the Black Stork.

Collision with Wind Turbines

In 2008–2009 (Carcamo et al. 2011) the first systematic research on bird mortality due to collision with wind turbines was carried out. Carcass searches were conducted around 127 wind turbines, out of the 163 operating at the time, once every 14 days. The research included trials in order to estimate carcass removal rates and observer detection bias. Four Griffon Vultures, one Booted Eagle, eleven birds of other species and eight bats were found dead during this period. Based on these findings, the total number of collision fatalities was calculated using the formula proposed by Everaert and Stienen (2007).

The total number of collision fatalities at the 127 operating wind turbines and for the duration of the research (one year) was estimated at 19.27 birds of prey, whereas for vultures it was estimated at 9.12. Consequently, the approximate mortality rate was calculated at 0.152 birds of prey/WT/year and 0.072 vultures/WT/year. By using this result and doing a simple extrapolation to the 163 wind turbines operating at the time, the mortality rates are estimated at 24.8 birds of prey/year and 11.7 vultures/year. Whereas for the 200 wind turbines currently operating the numbers are 30.4 birds of prey/year and 14.4 vultures/year. In 2009–2010 (Doutaou et al. 2011) a follow-up research was conducted in order to address considerations that the results of the previous research were altered due to the removal of a significant number of carcasses from scavenger species and/or humans. The research of 2009–2010 was conducted in a more systematic way, as every wind turbine was searched daily in comparison with every 14 days that was searched in 2008–2009. The search for carcasses around the 88 wind turbines searched in this period found two Short-toed Eagles, three Common Buzzards, one Marsh Harrier, one Eurasian Sparrowhawk, one Black Vulture, one unidentified species of the Accipiter genus, 73 birds of others species and 183 bats. Using the same formula the total number of collision fatalities was calculated at 15.265 raptors and 1.696 vultures, while the approximate mortality rate was estimated at 0.173 raptors/WT/year and 0.02 Black Vultures/WT/year.

In addition, the per turbine annual mortality rate was estimated using the formula of Erickson et al. (2003). The total number of collision fatalities was estimated at 13.64 raptors and 1.52 Black Vultures, whereas the approximate mortality rate was at 0.15 raptors/WT/year and 0.02 Black Vultures/WT/year. In this case a simple extrapolation to the 163 wind turbines operating at the time results in an estimated 24.45 raptor deaths/year and 3.2 Black Vulture deaths/year. However, for the 200 turbines currently operating these numbers become 30 raptor deaths/year and four Black Vulture deaths/year. In a recent study that combined data of the Black Vulture's space use with a collision risk model, it was found that 5–11% of Thrace's Black Vulture population will collide annually with the wind turbines, signifying an increased risk of population decline (Vasilakis et al. 2016).

Bird Space Use

Aside from carcass searches, WWF Greece monitored bird behavior in the vicinity of the wind farms in two distinct periods. The first was in 2004–2005 (Ruiz et al. 2005) and involved monitoring bird activities from vantage points adjacent to the existing at the time 117 wind turbines. The aim was to produce information regarding areas most commonly used by the birds when crossing the wind farms and the factors that frequently affect bird interactions with the wind turbines. During this first research it was found that the local territorial birds and the vultures exhibited different behavior when flying close to the wind farms. They crossed the wind farms in different points, while the vultures flew in the risk zone (a zone with a radius of 250 meters around the wind turbines) more frequently. Furthermore, the vultures when inside the risk zone almost always crossed the rows of wind turbines, while the local birds of prey only in half of all recorded flights.

The monitoring of bird behavior was repeated in 2008–2009, this time from vantage points covering bird interactions with a total of 122 wind turbines (Carcamo et al. 2011). Vulture species had an increased presence in comparison with the research conducted in 2004–2005 accounting for 50% of all birds monitored. During this research vultures were observed flying inside the risk zone 70% of their total recordings, while they were also the most common species crossing through the wind farms.

Spatial Use of the Black Vulture Population

The range use pattern of the Black Vulture population that breeds inside the Dadia National Park was mapped via two different methods (Figs. 2 and 3).

During the period from 2004 to 2007, the range use was studied using an integrated radio-tracking methodology. Radio-transmitters were attached to ten vultures from different age classes and their bearings were obtained with standard radio-tracking techniques. During 2007–2009 (Noidou and Vasilakis 2011) WWF Greece recorded the movement of the Black Vulture populations via satellite telemetry, by placing seven transmitters. The data analysis from both methods reveals that the area where the Black Vulture is active daily extends outside the borders of the Dadia National Park and occupies the central part of Evros RU, a significant part of Rodopi RU and part of southeast Bulgaria's historical nesting grounds (Skartsi et al. 2008; Hristov et al. 2012). It overlaps, to a large extent, with the areas that until now have been the focus of wind farm development (Figs. 2 and 3).

The Black Vulture does not utilize the entire area in the same way, as its behavior differentiates significantly between areas. High use areas for the birds include those of significance for breeding and foraging activities of the population, roosting sites and important surveillance sites. In addition, vultures use areas where

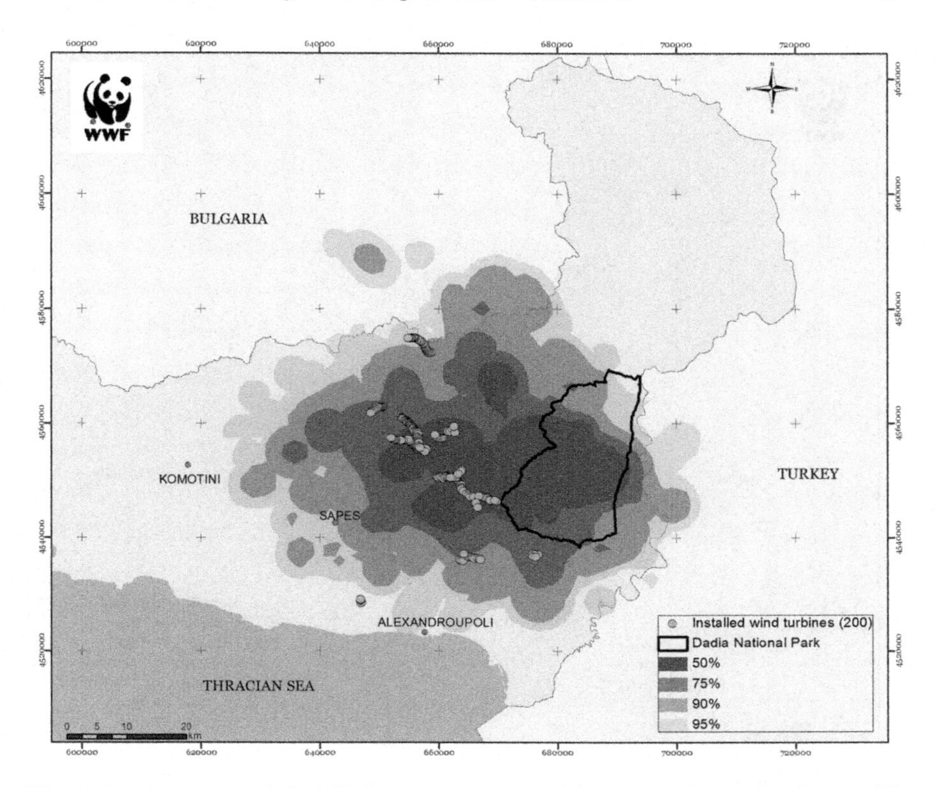

Fig. 2 Relative amount of Black Vulture presence based on the results of the radio telemetry. The *red* area (50%) signifies the core area of the population

they gain the necessary kinetic energy for their locomotion through suitable upward moving wind currents (Vasilakis et al. 2008). Equally important are the areas used as flight corridors by the Black Vulture population, which mostly link the foraging areas with the breeding grounds. The latter are found exclusively inside the Dadia National Park.

Nesting Sites of Territorial Birds of Prey

A survey was conducted, first in 2008 and then in 2010, to locate the nesting sites of territorial birds of prey outside the Dadia National Park (WWF Greece 2013). The results were used to map the nesting sites (Fig. 4) in the vicinity of the wind farms for the following species: Golden Eagle, White-tailed Eagle, Egyptian Vulture, Lesser Spotted Eagle, Booted Eagle, Peregrine Falcon, Long Legged Buzzard, the Eurasian Eagle Owl and the Black Stork. The latter was included even though it is not a raptor, but because it is a rare and endangered species.

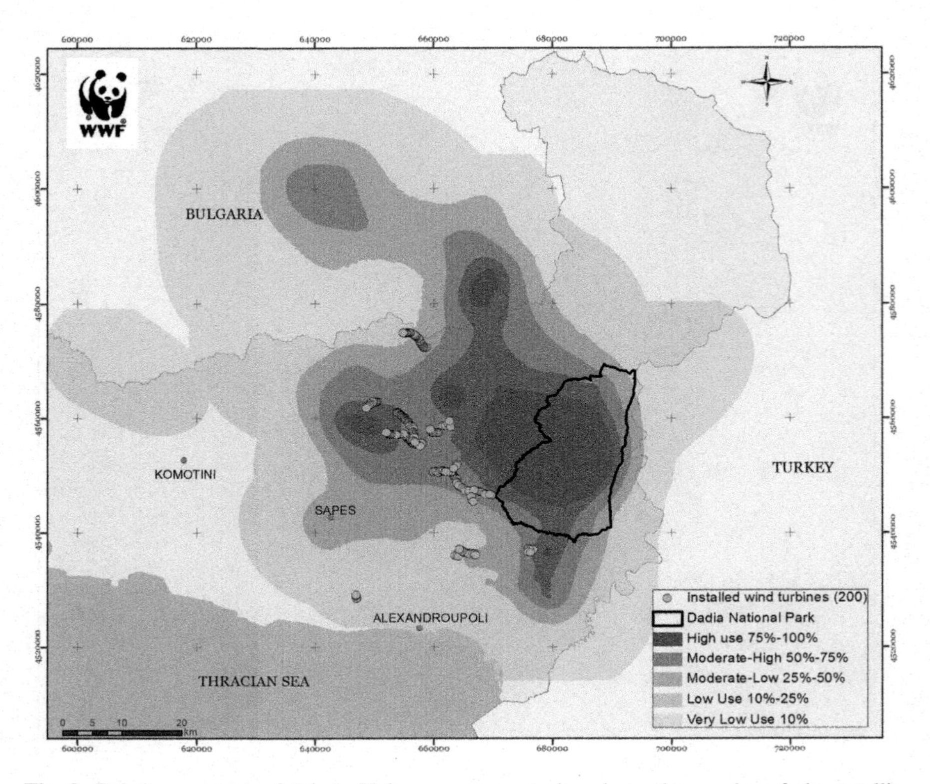

Fig. 3 Relative amount of Black Vulture movements based on the results of the satellite telemetry. The *red* area (75–100%) signifies the core areas, while the *orange* (50–75%) the most common flight movement corridors

Tools for a Proper Site Selection and Development of Wind Farms

The facts detailed above point to the undisputed conclusion that in order to protect the survival of the bird species in Thrace and safeguard the region's ecological integrity, the development of wind farms in the RUs of Evros and Rodopi must be drastically reconsidered. Any solution proposed must address the wind farm' impacts proactively and cumulatively and be based on sound scientific evidence that consider the links between seemingly distinct ecosystems. Solutions must also be based on the precautionary principle and be capable of assisting the decision-making process by creating a common ground for both sides of the dilemma.

Developing wildlife sensitivity maps has been identified as an appropriate tool that can assist in striking a balance at a strategic level between wind farm development and wildlife protection (EC 2010). In 2008, WWF Greece published its first

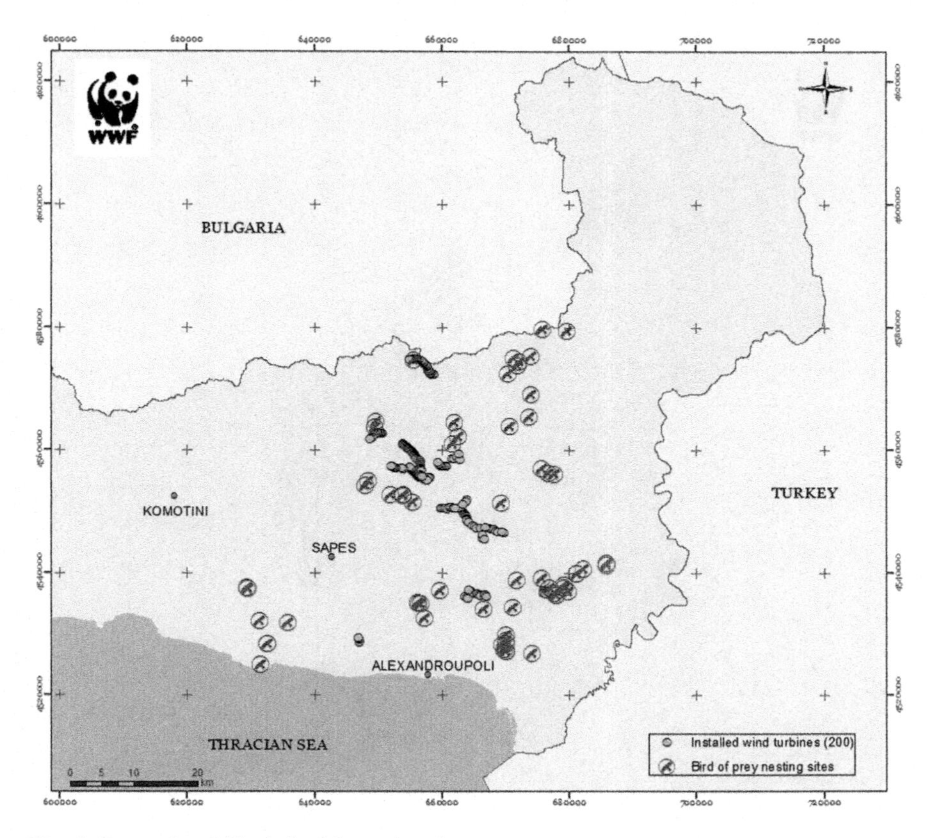

Fig. 4 Raptors' and Black Stork's nesting sites

spatially explicit proposal for the proper site selection of wind farms in Thrace (WWF Greece 2008). Based on available scientific data for the region, wind farms and specific rare species, Thrace was divided into two distinct zones: Exclusion Zones and Increased Protection Zones. Exclusion Zones are locations where wind farm installation should be excluded. The Increased Protection Zones are locations where wind farm installation could be realized based on certain prerequisites that minimise the negative impacts that wind farms could have on wildlife. Since then, new information has been produced on the wind farms' impacts and the most significant habitats of the birds of prey, highlighting the need to update the 2008 proposal.

In 2013, WWF Greece published the second proposal for the proper site selection of wind farms in Thrace (Fig. 5). The final sensitivity map adopted the same delineation of Exclusion and Increased Protection Zones of the first proposal. It was produced using information on the most crucial habitats for the breeding and the survival of the avifauna in the region, as presented previously (WWF Greece 2013).

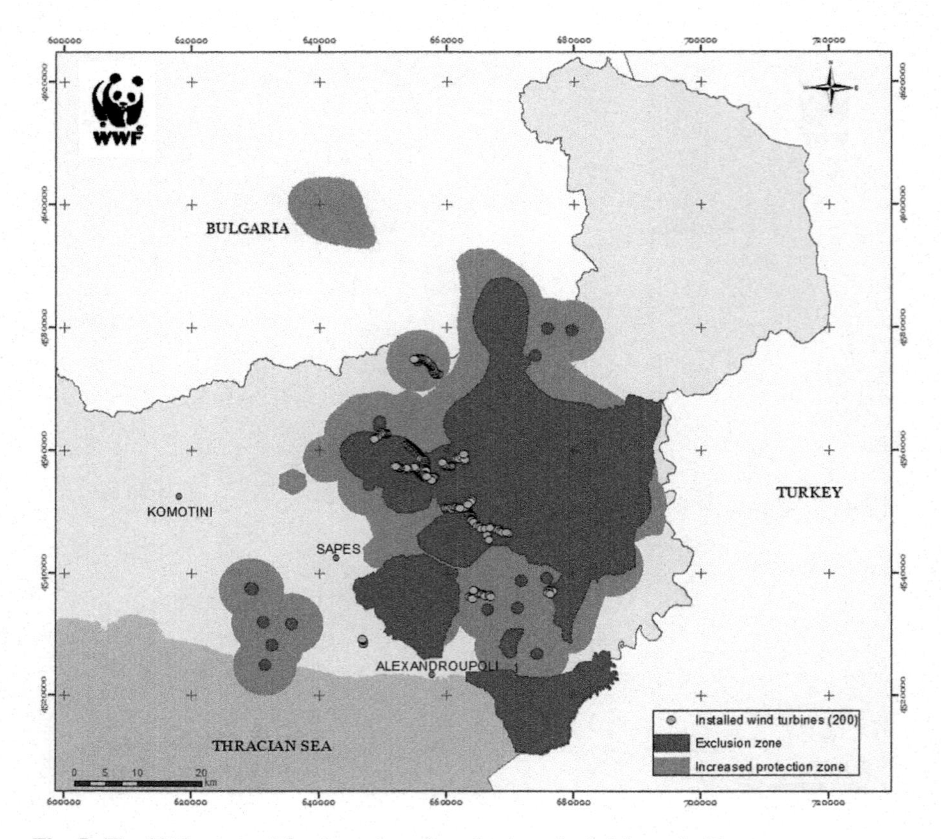

Fig. 5 The 2013 proposal for the proper site selection of wind farms in Thrace

The Exclusion Zone was created by combining: (i) areas of high and medium-high use of the Black Vulture population according to the results of the satellite telemetry, (ii) areas of high use of the Black Vulture according to the results of the radio-telemetry, (iii) the National Parks of Dadia and Evros Delta, (iv) the Griffon Vulture colony, (v) a very crucial breeding habitat for raptors and (vi) areas within a 1000 m radius of the nesting sites of birds of prey and the Black Stork. Similarly, the Increased Protection Zone emerged by combining: (i) areas of medium-low use of the Black Vulture according to the results of the satellite telemetry, (ii) areas of medium use of the Black Vulture according to the results of the radio-telemetry and, (iii) areas within a 5000 m radius of the nesting sites of birds of prey and the Black Stork.

In addition to creating zones, WWF's Greece proposal for the proper site selection of wind farms pinpoints suitable areas by superimposing the wind capacity of the region over the study area, as shown on Fig. 6. Even with the restrictions proposed it is clear that there is still room for wind farm development capable of making a noteworthy contribution to climate change mitigation. Figure 6

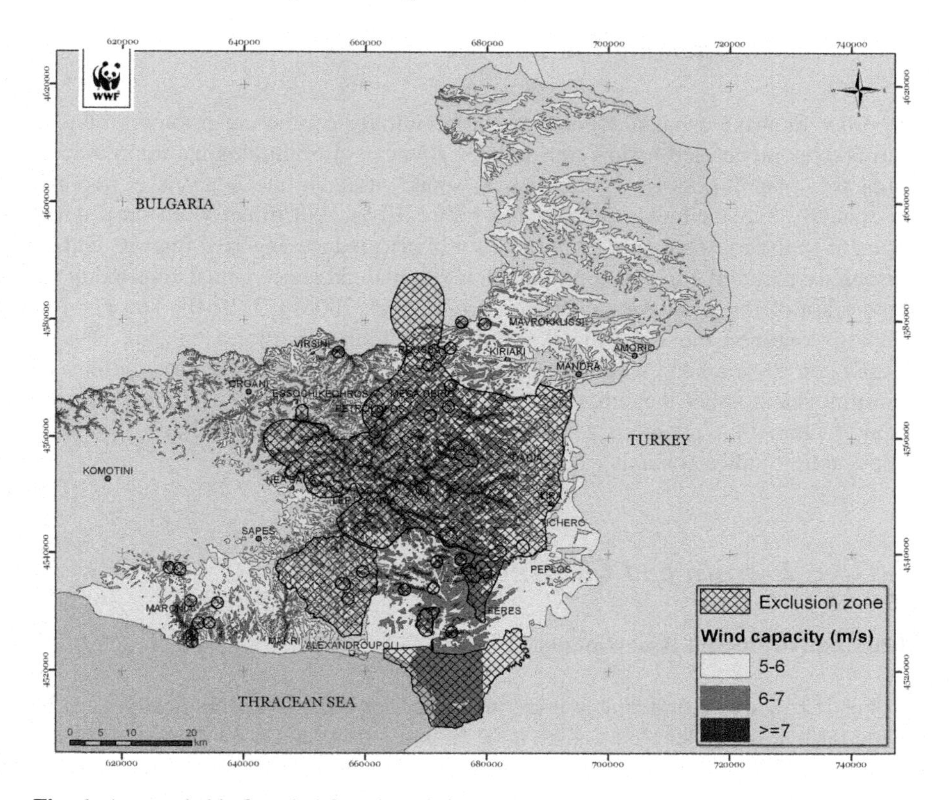

Fig. 6 Areas suitable for wind farm installation

illustrates the potential for WF development outside the proposed Exclusion Zone. This potential though is affected by additional parameters (e.g. grid limitations), which must be taken into consideration. When integrated with the proposed Exclusion Zone the final map will depict more accurately the potential for wind farm development in the region and highlight specific win-win locations.

In order to ensure that wind energy development is compatible with the provisions of the EU Habitats and Birds Directives (Directives 92/43/EEC and 2009/147/EC), apart from a strategic approach aiming to address cumulatively wind farms' impacts emphasis needs to be placed on enhancing the quality of Environmental Impact Assessment and Appropriate Assessments for individual projects. By using a hierarchical, multidimensional tool, known as the Lee et al. (1999) Review Package, WWF Greece evaluated 22 environmental impact assessments (EIAs) conducted for wind farms in Thrace. The evaluation pinpointed strong and weak elements and presented recommendations targeted at improving EIA quality for wind farms in the region (WWF Greece 2014).

Conclusions: Striking the Balance

In order to strike a balance between the seemingly conflicting priorities of wind farm development and nature conservation in areas of ornithological importance, a proactive planning system grounded on sound research and analysis is required. A spatially explicit tool, supported by robust EIAs, will direct wind farm development to the most suitable locations. It will also reduce negative impacts and the resources required for the approval of individual projects, whilst improving the reputation of the wind energy sector (Bright et al. 2006; EC 2010). The previous sections outlined the steps that WWF Greece took to strike a balance between nature conservation and wind farm development in Thrace. This section outlines the most prevalent issues that emerged during WWF Greece's efforts. While some stem from factors that relate to the Greek context, several observations should be applicable to other contexts.

Lessons Learned and Challenges Faced

Cumulative Impact Assessments

Robust EIAs are an undeniable prerequisite for achieving an ecologically sensitive wind farm development in areas of value for rare bird species. As mentioned above, the research has pinpointed significant quality gaps in the existing EIAs. One of the most important is in relation to the Cumulative Impact Assessments (CIAs). Though the CIAs are a cardinal part of EIAs, experience has shown that they are not suited to the project scale (Scott et al. 2014), especially when the environmental parameters under examination have an extensive range use, far wider than a single wind farm. As Therivel and Ross (2007) pinpoint, "most cumulative effects require cumulative solutions", so a CIA conducted independently for every single project is possibly more difficult and less reliable than it would be under a regional approach (Ferrer et al. 2011).

The need for such a study is further supported by the fact that wind farm development in Thrace is dominated by multiple, small-scale projects (ranging from two to a maximum of 12 wind turbines). Furthermore, a significant number of these projects are sited next to existing wind farms, thus expanding already extensive lines of wind turbines that occupy many of the long ridges characteristic of the region (WWF 2013), increasing the barriers birds must manoeuvre around.

Decision-Making Criteria and Priorities

Decisions on whether to approve or refuse individual wind farm proposals needs to be considered within the structural parameters of the licensing process in Greece.

First, while there is an effort to align the national legislation with the European directives in relation to wind farm development in areas crucial for nature conservation, it remains vague and at times provides scientifically questionable guidelines. Due to significant gaps, weaknesses and continuous changes in the legislation and provisions, the legal framework is rather complex and challenging to interpret and implement for all involved parties, authorities, wind farm developers and other stakeholders. The challenge becomes even greater for the small projects, such as those in Thrace, since it is the regional, rather than the central, authorities that provide the final environmental permit. The authorities lack capacity, staff and particularly know-how in assessing EIAs for wind farms. Moreover, during the decision-making process environmental issues are addressed as a secondary value in a system where national targets for renewable energy resources and broader economic priorities continue to prevail.

As the environmental impacts of each individual project are assessed at the latter stages of an invariably time-consuming process, which is linked strictly to a specific site, there is little possibility to reconsider the appropriate siting of a project. In particular, developers highlight the cost of not obtaining a permit, after significant prior investments. This tendency is further augmented due to the competitive nature of wind farm development in Thrace because the individual companies are entangled in a race to secure the best possible sites and connectivity to a limited capacity grid.

Conclusion

The experience of wind farm development in the region of Thrace, Greece, leads to the conclusion that a scientifically robust strategic approach would offer a long term solution, providing authorities, investors and other stakeholders with the tools required to take well informed decisions. Critical for the fruition of such a solution is the political will of public authorities, as well as all involved stakeholders to recognize and address overall the malpractices that characterize wind farm development in Thrace. Such a contribution will be catalytic for the resolution of the conflict between the priorities of wind exploitation and biodiversity conservation.

References

Bright JA, Langston RHW, Bullman R, Evans RJ, Gardner S, Pearce-Higgins J, Wilson E (2006) Bird sensitivity map to provide locational guidance for onshore wind farms in Scotland. Royal society for the protection of birds research report No 20. Available via http://www.rspb.org.uk/Images/sensitivitymapreport_tcm9-157990.pdf

Carcamo B, Kret E, Zografou C, Vasilakis D (2011) Assessing the impacts of 9 wind farms on the birds of prey of Thrace. (Greek). Technical report. WWF Greece. Available via http://www.wwf.gr/images/pdfs/WWF-aiolika-arpaktika2011.pdf

Dimalexis A, Kastritis T, Manolopoulos A, Korbeti M, Fric J, Saravia Mullin V, Xirouchakis S, Bousbouras D (2010) Identification and mapping of sensitive bird areas to wind farm development in Greece. (Greek). Hellenic Ornithological Society, Athens

Doutau B, Cafkaletou-Diaz A, Carcamo B, Vasilakis D, Kret E (2011) Impacts of wind farms on the birds of prey of Thrace. (Greek). Annual technical report August 2009-August 2010. WWF Greece. Available via http://www.wwf.gr/images/pdfs/WWF-aiolika-arpaktika2011-etisio.pdf

Erickson WP, Gritski B, Kronner K (2003) Nine canyon wind power project avian and bat monitoring report, September 2002—August 2003. Technical report submitted to energy Northwest and the Nine Canyon Technical Advisory Committee

European Commission (2010) Wind energy developments and natura 2000. Guidance document. Available via http://ec.europa.eu/environment/nature/natura2000/management/docs/Windfarms.pdf

Everaert J, Stienen E (2007) Impact of wind turbines on birds in Zeebrugge (Belgium). Significant effect on breeding tern colony due to collisions. Biodivers Conserv 16:3345–3359

Ferrer M, de Lucas M, Janss GFE, Casado E, Munoz AR, Bechard MJ, Calabuig CP (2011) Weak relationship between risk assessment studies and recorded mortality in wind farms. J Appl Ecol 49(1):38–46

Hristov H, Demerdzhiev D, Stoychev S (2012) The black vulture *Aegypius monachus* in Bulgaria. In: Berrios PMD, Gonzalez RA (eds) Proceedings of the first international symposium on the black vulture *Aegypius monachus*. Dirección General de Gestión del Medio Natural, Cordoba, pp 96–104

Langston RHW, Pullan JD (2003) Windfarms and Birds: an analysis of the effects of wind farms on birds, and guidance on environmental assessment criteria and site selection issues. Report by BirdLife international to the council of Europe. Available via http://www.birdlife.org/eu/pdfs/BirdLife_Bern_windfarms.pdf

Lee N, Colley R, Bonde J, Simpson J (1999) Reviewing the quality of environmental statements and environmental appraisals, 3rd edn. [online]. University of Manchester. Available via http://www.seed.manchester.ac.uk/medialibrary/Planning/working_papers/archive/eia/OP55.pdf. Accessed on 21 Mar 2016

Legakis A, Maragou P (2009) The red book of threatened animals in Greece. Hellenic Zoological Society, Athens

Martínez-Abraín A, Tavecchia G, Regan HM, Jiménez J, Surroca M, Oro D (2012) Effects of wind farms and food scarcity on a large scavenging bird species following an epidemic of bovine spongiform encephalopathy. J Appl Ecol 49:109–117

Noidou M, Vasilakis D (2011) Characterizing Eurasian black vulture's (*Aegypius monachus*) flight movement corridors in Thrace: a need for conservation on a landscape-level scale. Report of WWF Greece. Available via http://www.wwf.gr/images/pdfs/WWF-Flight-movement-corridors-2011.pdf

Poirazidis K, Schindler S, Kakalis E, Ruiz C, Bakaloudis D, Scandolara C, Eastham C, Hristov H, Catsadorakis G (2011) Population estimates for the diverse raptor assemblage of Dadia National Park, Greece. Ardeola 58:3–17

Ruiz C, Schindler S, Poirazidis K (2005) Impact of wind farms on birds in Thrace, Greece. Technical report. WWF Greece. Athens

Schindler S, Poirazidis K, Ruiz C, Scandolara C, Cárcamo B, Eastham C, Catsadorakis G (2014) At the crossroads from Asia to Europe: spring migration of raptors and black storks in Dadia national park (Greece). J Nat Hist 49:285–300

Scott A, Shannon P, Hardman M, Miller D (2014) Evaluating the cumulative problem in spatial planning: a case study of wind turbines in Aberdeenshier, UK. Town Plan Rev 85(4):457–487

Skartsi T, Vasilakis D (2008) Monitoring of the black vultures and the griffon vulture inside the National Park of the Forest of Dadia-Lefkimi-Soufliou. (Greek). Annual technical report 2007. WWF Greece. Athens

Therivel R, Ross B (2007) Cumulative effects assessment: does scale matter? Environ Impact Assess Rev 27:365–385

Treweek J (1996) Ecology and environmental impact assessment. J Appl Ecol 33:191–199

Vasilakis DP, Poirazidis KS, Elorriaga JN (2008) Range use of a Eurasian black vulture (*Aegypius monachus*) population in the Dadia-Lefkimi-Soufli National Park and the adjacent areas, Thrace, NE Greece. J Nat Hist 42(5–8):355–373

Vasilakis DP, Whitfield DP, Schindler S, Poirazidis KS, Kati V (2016) Reconciling endangered species conservation with wind farm development: cinereous vultures (*Aegypius monachus*) in South-Eastern Europe. Biol Conserv 196:10–17. doi:10.1016/j.biocon.2016.01.014

WWF Greece (2008) Wind farms in Thrace: recommendations on proper site selection. Position paper. Dadia-Athens

WWF Greece (2013) Wind farms in Thrace: updating the proposal for proper site selection. Dadia-Athens. Available via http://wwf.gr/images/pdfs/2013_ProperSite Selection.pdf

WWF Greece (2014) Wind farms in Thrace: quality assessment of environmental impact assessments, 2000–2010. (Greek). Dadia-Athens. Available via http://wwf.gr/images/pdfs/meletei-per-epiptoseon.pdf

Zakkak S, Babakas P (2015) Annual monitoring report for the species and habitats of european concern—2014. Management body of Dadia-Lefkimi-Soufli forest national park (unpublished data)

Zakkak S, Babakas P, Skartsi T (2014) Annual monitoring report for the species and habitats of european concern—2013. Management body of Dadia-Lefkimi-Soufli forest national park and WWF Greece (unpublished data)

Introducing a New Avian Sensitivity Mapping Tool to Support the Siting of Wind Farms and Power Lines in the Middle East and Northeast Africa

Tristram Allinson

Abstract The risk to certain soaring bird species posed by poorly-sited wind turbines and power lines is now well established, and badly planned operations can prove both environmentally and financially costly. Consequently, wind energy developers, governments and other stakeholder groups urgently need access to accurate ornithological information to better inform the planning process. This is particularly true in parts of the Middle East and Northeast Africa where a rapidly expanding wind energy sector coincides with a globally significant soaring bird migration route—the Rift Valley/Red Sea Flyway. Unfortunately, relevant data for this region has often been difficult to obtain and interpret. To address this problem, BirdLife International, working with its network of regional partners, has launched the *Soaring Bird Sensitivity Mapping Tool*, an online tool providing detailed information on the distribution of soaring bird species along the flyway. By providing and interpreting this information, it is hoped that the tool will become an essential instrument in the environmentally sound expansion of wind energy in the region.

Keywords Avian collision · Soaring birds · Raptors · Migration · Wind energy · Sensitivity mapping · Spatial planning · Rift valley/Red Sea flyway

Introduction

Across the Middle East and Northeast Africa, considerable potential exists for the expansion of the wind energy industry. Particularly rich wind resources exist on the Mediterranean and Red Sea coasts where wind speeds frequently exceed 7 m/s, whilst some areas such the Gulf of Suez in Egypt have average wind speeds of up to 10 m/s, reaching average annual full load-hours of 3000 (GWEC 2015).

T. Allinson (✉)
BirdLife International, David Attenborough Building, Pembroke Street, Cambridge, UK
e-mail: tris.allinson@birdlife.org

J. Köppel (ed.), *Wind Energy and Wildlife Interactions*,
DOI 10.1007/978-3-319-51272-3_12

As a result, a number of countries in the region have identified wind energy as an important means of increasing their energy independence, mitigating climate change and reducing air pollution. Egypt, for example, intends renewable energy to contribute 20% of its total electricity generation by 2020. Of this, 12%, roughly 7200 MW, will be provided by wind (IRENA 2014). Similarly, Jordan plans to generate 10% of its primary energy through renewables by 2020, with over 65% (1200 MW) coming from wind (IRENA 2014). Saudi Arabia meanwhile is aiming for 30% of electricity generation to come from renewables by 2032, with wind contributing 9000 MW (IRENA 2014).

The region is also home to one of the world's most important bird migration routes—known as the Rift Valley/Red Sea Flyway. Almost two million large soaring birds, such as eagles, hawks, cranes and storks, migrate along this route as they move biannually between Africa and Eurasia. To maintain their soaring flight, they rely on thermals and updrafts, and consequently concentrate in huge numbers above landscape features conducive for these conditions. Most significantly, they avoid lengthy sea crossings, preferring to circumvent areas of open water by following coastlines or to cross at narrow straits. These breath-taking avian aggregations constitute one of region's great natural spectacles. Key 'bottleneck' sites along the route include Suez, the Strait of Jubal, the Bab-el-Mandeb Straits and a number of sites in Israel including Eilat and the southern Arava valley (Zalles and Bildstein 2000).

Several studies have shown that soaring birds, especially raptors, are vulnerable to collision with wind turbines (e.g. Dahl et al. 2012; Smallwood and Thelander 2008). Soaring birds are often large-bodied, have limited manoeuvrability and have flight behaviours that heighten collision risk. For instance, when utilising thermals, soaring birds often adopt a spiralling flightpath. As such, the birds can find themselves rapidly ascending into a turbine rotor-swept zone whilst facing away from the direction of the hazard or suddenly turning sharply into an oncoming blade with little time to react. Some soaring birds, such as the *Gyps* vultures, are especially vulnerable as they have a limited visual field, designed to focus their attention fully on the ground whilst searching for carrion, and thus are often unaware of aerial hazards directly in front of them (Martin et al. 2012).

In addition, soaring birds are typically long-lived species that take several years to reach maturity and have high adult survival rates. There is, therefore, considerable potential for the mortality caused through wind farm collisions to cause population level declines (Carret et al. 2009; Martínez-Abraín et al. 2012). It is worth noting that migratory soaring birds are not only at risk on passage, but also on their wintering and breeding grounds. In fact, it is likely that birds engaging in activities such as foraging, displaying and roosting are more susceptible to collision than birds actively migrating, which are often travelling at altitudes well above the height of turbine blades (Barrios and Rodríguez 2004; de Lucas et al. 2008; Katzner et al. 2012).

Soaring birds using the Rift Valley/Red Sea Flyway already encounter numerous and escalating threats, including hunting, agricultural intensification and habitat loss and deterioration. The additional, cumulative impact caused by multiple, inappropriately sited wind energy developments would worsen an already perilous journey. To address these multiple threats, BirdLife International launched the Migratory Soaring Bird (MSB) project in 2010. The initiative, supported by the Global Environment Facility (GEF) and the United Nations Development Programme (UNDP), aims to mainstream soaring bird conservation across a range of sectors (see http://migratorysoaringbirds.undp.birdlife.org). One major outcome of the project's first phase has been the development of the *Soaring Bird Sensitivity Mapping Tool* to aid the appropriate siting of wind farms in the region (tinyurl.com/MSBmap).

It is now widely recognised that the most important way to ensure that wind farms do not have a detrimental impact on bird populations is to site them away from sensitive areas (EC 2010; Gove et al. 2013). To achieve this, developers, governments and funders need robust spatial planning tools interpreting available ornithological data so that they can identify and then avoid sensitive areas where the potential impacts on soaring birds could be high. BirdLife International is a world authority on the development of such tools. The first sensitivity maps were developed for Scotland and England by the RSPB, BirdLife's Partner organisation in the UK (Bright et al. 2008, 2009). Subsequently, a number of other BirdLife Partners around the world developed similar national tools, including in Bulgaria, Greece (Dimalexi et al. 2010), South Africa (Retief et al. 2010), Slovenia (Bordjan et al. 2012) and Ireland (McGuinness et al. 2015). Working with its extensive network of national partners across the Middle East and Northeast Africa, BirdLife's *Soaring Bird Sensitivity Mapping Tool* is the first regional tool of its kind. This paper describes the tool's intended practical application and the data and methodology that underpin it.

Tool Application

By the time that an Environmental Impact Assessment (EIA) is underway, plans for a wind farm or power line may already be at an advanced stage. Discovering at this juncture that the proposed development intersects with a major soaring bird migration route or the breeding grounds of a globally threatened raptor can prove financially, as well as environmentally, costly. Developers, funders and national planning authorities have a clear need for accurate, site-scale biodiversity data that can inform the earliest stages of the planning process when it is still relatively easy and inexpensive to make changes. Unfortunately, relevant information can be difficult to access and interpret, being frequently scattered across a disparate array of databases or buried within the scientific literature. The *Soaring Bird Sensitivity Mapping Tool* is an attempt to address this problem for the Middle East and Northeast Africa. Through the tool, users have unrestricted access to extensive spatial datasets relating to soaring birds. Most significantly, a simple, explicit formula is used to assign sensitivity categories, thus allowing for an objective assessment and comparison of prospective locations on the basis of available data.

Since its launch in early 2014, the tool has been extremely well received, generating considerable interest from across the wind energy sector and from the international financing community. Properly integrated into planning and safeguard procedures, the tool should help ensure that the expansion of wind energy in the region does not come at the expense of the area's unique and spectacular natural heritage.

Tool Overview

Geographical Scope

The *Soaring Bird Sensitivity Mapping Tool* focuses on the potential impact of terrestrial wind farms on soaring birds and is not intended to be used in the assessment of offshore developments. At present, the tool covers an area of 17.8 million km^2 centred on the Red Sea and extending from 37.523 decimal degrees north to 3.187 decimal degrees south and from 21.803 decimal degrees west to 60.232 decimal degrees east. This encompasses the majority of the Middle East and Northeast Africa. Geographical expansion of the tool is underway, most imminently, to cover additional parts of Northern Africa and the Mediterranean.

Platform

The tool is an open access, web-based mapping tool built using the Esri ArcGIS API for Flex. The tool is available in English, French and Arabic.

Species Covered

The tool contains information on 83 species of soaring bird from the following families—Pandionidae (Osprey), Accipitridae (Hawks, Eagles), Falconidae (Falcons, Caracaras), Gruidae (Cranes), Ciconiidae (Storks), Threskiornithidae (Ibises, Spoonbills) and Pelecanidae (Pelicans).

Mapping Interface

The tool employs a simple and intuitive mapping interface designed to be easily operated by non-GIS specialists. A retractable left-hand legend (see Fig. 1; A) shows the various spatial layers underpinning the tool and allows the user to choose which are displayed on the map. The sensitivity search instrument (Fig. 1; B) is used to carry out the site sensitivity assessments. The right-hand navigational instrument (Fig. 1; C) allows users to move around the map and zoom in and out.

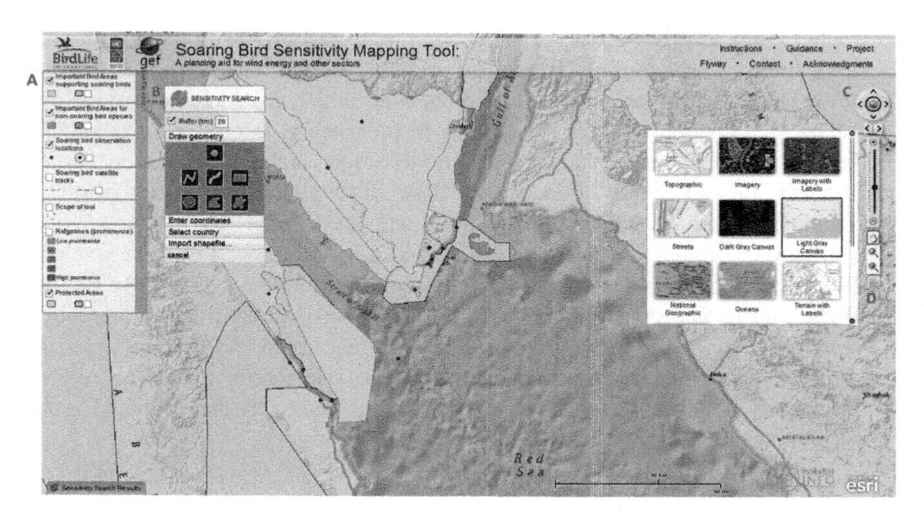

Fig. 1 Screenshot of mapping interface showing the legend (*A*) and instruments for conducting a sensitivity assessment (*B*), navigating the map (*C*) and changing the base map view (*D*)

The map view instrument (Fig. 1; D) allows users to choose from a range of base maps including satellite imagery.

Data Sources

The tool contains thousands of georeferenced records of soaring birds from across the region. The principal source of these is BirdLife's Important Bird and Biodiversity Areas (IBAs) database. Additional soaring bird records have been collated from a wide range of literature sources and data archives. These records are used to generate the sensitivity assessments. Additional, contextual, soaring bird information is provided through satellite tracking data and species' range maps. The tool also contains spatial data on protected areas and relevant topography.

Important Bird and Biodiversity Areas

IBAs are a set of sites of international significance for birds that are identified using a standardised set of data-driven criteria and thresholds, based on threat and irreplaceability. To date, more than 12,800 IBAs have been identified in nearly 200 countries and territories.

The global network of IBAs has been identified using four internationally agreed criteria. These are based upon globally threatened species, groups of species with a restricted range (defined as less than 50,000 km^2), species assemblages confined to a single biome, and significant congregations of one or more species. For each criterion, lists of 'trigger' species and associated thresholds have been developed, and IBA qualification requires the confirmed presence of one or more populations or sets of species that meet these thresholds under any criterion. The BirdLife partnership has identified global-level IBAs throughout the world (using the 'A-level' criteria); however, within Europe and the Middle East, IBAs have also been identified at the regional and sub-regional levels. A more detailed explanation of IBA criteria, and how they have been applied in different regions, can be found in Heath and Evans (2000), Fishpool and Evans (2001) and BirdLife International (2004); see also www.birdlife.org/datazone/info/ibacriteria. As well as population information on relevant trigger species, BirdLife also maintain data on all other significant bird populations within IBAs. This constitutes the principal dataset underpinning the tool.

Soaring Bird Observation Locations

Outside of IBAs, records of soaring bird have been collated from a wide range of sources, including scientific journals, trip reports, environmental impact assessments and from Worldbirds, a global archive of bird observation records.

Satellite Tracking Data

Increasingly, researchers are investigating the migratory patterns and behaviours of soaring bird species by fitting them with satellite transmitters. Through direct collaborations with researchers and with Movebank (www.movebank.org), an online database of animal tracking data hosted by the Max Planck Institute for Ornithology, the tool also contains information on the routes taken by hundreds of individual soaring birds of several species within the region.

Species Range Maps

BirdLife compiles and maintains digitized distribution maps for all of the world's bird species. Maps for 83 soaring bird species are incorporated into the tool and used to indicate which species may be present within a selected search area in addition to those for which observational records are available.

Protected Area Information

The tool contains spatial data on the region's protected areas courtesy of the World Database of Protected Areas (WDPA): a joint venture of UNEP and IUCN, produced by UNEP-WCMC and the IUCN World Commission on Protected Areas (IUCN-WCPA) working with governments and collaborating NGOs. The WDPA is compiled from multiple local and national sources and is the most comprehensive global dataset on marine and terrestrial protected areas available. It also includes the official set of protected areas submitted by national protected areas authorities (including protected areas classified using the IUCN protected area category system) and the secretariats of international conventions (e.g. Ramsar, World Heritage and Man & Biosphere) which is used to compile the United Nations List of Protected Areas.

Ridgelines

Topography has been identified as an important factor in the collision risk of soaring birds (Katzner et al. 2012). Consequently, the tool includes a layer showing all slopes with an angle greater than 14.5°, colour-coded to indicate the steepness of the incline. Such ridgelines are frequently used by soaring bird species which exploit rising air masses to maintain their flight.

Calculating Sensitivity

When a user conducts a sensitivity search of a location (e.g. a potential development site), the tool not only collates and displays the relevant soaring bird information, it also uses these data to derive a numeric measure of the site's sensitivity. This makes it much easier for non-ornithologists to judge the avian sensitivity of a site and compare it objectively with other locations. The tool evaluates all the soaring bird records intersecting with the buffered search area, and sums the sensitivity scores for each species' population at the site to get an overall Sensitivity Index (SI), following the equations:

$$\mathbf{SI = SSS_1 + SSS_2 + SSS_3 \ldots\ldots SSSn}$$

$$\mathbf{SSS = SSI \times (Site\ Population/Global\ Population)}$$

$$\mathbf{SSI = SVI \times ERI}$$

SI Sensitivity Index of site
SSS_1 Species' Sensitivity at Site for Species 1
SSI Species' Sensitivity Index
SVI Species' Vulnerability Index
ERI Extinction Risk Index

The sensitivity scores for each species (or Species Sensitivity at Site; SSS) take into account the proportion of the species' global population recorded at the site and the Species Sensitivity Index (SSI), a species-specific measure of sensitivity derived from two parameters. Firstly, the extinction risk of the species as indicated by its IUCN Red List category. Table 1 shows the values assigned to different Red List categories.

Secondly, the Species' Vulnerability Index (SVI), a measure of each species' collision susceptibility based on an assessment of its body mass, flight style, behaviour and documented incidents of collision (Table 2).

Sensitivity Categories

SI values correspond to six sensitivity categories, ranging from "Unknown", through "Potential" and "Medium", to "High", "Very High" and "Outstanding". If a site has no associated population records of soaring bird species, it is placed in the "Unknown" sensitivity category. However, although this means that there is insufficient soaring bird data on which to base a sensitivity score, this should not be

Table 1 Extinction risk index

IUCN red list category	ERI
Critically endangered	10
Endangered	8
Vulnerable	6
Near threatened	2
Least concern	1

Table 2 Species' vulnerability index scores

Species	SVI
Cranes, storks, vultures, pelicans, Secretarybird and *Haliaeetus* spp. sea eagles	10
Large eagles	9
Harriers, *Milvus spp.* kites, Scissor-tailed Kite, Eurasian Spoonbill and Northern Bald Ibis	8
Osprey, buzzards, honey-buzzards, snake-eagles, harrier-hawks and Bat Hawk	7
Accipiters, falcons, cuckoo-hawks and Black-winged Kite	6

interpreted as meaning that a site has no or low sensitivity. Information on soaring birds may be incomplete, out of date or lacking for this region. Alternatively, the site may be sensitive for another reason, such as the presence of other vulnerable wildlife such as Chiroptera (bats). Likewise, sites that fall within the "Very High" and "Outstanding" categories should not automatically be assumed to be unsuitable for all forms of development, and it may be possible that, with the right mitigation measures in place, development at these sites can take place without any negative impact on soaring birds.

Conducting a Sensitivity Search

The aim of the tool is to provide users with an authoritative, transparent and accurate assessment of the soaring bird sensitivity of a site in relation to wind farm or power line development. Searches are conducted using the Sensitivity Search facility (see Fig. 1). Users can choose to delineate the boundaries of a proposed wind farm or the route of a proposed power line, either by importing an existing shapefile, plotting a sequence of coordinates or by demarcating the boundaries using a range of geometry drawing tools. As soaring birds often range over considerable areas, and since this tool is designed for use at the very earliest stages of the planning process when the broadest and most precautionary assessment is warranted, the user-defined search area is automatically buffered to 20 km. If desired, users can set larger, more precautionary buffers.

The tool then displays all of the features intersecting with the buffered point, line or polygon and renders them on the map bordered in red (see Fig. 2). The associated data are presented in table arrays that open along the bottom of the map (Fig. 2; B1–6). An accompanying search summary (Fig. 2; A) provides the key information, namely the number of species observed and expected within the delineated area, the number of intersecting IBAs and protected areas, the number of additional soaring bird observation locations and the number of satellite tracks. Most importantly, the summary provides the overall sensitivity score and the sensitivity category with which this score is associated.

A more detailed breakdown of the data is provided in the table array:

- Table 1 (Fig. 2; B1) lists all of the species recorded within the intersecting features, with their known peak count and information of the source of the data. Species that have not been observed within the intersecting features, but whose range maps indicate that they are expected to occur, are also listed. A link is provided for each species to its factsheet on the BirdLife data zone website. Clicking on a species in the list highlights the map features with which it is associated and brings up additional information in the summary box (Fig. 2; C). This includes the name and type of feature in which it was recorded, the season in which it has been recorded and the year from which the record dates.

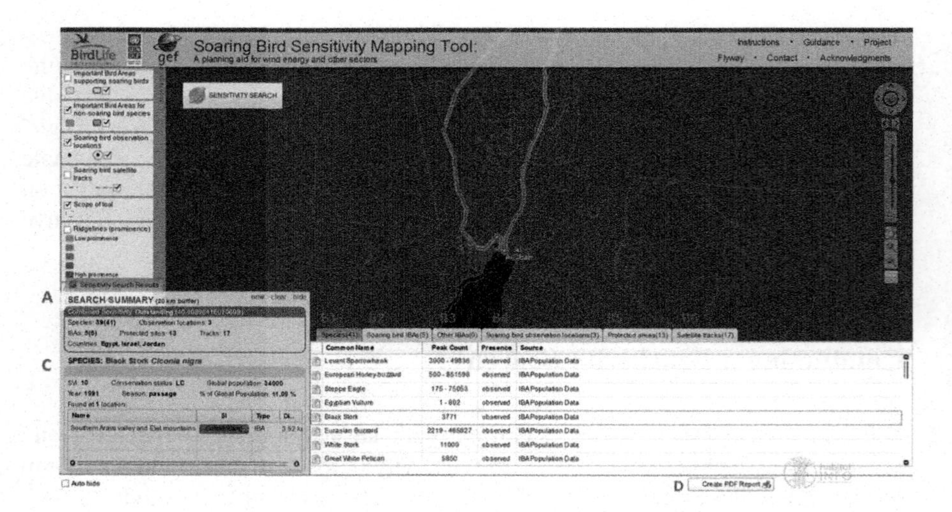

Fig. 2 Screenshot of search results showing the search summary (*A*), the table array (*B1–6*), additional information box (*C*) and the 'create PDF report' function (*D*)

- Table 2 (Fig. 2; B2) lists the intersecting IBAs at which soaring bird species have been documented, whether they fall within the delineated area or the distance from the delineated area with the buffer and their sensitivity category. A link is provided to the IBA factsheet in the BirdLife data zone website. Clicking on an IBA in the list highlights the site on the map and brings up additional information in the summary box. This includes a breakdown of the species present at the site, the peak numbers in which they occur, the percentage of the global population this represents, the season in which they occur and the year in which the data were collected.
- Table 3 (Fig. 2; B3) lists the intersecting IBAs for which notable populations of soaring bird species have not been documented. Although these sites do not contribute to the soaring bird sensitivity score, it is important that users are made aware of them as they could well support populations of other sensitive taxa.
- Table 4 (Fig. 2; B4) lists the supplementary soaring bird observations. Clicking on an observation in the list highlights the location on the map and brings up additional information in the summary box similar to that presented for soaring bird IBAs.
- Table 5 (Fig. 2; B5) lists the intersecting protected areas, providing their name, IUCN category, status, year of establishment and link to the official site information on the UNEP-WCMC/IUCN Protected Planet website (www.protectedplanet.net).
- Table 6 (Fig. 2; B6) lists the intersecting satellite tracks. Clicking on a track in the list brings up additional information in the summary box, including the species, the season in which it passed through the area and the source of the data.

All this information can be downloaded in a pdf, along with a screenshot of the map view and information on how to interpret and use the sensitivity scores.

Acknowledgements The *Soaring Bird Sensitivity Mapping Tool* has been developed as part of the Migratory Soaring Birds Project, an initiative supported by The Global Environment Facility (GEF) and the United Nations Development Programme (UNDP).

The project is coordinated by the Regional Flyway Facility (RFF) with support from BirdLife's regional offices in Amman (Jordan) and Nairobi (Kenya) and oversight by the BirdLife Global Secretariat in Cambridge (UK). Fundamental to this project are BirdLife's national partners in the region: Association Djibouti Nature, Nature Conservation Egypt, The Egyptian Environmental Affairs Agency, Ethiopian Wildlife and Natural History Society, Royal Society for the Conservation of Nature, Republic of Lebanon Ministry of Environment, Society for the Protection of Nature in Lebanon, Palestine Wildlife Society, Saudi Wildlife Authority, The Sudanese Wildlife Society, Syrian Society for Conservation of Wildlife, Foundation for Endangered Wildlife (Yemen).

The tool was built by Habitat INFO (www.habitatinfo.com) using the ESRI ArcGIS Server platform (www.esri.com).

References

Barrios L, Rodríguez A (2004) Behavioural and environmental correlates of soaring-bird mortality at on-shore wind turbines. J Appl Ecol 41:72–81

BirdLife International (2004) Important Bird Areas in Asia: key sites for conservation. BirdLife International, Cambridge, UK (BirdLife Conservation Series No. 13)

Bordjan D, Janča T, Mihelič T (2012) Karta občutljivih območij za ptice za umeščanja vetrnih elektrarn v Sloveniji Verzija 2.0. DOPPS (BirdLife Slovenia), Ljubljana, Slovenia

Bright JA, Langston R, Bullman R, Evans R, Gardner S, Pearce-Higgins J (2008) Map of bird sensitivities to wind farms in Scotland: a tool to aid planning and conservation. Biol Conserv 141:2342–2356

Bright JA, Langston RHW, Anthony S (2009) Mapped and written guidance in relation to birds and onshore wind energy development in England. RSPB Research Report No 35. A report by the Royal Society for the Protection of Birds, funded by the RSPB and Natural England. Royal Society for the Protection of Birds, Sandy, UK

Carret M, Sánchez-Zapata JA, Benítez JR, Lobón M, Donázar JA (2009) Large scale risk-assessment of wind-farms on population viability of a globally endangered long-lived raptor. Biol Conserv 142:2954–2961

Dahl EL, Bevanger K, Nygård T, Røskaft E, Stokke BG (2012) Reduced breeding success in white-tailed eagles at Smola windfarm, western Norway, is caused by mortality and displacement. Biol Conserv 145:79–85

de Lucas M, Janss GFE, Whitfield DP, Miguel F (2008) Collision fatality of raptors in wind farms does not depend on raptor abundance. J Appl Ecol 45:1695–1703

Dimalexi A, Kastritis T, Manolopoulo A, Korbeti M, Fric J, Saravia Mullin V, Xirouchakis S, Bousbouras D (2010) Identification and mapping of sensitive bird areas to wind farm development in Greece. Hellenic Ornithological Society, Athens

European Commission (EC) (2010) Wind energy developments and Natura 2000: EU guidance on wind energy development in accordance with the EU nature legislation. European Commission

Fishpool LDC, Evans MI (eds) (2001) Important Bird Areas in Africa and associated islands: priority sites for conservation. BirdLife International, Cambridge, UK (BirdLife Conservation Series No. 11)

Gove B, Langston RHW, McCluskie A, Pullan JD, Scrase I (2013) Windfarms and birds: an updated analysis of the effects of windfarms on birds, and best practice guidance on integrated planning and impact assessment. Report prepared by Birdlife International on behalf of the Bern Convention. RSPB/ BirdLife in the UK, Sandy, UK

GWEC (2015) Global Wind Report 2014: annual market update. Global Wind Energy Council

Heath MF, Evans MI (eds) (2000) Important Bird Areas in Europe: priority sites for conservation, vol 2. BirdLife International, Cambridge, UK (BirdLife Conservation Series No. 8)

IRENA (2014) Pan-Arab renewable energy strategy 2030: roadmap of actions for implementation

Katzner TE, Brandes D, Miller T, Lanzone M, Maisonneuve C, Tremblay JA, Mulvihill R, Merovich GT (2012) Topography drives migratory flight altitude of golden eagles: implications for on-shore wind energy development. J Appl Ecol 49:1178–1186

Martin GR, Portugal SJ, Murn CP (2012) Visual fields, foraging and collision vulnerability in Gyps vultures. Ibis 154:626–631

Martínez-Abraín A, Tavecchia G, Regan HM, Jiménez J, Surroca M, Oro D (2012) Effects of wind farms and food scarcity on a large scavenging bird species following an epidemic of bovine spongiform encephalopathy. J Appl Ecol 49:109–117

McGuinness S, Muldoon C, Tierney N, Cummins S, Murray A, Egan S, Crowe O (2015) Bird sensitivity mapping for wind energy developments and associated infrastructure in the Republic of Ireland. BirdWatch Ireland, Kilcoole, Wicklow

Retief EF, Diamond M, Anderson MD, Smit HA, Jenkins A, Brooks M, Simmons R (2010) Avian wind farm sensitivity map for South Africa: criteria and procedures used. BirdLife South Africa

Smallwood KS, Thelander C (2008) Bird mortality in the Altamont Pass Wind Resource Area, California. J Wildl Manage 72:215–223

Zalles JI, Bildstein KL (eds) (2000) Raptor Watch: A global directory of raptor migration sites. BirdLife International and Kempton, PA, Hawk Mountain Sanctuary, Cambridge, UK (BirdLife Conservation Series No. 9)

A Framework for Assessing Ecological and Cumulative Effects (FAECE) of Offshore Wind Farms on Birds, Bats and Marine Mammals in the Southern North Sea

Maarten Platteeuw, Joop Bakker, Inger van den Bosch, Aylin Erkman, Martine Graafland, Suzanne Lubbe and Marijke Warnas

Abstract The European Union's Birds and Habitats Directives and Marine Strategy Framework Directive demand that the proponent describes and assesses the potential cumulative effects on wildlife in environmental impact assessment reports for proposed actions. Based upon the DPSIR (**D**riving forces, **P**ressures, **S**tates, **I**mpacts, **R**esponses) approach (European Environment Agency 1999), a 6-step framework for undertaking a cumulative impacts assessment was developed to address this requirement:

1. Identify the relevant pressures the envisaged activities could cause.
2. Identify the habitats and species that may be affected by these pressures.
3. Describe all other activities that could affect the same species.
4. Describe the nature and scale of the cumulative effects of all the activities selected in Step 3 on the selected habitats and species.
5. Evaluate the significance of the effects on the selected habitats and species.
6. If necessary, adapt the activity by taking measures to prevent the activity from causing significant effects.

The first application of this approach was carried out for the initiative of the Dutch Social Economic Council, the *Energy Agreement for Sustainable Growth* (2013). The agreement proposed an offshore wind capacity of 4500 MW by 2030 for the Dutch part of the North Sea. The application of this cumulative impacts assessment framework showed that effective mitigation measures are required to achieve this agreement without endangering protected species and habitats, in line with European Union (EU) nature and environmental legislation. Mitigation measures can be implemented through conditions on permits for offshore wind farm projects, which include restrictions on the maximum underwater noise levels during

M. Platteeuw (✉) · J. Bakker · I. van den Bosch · A. Erkman ·
M. Graafland · S. Lubbe · M. Warnas
Rijkswaterstaat, Ministry of Infrastructure and Environment, Amsterdam, The Netherlands
e-mail: maarten.platteeuw@rws.nl

J. Köppel (ed.), *Wind Energy and Wildlife Interactions*,
DOI 10.1007/978-3-319-51272-3_13

construction, the minimum capacity of individual wind turbines, and measures to reduce bird and bat collisions and fatalities during seasons of major migration.

Keywords Offshore wind farms (OWFs) · Ecology · Nature legislation · Cumulative effects · Underwater noise · Collision · Displacement · Harbour porpoise · Birds · Bats

Introduction

There has been a growing recognition since the 1970s of the importance of describing and assessing the effects of human activities on natural ecosystems. Moreover, as early as the 1980s it was concluded that just describing and assessing specific proposals and activities was not enough. It is necessary to also examine whether the effects of various activities can accumulate to produce larger or more damaging ecological or environmental impacts (Dijkema et al. 1985). Despite the difficulties, the importance of properly describing and addressing the issue of cumulative effects was acknowledged and incorporated into international legislation, such as the EU Birds Directive 1979 and EU Habitats Directive 1982.

Since 2005, offshore wind farm initiatives have been booming in the North Sea region, following the positive experience of the Danish Horns Rev 1 Wind Farm (2002). Between 2005 and 2009, the Dutch government received 12 applications for offshore wind farms (OWFs). Since little was known about ecological effects, let alone cumulative effects of OWFs, a new scheme to address gaps in knowledge and assessment was launched. This was led by the precautionary principle and sound decision-making based on monitoring and science.

Four of the 12 OWF applications were approved on the condition that an ecological monitoring programme be prepared for benthos, fish, marine mammals, birds and recently bats. Parallel research on impacts, such as the impacts of piling sounds on fish (larvae), Harbour Porpoise and Common and Grey Seal, has helped in filling knowledge gaps.

Moreover, the Dutch Social Economic Council prepared the *Energy Agreement for Sustainable Growth* (SER-agreement 2013). The purpose of this agreement was to increase the share of renewable energy in the Netherlands to 14% by 2020 and to 16% by 2023. Therefore, a new phase for offshore wind power was heralded. In addition to the existing 1050 MW, an additional 3450 MW of operational offshore wind capacity is planned to be installed in the Dutch North Sea waters by 2023.

The recent research on ecological effects by construction and operation of OWFs allows for a new approach to address cumulative effects. This paper outlines a Framework for Assessing Ecological and Cumulative Effects (FAECE).[1] It

[1] Information and basic reports at http://www.noordzeeloket.nl/en/functions-and-use/Maritime_wind_energy/ecology/.

identifies the most important species, their relevant populations (size and distribution), the relevant activities, and shows how to identify and assess cumulative effects of these activities in Environmental Impact Assessments (EIAs) for OWFs. It includes generic information on the cumulative effects as well as more specific information on how cumulative effects of OWF activities should be incorporated into EIAs.

The OWF case study (shown in Fig. 1) in the southern North Sea region addresses both impacts during the construction phase (by underwater sound production through pile driving) and during the operational phase (due to habitat change and collision risk by wind turbines). Based on the outcomes of the FAECE, mitigation measures will be implemented in the permit process of OWFs applications.

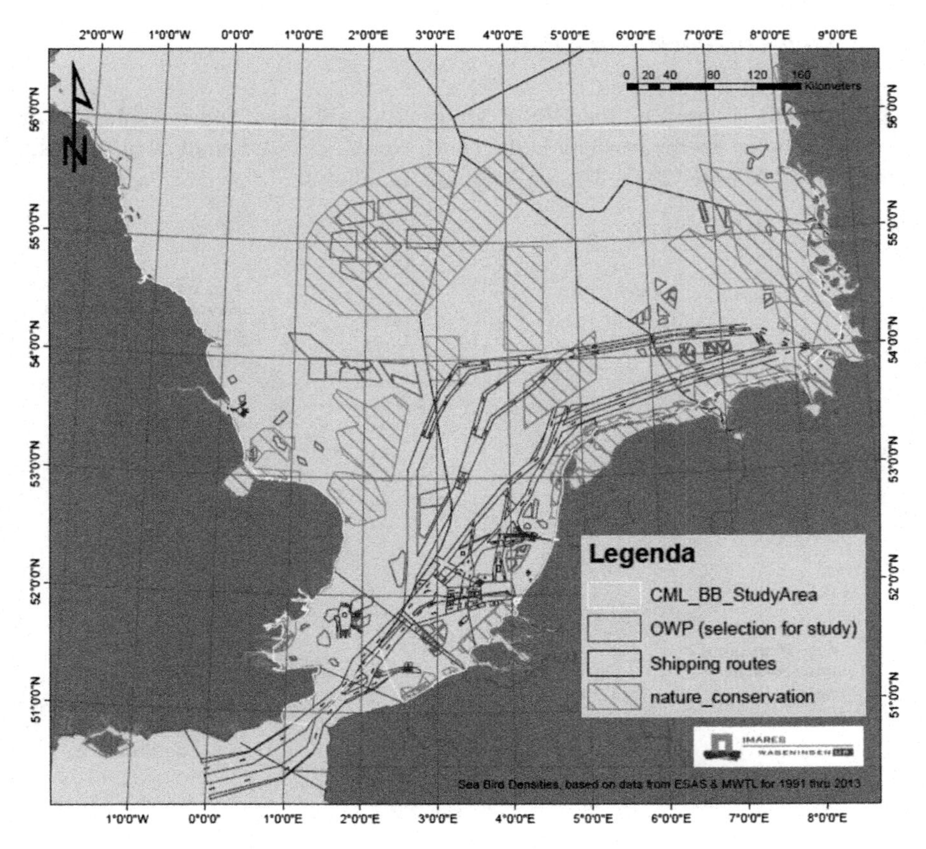

Fig. 1 Southern North Sea showing offshore wind farms (actual + projected), Natura 2000 sites and other (proposed) Marine Protected Areas (MPAs). The *yellow frame* is the case study area

Methods

Using a Common Language: The DPSIR Approach

In assessments of ecological cause-effect relationships, the DPSIR (Drivers, Pressures, State, Impact, Response) approach (European Environment Agency 1999) provides a systematic way of addressing these relationships and their interpretation in terms of drivers (and activities), pressures, impacts, state assessments and necessary responses. The DPSIR elements (Fig. 2) associated with planned and existing projects are assessed in the FAECE by a step-by-step procedure.

Case Study: Applying the Framework to Dutch Offshore Wind Plans

The FAECE is based on the DPSIR sequence of six steps, and is applied to the "Offshore wind energy capacity in the Dutch North Sea" case study, which aims to reach 4450 MW by 2023. The six steps are as follows:

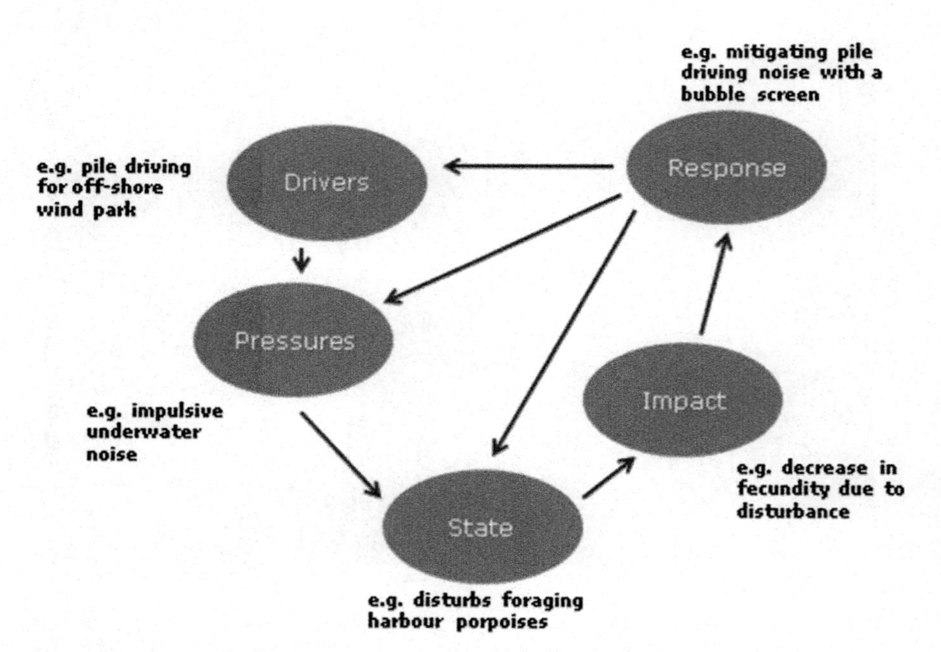

Fig. 2 Generic representation of the DPSIR approach (http://www.eea.europa.eu/publications/92-9167-059-6-sum) based upon the environmental indicators typology and overview (European Environment Agency 1999)

1. Identify the relevant pressures the envisaged activities could cause.
2. Identify the habitats and species that may be affected by these pressures.
3. Describe all other activities that could affect the same species.
4. Describe the nature and scale of the cumulative effects of all the activities selected in Step 3 on the selected habitats and species.
5. Evaluate the significance of the effects on the selected habitats and species.
6. If necessary, adapt the activity by taking measures to prevent the activity from causing significant effects.

Step 1: Pressures

The main pressures on wildlife caused by OWFs are underwater sound pulses from piling during construction, and habitat loss and collisions during operation. The main pressures on wildlife during decommissioning have not yet been included in the assessment.

The main pressures from OWFs on birds and bats may include:

- the loss of surface area and quality of habitat due to displacement of individuals who are intimidated by OWFs, and
- direct mortalities due to collisions with wind turbine blades and towers.

Step 2: Habitats and species that may be affected

This step deals with habitats and species that may be affected by these pressures. For the protection of the marine habitats mentioned in the Habitats Directive, it is necessary to avoid the Natura 2000 sites when planning the OWF activities and locations. Natura 2000 is a network of nature protection areas within the European Union.

The species groups (birds, marine mammals, fish, bats etc.) listed under the Birds and Habitats Directive and the Marine Strategy Framework Directive, however, do not only occur in these protected areas, but outside of these areas as well. Thus, the North Sea Region biogeographical populations of bird and marine mammal species were considered.

Fish were not considered in this step, since research indicates that there are behavioural effects found on fish but no permanent physical effects by impulsive noise such as piling sounds (Neo et al. 2014, 2015).

Among all North Sea dwelling species of marine mammals, the Harbour Porpoise (*Phocoena phocoena*) is considered to be the most sensitive to loud piling sounds during construction (SEAMARCO 2013a, b, 2014, 2015a, b). Other relevant species sensitive to piling sounds may include the Common Seal (*Phoca vitulina*) and the Grey Seal (*Halichoerus grypus*).

Broadly, the categories of birds that occur in the North Sea region are:

1. Seabirds, which spend all their time at sea outside the breeding season.
2. Coastal birds, which live year-round on the coast, for breeding and resting purposes included. They fly over the North Sea daily.
3. Migratory land and waterbirds, which migrate seasonally along the North Sea shores or cross the North Sea between the European continent and the British Isles.

All three of these categories should be taken into account when assessing OWF impacts. In recent years the presence of bats around OWFs has been monitored and found to be a much more regular and abundant than originally assumed. Bats are regularly present at over 85 km from the shore, which suggests seasonal migration across the North Sea (Lagerveld et al. 2014). Pressures from OWFs for birds and bats may include the loss of habitat area and quality due to displacement by OWFs, as well as direct mortality due to collisions with wind turbine blades and towers.

Step 3: Other activities that could affect the same species

Other pressures contributing to the disturbance of the marine environment are sounds by airguns, sonar, shooting and explosions, as well as mortality due to fishing activities, disturbance by vessels, pollution, and other human activities. Above-water disturbances to these species may come from collisions with other objects, disturbance by shipping, fishing, platforms, sand and shell extraction, military activities, oil spills, plastic waste, and bioaccumulation of micro-contaminants.

Step 4: Nature and scale of the cumulative effects

Wind farms negatively affect birds in three ways:

1. Loss of habitat: Some species tend to avoid OWF areas as they no longer 'recognise' an OWF as part of their habitat and are displaced from these areas.
2. Additional energy expenditure: Wind farms intersecting with foraging and migration routes may force some species to fly around them. However, this effect is believed to be minor in relation to the scale of the southern North Sea (Leopold et al. 2014).
3. Additional mortality by collision: Species flying through OWF areas run a higher risk of collision and mortality.

Two scenarios were modelled: all 2023 Dutch OWFs and all 2023 North Sea OWFs. The reason for modeling all 2023 OWFs was to gain a better understanding of the total cumulative impact of the *Energy Agreement for Sustainable Growth* (2013). The reason for distinguishing between the Dutch OWFs by that year and the total amount of foreseeable OWFs in the southern North Sea was to be able to identify which mitigation measures would be required for the Netherlands. In addition, this would help to identify the international contribution to the total cumulative impact. The combined effects were predicted for each species, which served as input for the specific population models used to predict effects on population level.

First of all, seabird and coastal bird densities were mapped for the entire southern North Sea region, based on available survey data (database of European Seabirds At Sea, ESAS), as presented by Leopold et al. (2014), in combination with available information on where operational OWFs are likely to be constructed by 2023. Then, a 10% mortality rate was included for seabirds likely to become displaced by OWFs (interpretation by Leopold et al. 2014 from assumptions by Bradbury et al. 2014). Then, the number of collision victims was predicted by calculation from the Extended Band Model SOSS (Band 2012), a model in which bird fatalities of OWFs are estimated. This estimation is based on site-specific information on bird fluxes (estimated from the bird densities as estimated from ESAS data and surveys of migrating birds), flight speeds and heights and information on the total rotor-swept area of the individual turbines in the OWFs. The numbers of fatalities calculated were then added to the estimated mortality caused by displacement from suitable habitats. In addition, an attempt was made to take into account the similar effects of other plans, projects and activities in the southern North Sea on the same species or groups.

No models currently exist to estimate the effects of OWFs on bats. Based on expert judgment, a worst-case scenario was developed whereby one bat fatality occurs per wind turbine per year (Leopold et al. 2014).

The critical sound contours of all OWF constructions were calculated with the sound propagation model AQUARIUS of TNO, which was developed to determine the underwater sound produced during the construction of wind farms and seismic surveys. The disturbance threshold of initially 136 dB (Fig. 3) and later 140 dB was used to determine the surface area (km^2) where the sound levels 1 m above the seabed were above this threshold. The density of Harbour Porpoise was determined based on aerial survey data for the Dutch continental shelf. For the EEZ of other North Sea countries, these densities were derived from several sources, such as the SCANS data, aerial counts and several environmental impact assessments (Heinis and de Jong 2015). There is a 50% uncertainty in the density estimates. Multiplying densities by the surface area within the critical sound contours determined the number of animals that will be disturbed by one piling event (piling of one foundation). Multiplying the number of disturbed animals by one piling event by the number of piling days for each modeled OWF yielded an estimate for the number of disturbance days.

The number of disturbance days is used as an input into the Interim Population Consequences of Disturbance (Interim PCoD) model (Fig. 4) to calculate the net population effect. The iPCoD model is a mathematical model developed by SMRU Marine (Harwood et al. 2014). This model is based on the conceptual model PCAD (Population Consequences of Acoustic Disturbance) of the US National Research Council's Committee on Characterizing Biologically Significant Marine Mammal Behavior in 2005 (National Research Council 2005). The iPCoD model calculates the effects of disturbance on vital rates (Stage specific survival, maturation and reproduction) through behavioral and physiological changes. The effects on vital

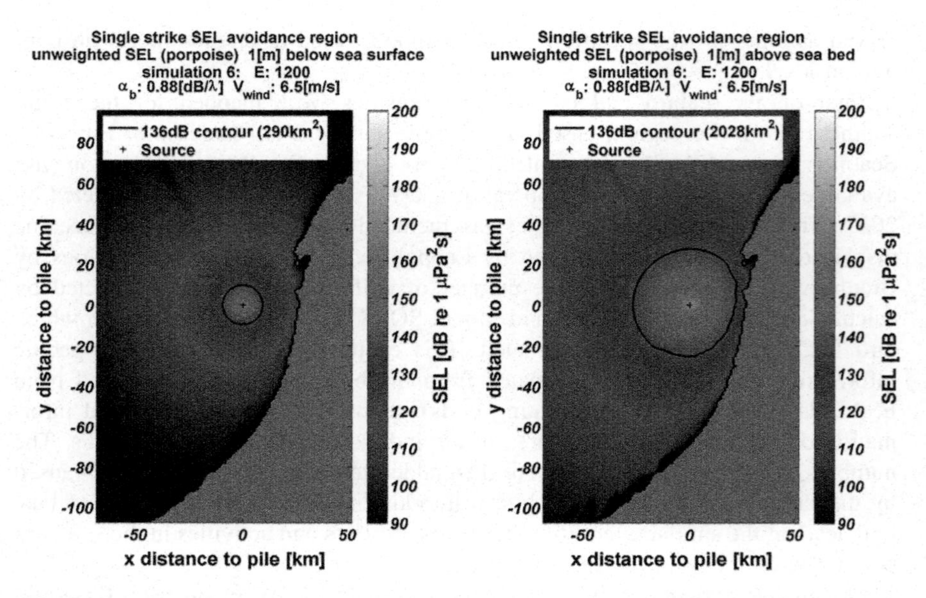

Fig. 3 Calculated distribution from Arends et al. (2013) of SEL_1 around a piling location (+) in the North Sea at a depth of 1 m below the surface (*left*) and 1 m above the seabed (*right*). Wind speed 6.5 m/s. The *black lines* show the contour within which the threshold value for avoidance is exceeded for Harbour Porpoises. The *grey area* shows the Dutch coast

rates due to behavioural and physiological changes were obtained through an expert elicitation process. The expert elicitation process is a formal technique that combines the opinions of several experts to arrive at a quantitative estimate of effects of an activity, where there is no data to quantify these effects directly. Harwood et al. (2014), describes in detail the process of the expert elicitation carried out for the iPCoD model. Behavioural change is defined as changes that are equivalent to 'Scale 5' or above on the severity scale for ranking observed behavioural responses of free-ranging marine mammals laboratory subjects to various types of anthropogenic sound compiled by Southall et al. (2007). The responses are scaled from '0 no observable reaction' to '9 outright panic'. The following behaviour is categorized as Scale 5:

- Extensive or prolonged changes in locomotion speed, direction, and/or dive profile but no avoidance of sound source
- Moderate shift in group distribution
- Change in inter-animal distance and/or group size (aggregation or separation)
- Prolonged cessation or modification of vocal behaviour (duration > duration of source operation).

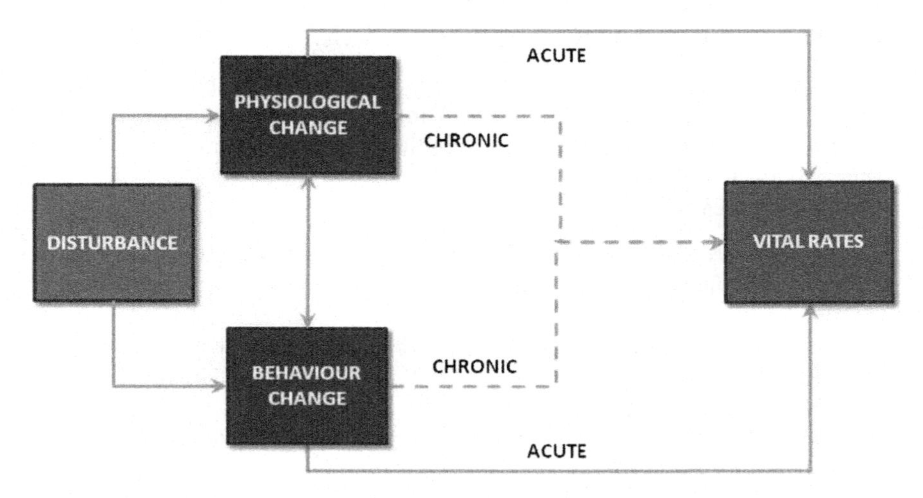

Fig. 4 This simplified version of the PCoD model was used in the interim PCoD model (Harwood et al. 2014)

Step 5: Evaluating the significance of the effects

The FAECE applies the concept of Potential Biological Removal (PBR) for assessing cumulative additional mortality of birds and bats due to OWFs. The PBR is "the maximum number of animals, not including natural mortalities, that may be removed from a marine mammal stock while allowing that stock to reach or maintain its optimum sustainable population" (defined by NOAA Fisheries 2014; marine mammals: Wade 1998; Dillingham and Fletcher 2008; Richard and Abraham 2013; short-lived: Milner-Gulland and Akçakaya 2001; Bellebaum and Wendeln 2011; bats: Lebreton 2005; Niel and Lebreton 2005). Therefore, it is assumed that as long as the cumulative additional mortality (due to the combined effects of all the pressures) does not exceed PBR, a population will not be at risk of decline due to cumulative impacts. If a population of a species is already in decline, the PBR value is calculated as low and will thus be exceeded. In comparison to bird estimations, bat population data is still rudimentary, causing this assessment to be 'indicative' at best.

The conservation status of the Harbour Porpoise, expressed as limits of 'acceptable change in population', is derived from the interim[2] objective of ASCOBANS (ASCOBANS Secretariat 1998). Effects by shipping, explosions and other anthropogenic sources could not be estimated and were not included.

[2]Since the available data on Harbour Porpoise densities and numbers in the North Sea are still considered insufficient, ASCOBANS has, for the time being, only established 'interim' objectives.

Step 6: Measures to prevent any significant effects

Prevention and mitigation measures are the proactive response to reducing impacts on targeted species. The extent of the response should effectively ensure that cumulative effects on species are not putting the species' conservation status or Natura 2000-site objectives at risk.

Mitigation measures may include:

- Sound screens and/or bubble curtains to reduce emitted construction sounds
- The implementation of large (and therefore lower numbers of) wind turbines per OWF
- Turning off or slowing down wind turbines during intensive bird or bat migration at rotor height
- Careful selection of OWF locations
- Enhancing habitat quality within the geographical range of the impacted species else or within the OWFs.

Effect Modelling

Suitable models and concepts of cumulative effect estimates are required to obtain net impact calculations for the targeted species. Data on spatial and temporal distribution of species (marine mammals, birds and bats) in the southern North Sea are also needed.

For marine mammals, critical sounds levels that disturb the Harbour Porpoise are needed to determine the geographical extent and thus the number of individuals potentially displaced, as explained in Step 4. This is then used to calculate the potential impacts on their vital rates. The FAECE applied the interim PCoD (Population Consequence of Disturbance) model (Harwood et al. 2014,[3] Fig. 3).

The model translates the duration of disturbance of a certain severity (i.e. Scale 5 or above) through changes in vital rates to changes in the population for a chosen amount of years. The model uses as input the amount of calculated disturbance days. This input has been generated as follows:

1. Calculate underwater sound propagation per blow and determine the distance where the disturbance threshold is reached
2. Estimate the surface area that will be avoided by porpoises during a piling day
3. Estimate number of porpoises and seals within this area
4. Calculate number of porpoise disturbance days (number of disturbed animals per day multiplied by number of days the piling sound is generated).

[3]iPCoD: https://www.st-andrews.ac.uk/news/archive/2014/title,248538,en.php.

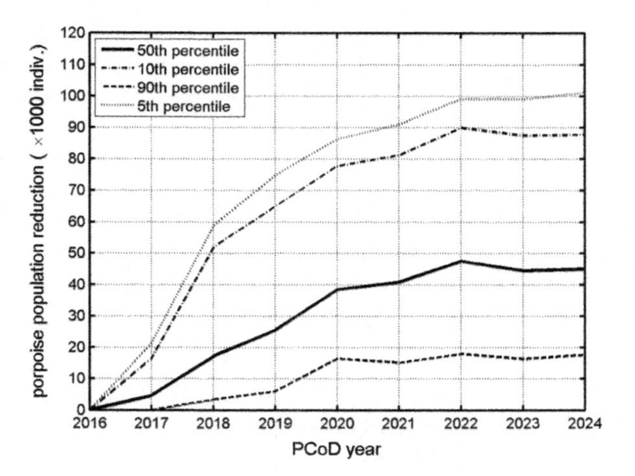

Fig. 5 The reduction in the Harbour Porpoise population in the North Sea calculated with interim PCoD and caused by piling sound in one of the chosen scenarios

Several scenarios were developed to determine the effects of construction under varying conditions such as season, number of simultaneous constructions and the presence or absence of maximum noise criterion. Some scenarios were defined where the duration of disturbance and vulnerable population were varied to test the sensitivity of the Interim PCoD model to variations in these parameters (Heinis and de Jong 2015). Parameters such as the duration of disturbance, which controls the number of disturbance days, had the biggest impact on the model outcome. The Interim PCoD model carries out 500 simulations. The result is a graph with the reduction of the population for the chosen number of years (Fig. 5). The graph shows the median as well as some chosen percentiles of the runs, in this case the 5th, the 10th and the 90th percentile. The 5th percentile of the runs is chosen to express the effects in. This means that 95% of the effects will be under this value and there is a 5% probability that the effects will exceed this value. In reality, the probability that the effects will exceed this value will be lower because the assumptions are all based on the worst-case scenario.

Some species of seabirds displaced by OWFs may experience diminishing foraging opportunities. The cumulative effects may lead to significant loss of food supply and impact on the population, unless managed properly. Bradbury et al. (2014) propose a vulnerability index that includes both disturbance by wind farm structures and disturbance by other human activities, as well as the amount of 'habitat specialisation' (the species' vulnerability to habitat loss and fragmentation and its dispersal ability). Leopold et al. (2014) arbitrarily assumed a maximum 10% mortality of the displaced birds, recognising the need for the precautionary principle, which requires that impacts are not to be underestimated.

Displacement scores due to OWFs have been estimated for all relevant seabird species in the southern North Sea, by comparing seasonal seabird distribution maps. This is based on all available survey data (European Seabirds At Sea database, ESAS), which includes maps of both existing OWFs as well as maps of projected OWFs up until the year 2023. Additional mortality due to habitat loss is expected to increase as the number of OWFs increase in the southern North Sea. It is assumed that the seabird species do not habituate to the presence of OWFs. The SOSS02 Band model (2012) was used to estimate the cumulative number of bird collisions with OWF turbines. This model is based upon the core theoretical model for collision risk described by Tucker (1996) and elaborated on by Band (2000) and Band et al. (2007). For modelling a collision probability, the number of collisions is calculated from the flux or density of birds within an OWF area. Seabird fluxes were derived from both boat surveys of birds and direct flux measurements by radar. The model calculates the species-specific number of collisions within an OWF, determined by type and number of turbines.

The Need for Species-Specific Threshold Levels for Acceptable Total Impacts

The common denominator of all human-induced pressures is the extent to which a species can sustain additional mortality, which is defined as Potential Biological Removal (PBR). The PBR approach was applied to birds colliding with wind turbines by Watts (2010), Poot et al. (2011), Sugimoto and Matsuda (2011), Bellebaum et al. (2013). These studies underlined the need for mitigation measures and monitoring of bird populations at risk.

In this study, the PBR of 61 (southern) North Sea bird species was calculated based on estimates of adult survival and age at first reproduction (Richard and Abraham 2013). These PBRs were compared with the calculated cumulative additional mortalities by displacement and collisions (Leopold et al. 2014).

The PBR approach has been applied with the Harbour Porpoise in the Dutch Exclusive Economic Zone (Scheidat et al. 2013), but does not fit in with the Interim PCoD model outcome because they deal with population dynamics and recovery factors in an incomparable way. As our studies currently heavily depend on the application of the Interim PCoD model, the interim ASCOBANS conservation objective “to restore populations to, or maintain them at, 80% or more of the carrying capacity” was selected for this paper. As ASCOBANS did not define which carrying capacity is meant by this objective (the current or a future maximum carrying capacity, where anthropogenic influences have been reduced to a minimum), the current Dutch EEZ population of the Harbour Porpoise was chosen to derive the maximum acceptable effect. This population was estimated to be at 51,000 (2010–2014, pers com Scheidat) and the North Sea population at 227,298 (Hammond et al. 2013). The maximum acceptable effect on this population was set at 5% so as not to be inconsistent with the interim objective of ASCOBANS. This means that a maximum reduction of the population by 2550 individuals is deemed acceptable for all the wind farms to be built until 2023 in the Netherlands.

Results of the Case Study

Underwater Sound During Pile Driving Impacts on the Harbour Porpoise

Results from the iPCoD model showed that pile driving of two wind farms per year could lead to a maximum population reduction of 19,000 Harbour Porpoises. Using ASCOBANS, the maximum annual population reduction for the Harbour Porpoise should only be 6375. Scenarios in which underwater noise was mitigated positively, showed a significant decrease in population reduction (up to an 86% decrease).

It is clear that the construction of OWFs in the North Sea will impose considerable pressure on the Harbour Porpoise population. Even if only the effects of the construction phase of the Dutch OWFs are taken into account, in only two scenarios do they remain within the ASCOBANS limits. It is also evident that a considerable proportion of the impacts is caused by foreign wind farms if the planned expansion of capacity by 2023 is achieved. This is without mitigation measures and with the exception of Germany, where the currently applicable sound standards have been taken into account.

On an international level, when no mitigation measures are taken to reduce underwater noise, population reduction of the international Harbour Porpoise population may be as high as 100,000. This is because it is not yet known which mitigation measures will be compulsory for the OWFs. The only exception to this is Germany, where a noise threshold for sound produced during construction is compulsory, which has been taken into account. In the international scenario the ASCOBANS threshold is exceeded at both the national and international levels and it is assumed that the Dutch share of the pressure on the total population of Harbour Porpoise in the North Sea is around 20%. The numbers clearly show that mitigation measures are necessary. The contribution to these numbers by the German OWFs, where the noise threshold is operational, is much smaller than that of other countries.

In regards to the effect of the construction of OWFs on the Harbour Porpoise population, the thresholds are at risk of being exceeded both at the national and international levels. It is therefore concluded that for the Harbour Porpoise significant adverse effects can only be avoided by taking effective mitigation measures to reduce the sounds emitted during construction.

Habitat Loss for Seabirds

Some species of seabirds, e.g. Common Guillemots (*Uria aalge*), Northern Gannet (*Morus bassanus*), Northern Fulmar (*Fulmarus glacialis*) and Red-throated and Black-throated Divers (*Gavia stellata* and *Gavia arctica*), tend to avoid operational wind farms. Therefore they may experience structural displacement and suffer loss of their marine habitat. The initial results of the work by Leopold et al. (2014) indicate that the plans for OWFs in the southern North Sea will not likely lead to unacceptable levels of extra annual mortality for the Common Guillemot (13% of PBR), Northern Gannet, Northern Fulmar (both 2% of PBR) or Red- and Black-throated Diver (8 and 7% of PBR, respectively). Nonetheless, the potential impacts might increase in the further future as the total surface area of the southern North Sea is increasingly occupied by OWFs, unless the birds become accustomed to the presence of turbines.

Collision Rates in Birds

Bird collision rates with wind turbines built by 2023 were modelled with additional annual mortality for the Lesser Black-backed Gull (*Larus fuscus*) at 160% PBR, Great Black-backed Gull (*L. marinus*) at 100% PBR and Herring Gull (*L. argentatus*) at 105% PBR (Leopold et al. 2014, 2015; van der Wal et al. 2015). The results indicate that OWFs consisting of less but larger wind turbines should be considered in order to reduce impacts on these species. It was also found that measuring real collision rates for gulls is critical.

For migratory birds such as the Eurasian Curlew (*Numenius arquata*), Black Tern (*Chlidonias niger*) and Bewick's Swan (*Cygnus bewickii*), mortality due to collisions was predicted for the year 2023 at about 40% of PBR (Leopold et al. 2014). Other sources of additional annual mortality are not considered in this already high rate of additional mortality. Therefore, the study found OWFs to be a significant contributor to mortalities of these species, which emphasizes the need to develop preventive and mitigating measures.

Bats at Sea

The distribution and behaviour of bats at sea and their sensitivity to OWFs are still largely unknown. Research has shown that at least one migratory species, the Nathusius's Pipistrelle (*Pipistrellus nathusii*), and possibly up to three species (also Parti-coloured Bat *Vespertilio murinus* and Common Noctule *Nyctalus noctula*) migrate across the southern North Sea (Lagerveld et al. 2014; Leopold et al. 2014). Lagerveld et al. (2014) recorded bat call activity at OWFs and meteorological masts in the North Sea, almost exclusively at night during spring and autumn when wind speeds are low.

Given the lack of data, a worst-case additional mortality was assumed of one bat per turbine per year (Leopold et al. 2014). This would mean 8000 bats, mostly Nathusius's Pipistrelle, killed by OWF turbines per year by 2023, a figure that might exceed the population's PBR (Leopold et al. 2014).

It is concluded that a research programme must be set up to gain a better knowledge base on bird and bat behaviour and to better devise prevention and mitigation measures.

Implementing FAECE

Both the Dutch spatial planning strategy on offshore wind power (Rijksstructuurvisie Windenergie op Zee, 26 September 2014) and the Second National Water Plan (2016–2021) designate the FAECE approach as a compulsory step in decision-making in relation to the boundaries and accumulation of OWFs within the already designated sites. The FAECE should be undertaken during the process of siting, determining the criteria by which an OWF may be positioned, configured, equipped and managed. This should assist in avoiding and/or minimizing the impacts of each OWF as well as the cumulative effects of OWFs and other activities.

The purpose of using the FAECE is to determine the parameters of OWFs and other activities, in order to establish sustainable development in the southern North Sea. Understanding the causes and impacts of cumulative effects by OWFs as well as of other activities may positively influence the siting of OWFs and help to achieve ecosystem based management, as required by the EU Marine Strategy Framework Directive 2008/56/EC. The planning includes aspects such as location, size, turbine characteristics, spatial configuration, operation management, etc. International collaboration is crucial, since energy companies as well as marine mammals, birds or (migratory) bats do not recognize national borders.

Already, the FAECE has resulted in the inclusion of mitigation measures within the approval procedure of OWFs at the Borssele site, close to the Belgian border. This will also happen for all new OWF approvals, in order to achieve the OWF objectives for the Netherlands by 2023 without hampering the conservation targets for nature and ecology at sea.

Mitigation measures include:

1. Disturbances to porpoises, seals and fish are limited through the introduction of a piling-sound criterion of 160–172 dB re $\mu Pa^2 s$ SEL at 750 m from the source depending on season and number of pilings.
2. Establishing a maximum number of turbines to become operational per OWF (supposing each OWF to provide a capacity of 380 MW), thus reducing collision risk to well below PBR for Herring Gull, Lesser Black-backed Gull and Great Black-backed Gull.
3. The cut-in wind speed of turbines must be at least 5.0 m s^{-1} at axis height during the period from August 15th until September 30th (both the period and the wind condition in which most bat migration is likely to occur; Lagerveld et al. 2014; Leopold et al. 2014) between 1 h after sunset to 2 h before sunrise.

When wind speeds are less than 5.0 m s^{-1}, the number of rotations per minute for each wind turbine must be reduced to less than 1.
4. In order to decrease collision risk during bird migration with more than 500 individuals km^{-1} hr^{-1} (spring and autumn mass migration), the number of rotations per minute for each wind turbine should also be less than 1. This will assist in minimizing the contribution to PBR for migratory (land) birds.

Discussion

Assumptions, Gaps in Knowledge, Monitoring and Evaluation

There are still a considerable amount of knowledge gaps, both methodological (process, legal) and ecological. Some of these gaps have been filled by assumptions based on expert judgment, and others have been filled by pragmatic assumptions. However, in due course these assumptions will need to be validated as best as possible. Research on ecological effects by OWFs is gathering pace, both in the Netherlands and elsewhere. Moreover, the Dutch government is in the process of establishing a nationally coordinated monitoring, research and evaluation plan to address, among other things, the knowledge gaps that are identified during the application of the FAECE.

Future Developments of FAECE

The development of OWFs is relatively new, and many effects are still largely unknown. The knowledge base and understanding of OWFs activities and their ecological impacts are expanding. For this reason, periodical updating of the framework is essential. A more connected, cooperative and international focus on cumulative effects is needed because wind, species and habitats do not obey borders and may lead to transboundary problems when not addressed appropriately. New and more detailed legislation might also lead to the necessity of a revision of the FAECE. In the upcoming years the following actions will be important:

- Greater international coordination
- Regular updates of the FAECE as new research results become available
- Keeping underlying models up-to-date (with options for substituting old models with new ones that become available if they are deemed to be more accurate and effective)
- Extending the scope to ecological targets from other policies and guidelines, e.g. from the EU Marine Strategy Framework Directive

- Including more human activities (drivers) and their associated pressures in the assessments.

In short, the concept of the FAECE is likely to be sound, but further shaping and extending its applicability, particularly on larger and more relevant spatial scales and for more Drivers and Pressures, seems to be inevitable.

Acknowledgements The framework tool presented in this paper, as well as its first application as described, has been made possible thanks to the dedicated work of the following persons and institutions: Mardik Leopold, Jan Tjalling van der Wal, Ruud Jongbloed, Michaela Scholl, Nara Davaasuren and Sander Lagerveld (all IMARES), Martin Poot, Ruben Fijn, Karen Krijgsveld, Mark Collier, Abel Gyimesi, Job de Jong and Martijn Boonman (all Bureau Waardenburg), Bob Jonge Poerink (The Fieldwork Company), Christ de Jong (TNO) and Floor Heinis (HWE). We want to express our gratitude to all of these professional researches as well as to Rob Gerits (Rijkswaterstaat) and Eeke Landman-Sinnema and Jeroen Vis (both ministry of Economic Affairs) for their keen interest in the matter of assessing cumulative effects of offshore windfarms.

References

Arends E, Jaspers Faijer M, van der Bilt S (2013) Passende Beoordeling Windpark Q4 west, Pondera Consult

ASCOBANS Secreatriat (1998) Progress report to the IWC on the agreement on the conservation of small cetaceans of the Baltic and North Seas (ASCOBANS), Oct 1997–May 1998

Band W (2000) Windfarms and birds: calculating a theoretical collision risk assuming no avoiding action. Guidance Notes Series, Scottish Natural Heritage

Band W (2012) Using a collision risk model to assess bird collision risks for offshore windfarms. SOSS, The Crown Estate, London, UK. Available via www.bto.org/science/wetland-and-marine/soss/projects. http://www.bto.org/sites/default/files/u28/downloads/Projects/Final_Report_SOSS02_Band1ModelGuidance.pdf

Band W, Madders M, Whitfield DP (2007) Developing field and analytical methods to assess avian collision risk at wind farms. In: de Lucas M, Janss GFE, Ferrer M (eds) Birds and wind farms: risk assessment and mitigation. Quercus, Madrid, pp 259–275

Bellebaum J, Wendeln H (2011) Identifying limits to wind farm-related mortality in migratory bird populations. In: Poster presented at the CWW conference, 2011 in Trondheim

Bellebaum J, Korner-Nievergelt F, Dürr T, Mammen U (2013) Wind turbine fatalities approach a level of concern in a raptor population. J Nat Conserv 21:394–400

Bradbury G, Trinder M, Furness B, Banks AN, Caldow RWG, Hume D (2014) Mapping seabird sensitivity to offshore wind farms. PLoS ONE 9(9):e106366. doi:10.1371/journal.pone.0106366

Dijkema KS, Dankers N, Wolff WJ (1985) Cumulatie van ecologische effecten in de Waddenzee. RIN-rapport 85/13. Rijksinstituut voor Natuurbeheer, Texel: 105

Dillingham P, Fletcher D (2008) Estimating the ability of birds to sustain additional human-caused mortalities using a simple decision rule and allometric relationships. Biol Conserv 141:1783–1792

European Environment Agency (1999) Environmental indicators: typology and overview. Technical report No 25/1999. Available via http://www.eea.europa.eu/publications/TEC25

Hammond PS, MacLeod K, Berggren P, Borchers DL, Burt ML, Cañadas A, Desportes G, Donovan GP, Gilles A, Gillespie D, Gordon J, Hiby L, Kuklik I, Leaper R, Lehnert K, Leopold M, Lovell P, Øien N, Paxton CGM, Ridoux V, Rogan E, Samarra F, Scheidat M, Sequeira M, Siebert U, Skov H, Swift R, Tasker ML, Teilmann J, van Canneyt O, Vázquez JA

(2013) Cetacean abundance and distribution in European Atlantic shelf waters to inform conservation and management. Biol Conserv 164:107–122

Harwood J, King S, Schick R, Donovan C, Booth C (2014) A protocol for implementing the interim population consequences of disturbance (PCOD) approach: quantifying and assessing the effects of UK offshore renewable energy developments on marine mammal populations. Report number SMRUL-TCE-2013–014. Scott Mar Freshw Sci 5(2)

Heinis F, de Jong CAF, RWS Werkgroep Onderwatergeluid (2015) Cumulatieve effecten van impulsief onderwatergeluid op zeezoogdieren. TNO-rapport, TNO 2015 R10335, Den Haag. Available via https://www.noordzeeloket.nl/en/Images/Frameworkfor%20assessing%20ecological%20and%20cumulative%20effects%20of%20offshore%20wind%20farms%20-%20Cumulative%20effects%20of%20impulsive%20underwater%20sound%20on%20marine%20mammals_4646.pdf

Lagerveld S, Jonge Poerink B, Verdaat H (2014) Monitoring bat activity in offshore wind farms OWEZ and PAWP in 2013. IMARES rapport C165/14

Lebreton JD (2005) Dynamical and statistical models for exploited populations. Aust N Z J Stat 47:49–63

Leopold MF, Boonman M, Collier MP, Davaasuren N, Fijn RC, Gyimesi A, de Jong J, Jongbloed RH, Jonge Poerink B, Kleyheeg-Hartman JC, Krijgsveld KL, Lagerveld S, Lensink R, Poot MJM, van der Wal JT, Scholl M (2014) A first approach to deal with cumulative effects on birds and bats of offshore wind farms and other human activities in the Southern North Sea. IMARES report C166/14. Available via https://www.noordzeeloket.nl/en/Images/Frameworkfor%20assessing%20ecological%20and%20cumulative%20effects%20of%20offshore%20wind%20farms%20-%20A%20first%20approach%20to%20deal%20with%20cumulative%20effects_4766.pdf

Leopold MF, Collier MP, Gyimesi A, Jongbloed RH, Poot MJM, van der Wal JT, Scholl M (2015) Iteration cycle: dealing with peaks in counts of birds following active fishing vessels when assessing cumulative effects of offshore wind farms and other human activities in the Southern North Sea. Additional note to IMARES report number C166/14

Milner-Gulland EJ, Akçakaya HR (2001) Sustainability indices for exploited populations under uncertainty. Trends Ecol Evol 16(12):686–692

National Research Council (2005) Marine mammal populations and ocean noise: determining when noise causes biologically significant effects. The National Academy Press, Washington D.C

Neo YY, Seitz J, Kastelein RA, Winter HV, Ten Cate C (2014) Temporal structure of sound affects behavioural recovery from noise impact in European seabass. Biol Conserv 178:65–73. doi:10.1016/j.biocon.2014.07.012

Neo YY, Parie L, Bakker F, Snelderwaard P, Tudorache C, Schaaf M, Slabbekoorn H (2015) Behavioral changes in response to sound exposure and no spatial avoidance of noisy conditions in captive zebrafish. Front Behav Neurosci, 17 Feb 2015. http://dx.doi.org/10.3389/fnbeh.2015.00028

Niel C, Lebreton JD (2005) Using demographic invariants to detect overharvested bird populations from incomplete data. Conserv Biol 19:826–835

NOAA Fisheries (2014) Protected resources glossary. Available via http://www.nmfs.noaa.gov/pr/glossary.htm

Poot MJM, van Horssen PW, Collier MP, Lensink R, Dirksen S (2011) Effect studies offshore Wind Egmond aan Zee: cumulative effects on seabirds a modelling approach to estimate effects on population levels in seabirds. Bureau Waardenburg report nr: 11-026, OWEZ_R_212_T1_20111118_Cumulative effects

Richard Y, Abraham ER (2013) Application of potential biological removal methods to seabird populations. New Zealand Aquatic environment and biodiversity report no. 108. Ministry for Primary Industries

Scheidat M, Leaper R, van den Heuvel-Greve M, Winship A (2013) Setting maximum mortality limits for harbour porpoises in Dutch waters to achieve conservation objectives. Open J Mar Sci, vol 3. Available via http://www.scirp.org/journal/ojms

SEAMARCO (2013a) Hearing thresholds of a harbor porpoise (*Phocoena phocoena*) for playbacks of multiple pile driving strike sounds. Report no. 2013-01

SEAMARCO (2013b) Behavioral responses of a harbor porpoise (*Phocoena phocoena*) to playbacks of broadband pile driving sounds. Report no. 2013.04

SEAMARCO (2014) Hearing frequencies of a harbor porpoise (*Phocoena phocoena*) temporarily affected by played back offshore pile driving sounds. Report no. 2014-05

SEAMARCO (2015a) Effect of pile driving sounds' exposure duration on temporary hearing threshold shift in harbor porpoises (*Phocoena phocoena*). Report no. 2015-09

SEAMARCO (2015b) Hearing thresholds of a harbor porpoise (*Phocoena phocoena*) for narrow-band sweeps (0.125–150 kHz). Report no. 2015–02

SER-agreement (2013) Energy Agreement for sustainable growth: implementation of the energy agreement. Available via https://www.ser.nl/en/publications/publications/2013/energy-agreement-sustainable-growth.aspx

Southall BL, Bowles AE, Ellison WT, Finneran JJ, Gentry RL, Greene CR Jr, Kastak D, Ketten DR, Miller JH, Nachtigall PE, Richardson WJ, Thomas JA, Tyack PL (2007) Marine mammal noise exposure criteria: initial scientific recommendations. Aquatic Mamm 33(4):411–509

Sugimoto H, Matsuda H (2011) Collision risk of White-fronted Geese with wind turbines. Ornithological Sci 10:61–71

Tucker VA (1996) A mathematical model of bird collisions with wind turbine rotors. J SolEnergy Eng 118:253–262

van der Wal JT, Fijn R, Gyimesi A, Scholl M (2015) 2nd Iteration: effect of turbine capacity on collision numbers for three large gull species, based on revised density data, when assessing cumulative effects of offshore wind farms on birds in the southern North Sea. Additonal note to IMARES Report C166/14

Wade PR (1998) Calculating limits to the allowable human-caused mortality of Cetaceans and Pinnipeds. Mar Mamm Sci 14(1):1–37

Watts BD (2010) Wind and waterbirds: Establishing sustainanle mortality limits within the Atlantic Flyway. Center for conservation biology technical report series, CCBTR-10-05. College of William and Mary/Virginia Commonwealth University, Williamsburg

Wind Turbines and Birds in Germany—Examples of Current Knowledge, New Insights and Remaining Gaps

Marc Reichenbach

Abstract The impacts of wind turbines on birds have been discussed in Germany for almost 25 years now. The current practices in the planning process for wind farms can be characterized as a mixture of scientifically based knowledge and precautionary assumptions. The basic principle of mitigating the impacts of wind farms on certain species of high conservation concern is to maintain a sufficient distance between wind turbines and breeding sites or important roosting areas. This paper uses examples of certain species of concern to illustrate the current knowledge and practice of mitigating impacts on birds in Germany. The results demonstrate that, in some cases, the impact is not as severe as originally assumed and that micrositing of turbines can be a powerful tool to minimize possible effects. In view of the growing number of wind turbines operating in Germany, the adequate assessment of cumulative impacts is a key issue. In addition, there is growing pressure to improve and demonstrate the effectiveness of mitigation measures. Future research should be focused on long-term effects and on population-level impacts.

Keywords Birds · Wind turbines · Current knowledge · Mitigating impacts · Micrositing

Introduction

Possible impacts of wind farms on birds have been a cause of concern in Germany for more than 25 years, with the first study published in 1990 (Böttger et al. 1990). A first systematic interpretation of data from six study areas using the Impact-Gradient-Design (survey of breeding or resting birds along a gradient of growing distance to the turbines) was given by Bach et al. (1999). This was a

M. Reichenbach (✉)
ARSU GmbH, Oldenburg, Germany
e-mail: reichenbach@arsu.de

J. Köppel (ed.), *Wind Energy and Wildlife Interactions*,
DOI 10.1007/978-3-319-51272-3_14

follow-up of the "Birds and wind energy in northern Germany" workshop in 1997, which was the first of its kind in Germany.

The discussion about bird impacts during the first 10–15 years was mainly focused on impacts resulting from disturbance and displacement (Reichenbach 2003; Reichenbach et al. 2004), but then it shifted gradually towards a stronger emphasis on the collision risk (Hötker et al. 2004; Dürr and Langgemach 2006; LAG VSW 2007; Hötker 2011; Hötker et al. 2013; LAG VSW 2015; Langgemach and Dürr 2015). Nowadays, both types of impacts play a major role in approval procedures, especially the collision risk of certain bird species, which has the potential to lead to a refusal of proposed wind farms.

Currently, there are more than 25,000 onshore wind turbines installed in Germany (Windguard 2015). Many Federal States aim at providing up to 1.5% of their land area for onshore wind energy, which will result in more than a doubling of the capacity already installed.[1] As a result, pressure on birds and their habitats continually grows and unproblematic locations become increasingly rare. Consequently, many new wind farm proposals face challenges in dealing with potential impacts on birds. Bats are, in contrast, much easier to handle in the approval procedure due to curtailment possibilities, for which Brinkmann et al. (2011) have laid the standards in Germany.

The aim of this paper is to provide an overview of the current German practice of dealing with certain bird species in the permit application procedure for wind farm projects. Another aim is to demonstrate that micro-siting and determining flight path behaviour can be effective and accurate and might modify general recommended radial distance buffers.

Collision Risk

Principles, Jurisdiction and the Problem of Significance Thresholds

In Germany the handling of collision risks of certain species is mainly guided by the distance recommendations of LAG VSW (2015). They define two types of buffer zones around nests or important roosting sites: the recommended minimum distance to the next wind turbine, as well as an additional assessment area, within which regularly used flight paths, foraging grounds or roosting sites must be identified. The minimum distances are justified by the general finding that the level of flight activity is significantly higher near the nest site, which has been shown in several studies (Hötker et al. 2013). Eichhorn et al. (2012) found that the collision

[1] http://windmonitor.iwes.fraunhofer.de/windmonitor_de/1_wind-im-strommix/1_energiewende-in-deutschland/6_Ausbaustand_der_Bundeslaender/

risk declines exponentially with increasing distance between a bird's nest and the wind turbines.

The recommended minimum buffer zones around nest sites to be kept clear from wind farms range from:

- 500 m (e.g. Curlew, Nightjar, Woodcock, Corncrake and Hobby),
- 1000 m (e.g. most raptor species such as Harriers, Black Kite, Honey Buzzard, Osprey and Eagle Owl, White Stork, Golden Plover),
- 1500 m (Red Kite) 3000 m (e.g. Golden Eagle, Sea Eagle, Black Stork),
- up to 6000 m (e.g. Lesser Spotted Eagle).

The additional assessment areas range from:

- 3000 m (Eagle Owl, Hobby and Hen Harrier),
- 4000 m (e.g. Red Kite, Osprey),
- 6000 m (e.g. Sea Eagle, Golden Eagle and Golden Plover),
- up to 10,000 m (e.g. Black Stork).

Moreover, it is recommended to exclude core areas with particularly high breeding densities from the development of wind power plants (LAG VSW 2015).

Under Article 12 of the Council Directive 92/43/EEC of 21 May 1992 (EU habitats directive), section 44 of the German Nature Conservation Act prohibits the intentional killing of all European bird species (among others). With respect to bird deaths by road traffic and wind turbines, the Federal Administrative Court has ruled that this regulation relates only to a 'significant increase' of bird deaths, not a single incidental killing. "Significant increase" is only vaguely described by the Court. The bird species must be one that is strongly affected by collisions, and the species must occur frequently at that location. The scale for assessing a "significant increase" has to be the risk for the affected individuals, not for the population.

Both LAG VSW (2015) and the Federal Administrative Court assume that a species' risk of being killed will significantly increase if the proposed wind farm is located within the recommended minimum distance. However, it is possible to show in individual cases that the implementation of a (circular) buffer radius is not appropriate in view of the actual flight behaviour of the individual birds. This flight behaviour can be determined through extensive flight observations comparable to the so-called 'vantage point watches' recommended by Scottish Natural Heritage (SNH 2014).

If a species' risk of being killed is deemed as a 'significant increase', this may have serious legal implications for a wind farm proposal, up to its rejection. Consequently, there is a major debate among wind farm planners, biologists, consultants and legal experts about how to deal with the threshold of significance. One possibility discussed in this context is the use of quantitative collision risk models. The PROGRESS project (Grünkorn 2015) has tested whether the well-known 'Band' collision risk model (Band et al. 2007) is able to predict collision risk numbers reliably. This model was assessed in comparison to fatality searches, including the necessary correction factors (Weitekamp et al. 2015). Quantitative thresholds can only be applicable with a predictive tool to assess whether the threshold will be met or not. The result was a complete mismatch between the predictions of the Band-model—based on 36 h of flight observations in each of the 55 wind farms—and the estimated numbers yielded by the fatality searches (Weitekamp et al. 2015). Therefore, the assessment of a significant increase of being killed because of collision with a proposed wind turbine has to be predominantly based on a qualitative evaluation rather than on quantitative collision risk modelling alone (i.e. not relying just on numbers but also taking the influence of certain behaviour patterns or topographical factors into account).

Species of Concern

White Tailed Sea Eagle

Currently, there are about 700 breeding pairs of Sea Eagles in Germany (Gedeon et al. 2015) and 119 fatalities due to wind farms have been documented so far (Langgemach and Dürr 2015). Consequently, in proportion to population size the Sea Eagle is the species most affected by collisions at wind turbines in Germany. Nevertheless, this anthropogenic mortality has not impeded the strong population growth and the recolonization of large parts of Germany since 1990 (Hötker et al. 2013).

Mitigation is mainly handled by implementation of the aforementioned turbine distance recommendations, whereby the minimum buffer around nest sites for Sea Eagles is 3 km. From 3 to 6 km extensive flight observations have to be carried out to check for frequently used flight paths and foraging areas during at least 30 days with 8 h per day, covering the whole breeding season. The validity of the data strongly depends on the breeding success, for example the recorded flight activity might not be representative in the absence of chicks that the parents need to supply with food (MELUR and LLUR 2013; NLT 2014). If a proposed wind farm is located within the 3 km radius, a multi-year survey of at least 70 days the whole year round is needed (MELUR and LLUR 2013).

Sea eagles show high breeding site fidelity (Bauer et al. 2005) and their preferred foraging areas can be assumed to be relatively constant (lakes, rivers, coast line, wetlands). Thus, patterns of flight activity can be expected to remain fairly stable which reduces the uncertainty in the assessment of collision risk. On the other hand in some parts of northern Germany the Sea Eagles are so common now that flight activity—especially of non-breeders—can be expected more or less everywhere. As a consequence, results of the respective survey (example in Fig. 1) and the assessment of a significant increase of the risk of being killed become less and less precise.

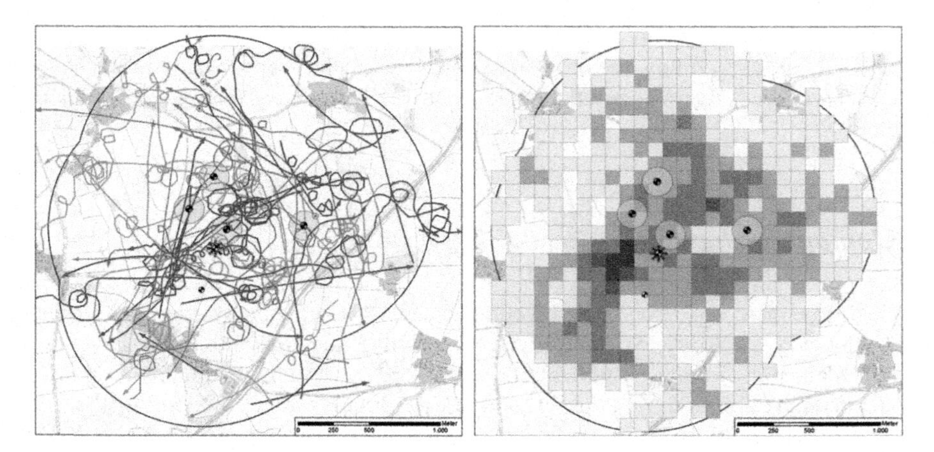

Fig. 1 Example of flight activity of sea eagles sampled during 70 days in an area with four proposed wind turbines—no nest site within 3 km, no attractive foraging habitat (arable land only), but still quite frequent sightings of sea eagles (own data)

Red Kite

Currently there are about 15,000 breeding pairs of Red Kites in Germany (Gedeon et al. 2015), which is about half of the European population (25,200–33,400 BirdLife International 2015). So far 301 fatalities due to windfarms have been documented in Germany (Langgemach and Dürr 2015). Considering the necessary correction factors, it is estimated that about 300 fatalities of red kites have occurred per year in the Federal State of Brandenburg alone, which is already close to a population relevant threshold (Bellebaum et al. 2013). On the basis of the results of the fatality searches within the PROGRESS project, Potiek and Krüger (2015) show with age- and stage-structured matrix models that the estimated mortality rates derived from the numbers of fatalities found lead to a population decline in the model.

Mitigation practice mainly centres on the distance recommendations (minimum buffer around nest sites = 1.5 km). However this measure is much less effective for Red Kites because of lower breeding site fidelity (alternating use of different nest sites) and higher variability of foraging areas due to changes in agricultural use (Hötker et al. 2013). Consequently, the results of flight activity measurements are less reliable for the assessment of collision risks than in the case of the sea eagle.

Further mitigation measures aim at luring the birds away from wind farms by creation of attractive habitat at a greater distance. Another example is reducing the collision risk inside the wind farm by temporarily shutting down turbines during agricultural harvest.

In the course of ornithological surveys for the growing number of repowering projects, it is not uncommon to find red kites nesting very close to existing wind farms. These birds obviously face an increased collision risk (see also Hötker et al.

2013). So far there is no established practice in Germany to deal with such cases, since all efforts to follow the provisions of the strict species protection in the Nature Conservation Act are focused on the approval procedure, not on the operating phase.

Montagu's Harrier

Currently there are about 500 breeding pairs of Montagu's Harrier in Germany (Gedeon et al. 2015). So far only five fatalities due to wind farms have been documented in Germany, but there have not been any systematic fatality searches in the relevant habitats during the breeding season (Langgemach and Dürr 2015). This is mainly due to vegetation height, especially of winter cereals. Collision risk is generally classified as high in closer proximity to nest sites because of the higher proportion of flight activity around rotor height, and as low during hunting (Hötker et al. 2013; Langgemach and Dürr 2015).

The implementation of buffer zones around known nest sites of Montagu′s Harrier does not make sense within arable land, because nest locations change from year to year following agricultural crop rotation. Correspondingly, extensive flight activity surveys are also meaningless because of the variability of flight patterns according to changes in agricultural use and food availability. So there are only two major mitigation strategies: generally avoid core areas with high breeding density of Montagu's Harriers; and/or follow an adaptive management approach (Köppel et al. 2014) based on monitoring of nest locations during operation of the wind farm. The latter is already established practice in some parts of northern Germany and can be characterized by a combination of measures to keep the Harriers out of the wind farm. The objectives are to enhance breeding success and to avoid obvious high collision risks (Hötker et al. 2013), and the measures include:

- Management of crop selection: discontinue the cropping of winter cereals (preferred breeding habitats) within the wind farm, and crop winter cereals in areas preferably outside the warm farm,
- Management of prey availability: avoid the location of attractive foraging habitats within the wind farm (no grassland, no fallow margins along the roads or around the turbines, and no mowing or harvesting inside the wind farm before outside), and creation of attractive foraging habitat outside the wind farm,
- Monitoring of nest site locations,
- Protection of nest sites against mowing and predators in order to enhance breeding success,
- Temporary shutdown of individual turbines if nest sites are closer than a certain distance, which is to be negotiated with the responsible nature conservation authority.

Common Buzzard

Currently there are about 107,000 breeding pairs of common Buzzards in Germany (Gedeon et al. 2015) and 337 fatalities due to wind farms have been documented so far (Dürr 2015). According to the results of the PROGRESS project the fatality rate of this species is estimated at about 0.5 collisions per turbine per year in the northern half of Germany (Grünkorn 2015). In view of its high breeding density in Germany and its nationwide distribution Buzzard collisions can be expected in more or less any onshore wind farm.

Mitigation for this common species with its unspecific preferences for breeding and foraging habitats is rather difficult, and there is no established practice for dealing with it in Germany. But it has to be expected that the collision rate identified by Grünkorn (2015) leads to population consequences, which might end up in a serious decline (Potiek and Krüger 2015). This is predominantly due to the cumulative effects of the high number of wind turbines in Germany. Distance recommendations with regard to nest sites are not feasible as they would lead to large exclusion zones, which would seriously impede the political objectives for the development of the wind energy in Germany. Moreover they are not feasible due to the high variability in the distribution of nest sites.

The first provisions in some approval documents for wind farms demand the avoidance of attractive foraging habitats within the wind farm, as well as the removal of Buzzard nests (which is controversial) located in close vicinity of the proposed wind farm. The idea is to allow the Buzzard to build a new nest anywhere else rather than run the high risk of it being killed at the wind farm. Temporary shutdowns during certain weather conditions favourable for thermic soaring are currently being discussed.

Skylark

Like the Buzzard, the Skylark is another species which is still common—albeit declining (Sudfeldt et al. 2013)—and affected to a greater or lesser extent by any onshore wind farm in Germany, at least on agricultural land. The number of breeding pairs in Germany is estimated between 1.3 and 2 million (Gedeon et al. 2015). The actual number of recorded fatalities is 87 (Dürr 2015), but this number is highly underestimated due to tall vegetation in the breeding period, the general low detectability of dead Skylarks and the lesser probability of being reported to the national database compared to birds of prey. The Skylark mainly collides during the song flights of the males ranging up into the rotor height (Morinha et al. 2014).

Mitigation mainly consists of trials to lure the larks out of the wind farm by creating attractive habitat, such as so-called "lark windows", which aim to open several patches within extremely dense cereal fields. The problem is that it is not actually possible to prevent skylarks from nesting close to wind turbines in agricultural land. Temporary shutdowns during weather conditions favourable for song flights are currently being discussed in Germany. But the question still remains: Up to which level can the number of collisions of skylarks still be regarded as not significant in light of the current jurisdiction?

Displacement—Species of Concern

Curlew and Lapwing

Among breeding birds it is mainly the group of meadow birds which are thought to be especially affected by displacement. In the case of the Lapwing, extensive studies showed that a significant displacement effect could only be found up to a range of 100 m and that other parameters describing habitat quality have much more influence on the spatial distribution of the breeding pairs. For the Curlew no significant displacement effects could be detected (Reichenbach and Steinborn 2011; Steinborn and Reichenbach 2011; Steinborn et al. 2011). New data from 13 years after the beginning of the study showed that the number of breeding pairs of Curlew still remained constant despite a strong increase in the number of turbines in the study area. In contrast, the number of Lapwings declined by more than 60% (Fig. 2). In view of the results mentioned above this decline is probably caused by habitat deterioration and low breeding success (which was shown not to be influenced by the wind turbines).

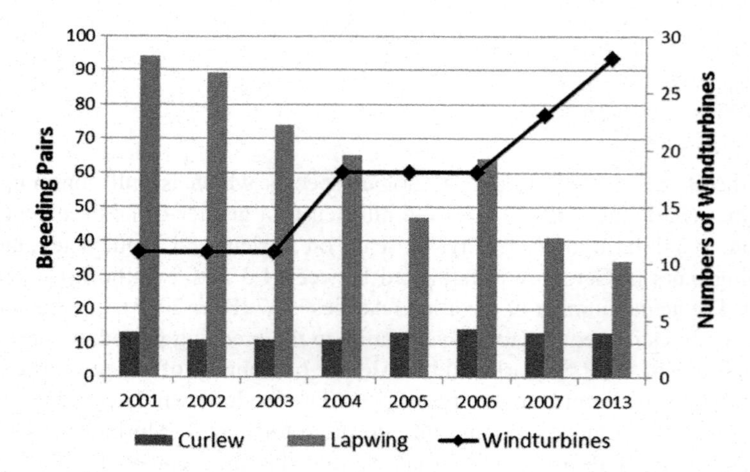

Fig. 2 Breeding population of curlew and lapwing in relation to number of wind turbines in a study area in Eastern Frisia, north-west Germany, over a period of 13 years (own data)

Mitigation measures include avoiding core areas with high breeding density and creating new attractive habitat in the surrounding area. This is undertaken before the installation of turbines, in order to facilitate more effective relocation of affected breeding pairs.

Roosting Geese

About 1.2 million geese come to Germany each year to roost or overwinter (ca. 450,000 greater White-Fronted Geese, ca. 450,000 Bean Geese, ca. 130,000 Greylag and about 200,000 Barnacle geese) (Kruckenberg et al. 2013). These geese are hardly affected by collisions (Douse 2013). However, a range of displacement effects occur up to more than 500 m, depending on flock size (Langgemach and Dürr 2015), which results in a corresponding habitat loss.

Mitigation measures include the avoidance of particularly valuable feeding and roosting areas and maintaining sufficient distances to overnight roost sites. Since wind farms also act as barriers it is important to identify the main flight paths between overnight roosts and the daily feeding areas and keep them free from wind turbines. A BACI-study was conducted for a wind farm in north-west Germany, at a distance of about 1 km from a roost site of more than 20,000 Bean Beese. This was conducted two years before and two years after construction of the wind farm, and it showed that the reduction of the area of the wind farm for the purpose of maintaining a major flight path free from turbines was successful. The geese recognise the wind farm as an obstacle and reorient their flight paths, and some of them do this right from the start—passing north of the turbines towards their feeding grounds (Fig. 3). No negative impact on the number of roosting Bean Geese was detected, as their numbers increased from 20,000 to about 30,000 during the operation of the wind farm.

Roosting Cranes

In the same way as geese, the impact on roosting Cranes due to collisions is negligible with an avoidance distance of several hundred meters to a wind farm, at least in larger flocks (Langgemach and Dürr 2015). In the fen wetland area around Diepholz in north-west Germany, where about 100,000 roosting cranes congregate in the period from October until December, there is growing concern about the cumulative impact of the increasing number of wind turbines. Apart from securing sufficient feeding grounds, which are not a limiting factor due to extensive maize cultivation in the region, the accessibility of the overnight roosts (wetlands and flooded fens) is the major issue. In an impact study for a proposed wind farm close to several other existing wind farms, the daily flight intensity of cranes was quantified using a heat map methodology. It could be shown that not only wind

Fig. 3 Morning flight paths of bean geese from their overnight roost in relation to an existing wind farm in the winter 2005–2006 (own data)

farms but also woodlands act as barriers to the low flying flocks. Gaps between any type of barrier need an effective width of about 1 km to be used as a flight path by bigger flocks of cranes (Fig. 4).

Thus, a major aspect of mitigation is to secure the routes between overnight roosts and feeding areas. Reduction of feeding grounds (usually maize fields) due to displacement effects is relatively straightforward to mitigate by improving the food availability on adjacent fields (e.g. no ploughing after harvest, no scaring off, and shallow flooding of grassland).

The Way Ahead

The handling of possible impacts of wind turbines on birds on different planning levels (regional, local, individual project) is generally guided by the internationally accepted hierarchy of mitigation: avoid—minimize—compensate (May et al. 2015). Therefore, the central question is: What do we need to know in order to be able to

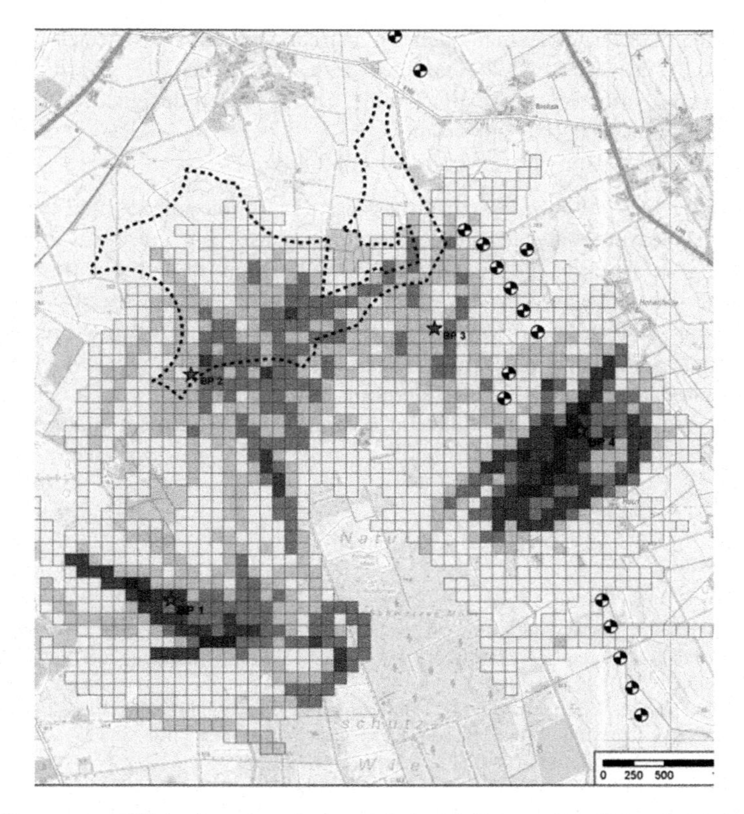

Fig. 4 Heat map of flight intensity of cranes during daily morning flights from the central overnight roost towards surrounding feeding grounds in autumn 2014 (own data)

successfully avoid, minimize and compensate possible impacts on birds and to assess reliably when these measures are really necessary? The concept of the SSS-specificity (site-species-season) by Bevanger (2011) acknowledges the need for a case-by-case approach for each individual wind farm project (Schuster et al. 2015).

In view of the growing numbers of wind turbines in Germany, the adequate assessment of cumulative impacts is a key issue. In addition, there is growing pressure to improve and demonstrate the effectiveness of mitigation measures (Gartman et al. 2016a, b). In this context, the lack of studies on long-term effects, especially on long-lived species, is still a major shortfall.

Consequently, research efforts should be focused on:

- cumulative impacts of multiple wind farms (alone and in combination with other anthropogenic changes) on different geographical levels,
- the effectiveness of mitigation and compensation measures including development of innovate approaches,

- the effectiveness of spatial planning in reducing risks to species of concern, especially in view of ecological variability,
- long-term effects of displacement, using well-designed Before-After-Control-Impact (BACI) approaches,
- Population-level impacts of collision mortality.

Currently, it is common practice to assess the possible impacts and the needs for mitigation measures in the pre-construction phase, but further validation and monitoring is lacking. Post-construction monitoring in combination with any additional measures based on the actual impact and changes in species composition and distribution is still very rare. Therefore, the concept of adaptive planning and management (Köppel et al. 2014) should be evaluated with regard to better mitigation strategies especially in consideration of the long lifetime of a windfarm (ca. 25 years) and the ecological and land-use variability to be expected during this time span.

References

Bach L, Handke K, Sinning F (1999) Einfluss von Windenergieanlagen auf die Verteilung von Brut- und Rastvögeln in Nordwest-Deutschland. Bremer Beiträge für Naturkunde und Naturschutz 4:107–122

Band B, Madders M, Whitfield DP (2007) Developing field and analytical methods to assess avian collision risk at wind farms. In: de lucas castellanos, m.: birds and wind farms—risk assessment and mitigation, Madrid

Bauer HG, Bezzel E, Fiedler W (2005) Das Kompendium der Vögel Mitteleuropas. Gefährdung und Schutz—Nonpasseriformes—Nichtsperlingsvögel, Aula-verlag, Wiebelsheim, Alles über Biologie

Bellebaum J, Korner-Niervergelt F, Dürr T, Mammen U (2013) Wind turbine fatalities approach a level of concern in a raptor population. J Nat Conserv 21(6):394–400

Bevanger K (2011) Wind energy and wildlife impacts—lessons learned from smøla conference on wind energy and wildlife impacts 2–5 May 2011, Trondheim Norway. Available via http://cww2011.nina.no/portals/cww2011/presentations/wind%20energy%20and%20wildlife%20impacts%20-bevanger.pdf?ver=2012-08-01-195453-807

Birdlife international (2015) European red list assessment—milvus milvus. Available via http://www.birdlife.org/datazone/userfiles/file/species/erlob/summarypdfs/22695072_milvus_milvus.pdf. Accessed 19 Oct 2015

Böttger M, Clemens T, Grote G, Hartmann G, Hartwig E, Lammen C, Vauk-hentzelt E, Vauk G (1990) Biologisch-ökologische Begleituntersuchungen zum Bau und Betrieb von Windkraftanlagen. Nna-Berichte 3 (Sonderheft)

Brinkman R, Behr O, Niermann I, Reich M (2011) Entwicklung von Methoden zur Untersuchung und Reduktion des Kollisionsrisikos von Fledermäusen an Onshore-Windenergieanlagen. Umwelt und Raum Band 4, Cuvillier Verlag, Göttingen

Douse A (2013) Guidance: avoidance rates for wintering species of geese in Scotland at onshore wind farms. Scott Nat Heritage 2013:20

Dürr T (2015) Auswirkungen von Windenergieanlagen auf Vögel in Feutschland: Stand 01.06.2015. Available via http://www.lugv.brandenburg.de/cms/media.php/lbm1.a.3310.de/wka_voegel_de.xls. Accessed 11 Sept 2011

Dürr T, Langgemach T (2006) Greifvögel als Opfer von Windkraftanlagen. Populationsökologie Greifvogel- und Eulenarten 5:483–490

Eichhorn M, Johst K, Seppelt R, Drechsler M (2012) Model-based estimation of collision risks of predatory birds with wind turbines. Ecol Soc 17(2):12

Gartman V, Bulling L, Dahmen M, Geißler G, Köppel J (2016a) Mitigation measures for wildlife in wind energy development, consolidating the state of knowledge. J Environ Assess Policy Manage 18(3):1650013-1-45. doi:10.1142/S1464333216500137

Gartman V, Bulling L, Dahmen M, Geißler G, Köppel J (2016b) Mitigation measures for wildlife in wind energy development, consolidating the state of knowledge—Part 2: operation, decommissioning. J Environ Assessment Policy Manage 18(2):1650014-1-1650014-31. doi:10.1142/S1464333216500149

Gedeon K, Grüneberg C, Mitschke A, Sudfeldt C, Eickhorst W, Fischer S, Flade M, Frick S, Geiersberger I, Koop BB, Kramer M, Krüger T, Roth N, Ryslavy T, Stübing S, Sudmann SR, Steffens R, Vökler F, Witt K (2015) Atlas deutscher Brutvogelarten—Atlas of german breeding birds. Herausgegeben von der Stiftung Vogelmonitoring und dem Dachverband deutscher Avifaunisten, Münster

Grünkorn T (2015) A large-scale, multispecies assessment of avian mortality rates at onshore wind turbines in northern Germany (Progress). Conference on wind energy and wildlife impacts (CWW), Berlin

Hötker H (2011) Vögel und regenerative Energiegewinnung. Der Falke 58:484–489

Hötker H, Krone O, Nehls G (2013) Greifvögel und Windkraftanlagen: Problemanalyse und Lösungsvorschläge. Schlussbericht für das Bundesministerium für Umwelt, Naturschutz und Reaktorsicherheit. Michael-Otto-Institut im Nabu, Leibnitz-Institut für Zoo- und Wildtierforschung, Bioconsult SH, Bergenhusen, Berlin, Husum

Hötker H, Thomsen KM, Köster H (2004) Auswirkungen regenerativer Energiegewinnung auf die biologische Vielfalt am Beispiel der Vögel und der Fledermäuse—Fakten, Wissenslücken, Anforderungen an die Forschung, ornithologische Kriterien zum Ausbau von regenerativen Energiegewinnungsformen. Nabu, gefördert vom Bundesamt für Naturschutz 80

Köppel J, Dahmen M, Helfrich J, Schuster E, Bulling L (2014) Cautious but committed: moving toward adaptive planning and operation strategies for renewable energy's wildlife implications. Environ Manage 54:744–755

Kruckenberg H, Mooij JH, Südbeck P, Heinicke T (2013) Die internationale Verantwortung Deutschlands für den Schutz arktischer und nordischer Wildgänse, Teil 1: Verbreitung der Arten in Deutschland. Naturschutz und Landschaftsplanung 43(11):334–342

LAG VSW (Länder-Arbeitsgemeinschaft der Vogelschutzwarten lag-vsw) (2007) Abstandsregelungen für Windenergieanlagen zu bedeutsamen Vogellebensräumen sowie Brutplätzen ausgewählter Bogelarten. Berichte zum Vogelschutz 44:151–153

LAG VSW (Länderarbeitsgemeinschaft der staatlichen Vogelschutzwarten in Deutschland) (2015) Abstandsempfehlungen für Windenergieanlagen zu bedeutsamen Vogellebensräumen sowie Brutplätzen ausgewählter Vogelarten in der Überarbeitung vom 155:29

Langgemach T, Dürr T (2015) Informationen über Winflüsse der Windenergienutzung auf Vögel - Stand 01. Juni 2015. Landesamt für Umwelt, Gesundheit und Verbraucherschutz, Staatliche Vogelschutzwarte

May R, Reitan O, Bevanger K, Lorentsen SH, Nygård T (2015) Mitigating wind-turbine induced avian mortality: sensory, aerodynamic and cognitive constraints and options. Renew Sustain Energy Rev 42:170–181

MELUR & LLUR (Ministerium für Energiewende, Landwirtschaft, Umwelt und ländliche Räume des Landes Schleswig-Holstein, Landesamt für Landwirtschaft, Umwelt und ländliche Räume des Landes Schleswig-Holstein) (2013) Errichung von Windenergieanlagen (WEA) innerhalb der Abstandsgrenzen der sogenannten potenziellen Beeinträchtigungsbereiche bei einigen sensiblen Großvogelarten—Empfehlungen für artenschutzfachliche Beiträge im Rahmen der Errichtung von WEA in Windeignungsräumen mit entsprechenden artenschutzrechtlichen Vorbehalten

Morinha F, Travassos P, Seixas F, Martins A, Bastos R, Carvalho D, Magalhaes, Santos PM, Bastos E, Cabral JA (2014) Differential mortality of birds killed at wind farms in northern portugal. Bird study 1–5. doi:10.1080/00063657.2014.883357

NLT (Niedersächsischer Landkreistag) (2014) Naturschutz und Windenergie—Hinweise zur Berücksichtigung des Naturschutzes und der Landschaftspflege bei Standortplanung und Zulassung von Windenergieanlagen (Stand: Oktober 2014). Hannover, 37 s. Available via http://www.nlt.de/pics/medien/1_1414133175/2014_10_01_arbeitshilfe_naturschutz_und_windenergie_5_auflage_stand_oktober_2014_arbeitshilfe.pdf

Potiek A, Krüger O (2015) Effects of collisions with wind turbines for population trends of three long-lived raptor species poster presentation at the CWW 2015—Conference on wind energy and wildlife impacts, Berlin

Reichenbach M (2003) Auswirkungen von Windenergieanlagen auf Vögel—Ausmaß und planerische Bewältigung. Dissertation. Landschaftsentwicklung und umweltforschung—Schriftenreihe der Fakultät Architektur Umwelt Gesellschaft, Technische Universität, Berlin

Reichenbach M, Handke K, Sinning F (2004) Der Stand des Wissens zur Empfindlichkeit von Vogelarten gegenüber Störungswirkungen von Windenergieanlagen. Bremer Beiträge für Naturkunde und Naturschutz 7:229–243

Reichenbach M, Steinborn H (2011) Wind turbines and meadow birds in germany—results of a 7 year baci-study and a literature review. Presentation held at the "conference on wind energy and wildlife impacts", Trondheim, Norway, 2–5 May 2011. Available via http://www.cww2011.nina.no/linkclick.aspx?fileticket=5vx5zl4qyqa%3d&tabid=3989

Schuster E, Bulling L, Köppel J (2015) Consolidating the state of knowledge: a synoptical review of wind endergy′s wildlife effects. Environ Manage 56(2):300–331

SNH (2014) Guidance: Recommended bird survey methods to inform impact assessment of onshore wind farms. 27 S. Available via http://www.snh.gov.uk/docs/c278917.pdf

Steinborn H, Reichenbach M (2011) Kiebitz und Windkraftanlagen—Ergebnisse aus einer siebenjährigen Studie im südlichen Ostfriesland. Naturschutz und Landschaftsplanung 43 (9):261–270

Steinborn H, Reichenbach H, Timmermann H (2011) Windkraft—Vögel—Lebensräume Ergebnisse einer siebenjährigen Studie zum Einfluss von Windkraftanlagen und Habitatparametern auf Wiesenvögel. Books on Demand, Norderstedt

Sudfeldt C, Dröschmeister R, Frederking W, Gedeon K, Gerlach B, Grüneberg C, Karthäuser J, Langgemach T, Schuster B, Trautmann S, Wahl J (2013) Vögel in Deutschland 2013

Weitekamp S, Timmermann H, Reichenbach H (2015) Progress—predictive modelling versus empirical data—collision numbers in relation to flight activity in 55 german wind farm seasons. Conference on wind energy and wildlife impacts (CWW), Berlin

Windguard (2015) Status des Windenergieausbaus an Land in Deutschland. 1. Halbjahr 2015 (30.06.2015). Available via http://www.windguard.de/_resources/persistent/b6ff13ecabb86fbbdd45851e498d686432a81a2c/factsheet-status-windenergieausbau-an-land-1.-halbj.-2015.pdf. Accessed on 11 Oct 2015

Part VII
Future Research and Knowledge Platforms

Future Research Directions to Reconcile Wind Turbine–Wildlife Interactions

Roel May, Andrew B. Gill, Johann Köppel, Rowena H.W. Langston, Marc Reichenbach, Meike Scheidat, Shawn Smallwood, Christian C. Voigt, Ommo Hüppop and Michelle Portman

Abstract Concurrent with the development of wind energy, research activity on wind energy generation and wildlife has evolved significantly during the last decade. This chapter presents an overview of remaining key knowledge gaps, consequent future research directions and their significance for management and planning for wind energy generation. The impacts of wind farms on wildlife are generally site-, species- and season-specific and related management strategies and practices may differ considerably between countries. These differences acknowledge the need to consider potential wildlife impacts for each wind farm project. Still, the ecological mechanisms guiding species' responses and potential

R. May (✉)
Norwegian Institute for Nature Research, Trondheim, Norway
e-mail: roel.may@nina.no

A.B. Gill
COREE, Cranfield University, Cranfield, UK

J. Köppel
Berlin Institute of Technology, Berlin, Germany

R.H.W. Langston
RSPB Centre for Conservation Science, Royal Society for the Protection of Birds, Sandy, UK

M. Reichenbach
ARSU GmbH, Oldenburg, Germany

M. Scheidat
IMARES Wageningen UR, IJmuiden, The Netherlands

S. Smallwood
Davis, CA, USA

C.C. Voigt
Leibniz Institute for Zoo and Wildlife Research, Berlin, Germany

O. Hüppop
Institute of Avian Research, Wilhelmshaven, Germany

M. Portman
Technion Israel Institute for Technology, Haifa, Israel

J. Köppel (ed.), *Wind Energy and Wildlife Interactions*,
DOI 10.1007/978-3-319-51272-3_15

vulnerability to wind farms can be expected to be fundamental in nature. A more cohesive understanding of the causes, patterns, mechanisms, and consequences of animal movement decisions will thereby facilitate successful mitigation of impacts. This requires planning approaches that implement the mitigation hierarchy effectively to reduce risks to species of concern. At larger geographical scales, population-level and cumulative impacts of multiple wind farms (and other anthropogenic activity) need to be addressed. This requires longitudinal and multiple-site studies to identify species-specific traits that influence risk of mortality, notably from collision with wind turbines, disturbance or barrier effects. In addition, appropriate pre- and post-construction monitoring techniques must be utilized. Predictive modelling to forecast risk, while tackling spatio-temporal variability, can guide the mitigation of wildlife impacts at wind farms.

Keywords Future research directions · Impacts of wind farms · Wildlife · Animal movement decisions · Mitigation hierarchy

Introduction

Reducing emissions of greenhouse gases to prevent anthropogenic climate change has boosted the innovation, development and application of renewable energy sources such as wind. At the same time, environmental and social issues will affect wind energy development opportunities (IPCC 2011). As wind energy development increases and larger wind farms are considered, existing concerns become more acute and new concerns may arise. Depending on the planned level of development, a need to reconcile renewable energy targets and biodiversity conservation will emerge, to ensure the lowest possible environmental costs per kWh (cf. van Kuik et al. 2016). This in turn requires comprehensive insight into potential effects of wind farms on wildlife such as disturbance, habitat loss and mortality, impacts on a population level (cf. Boehlert and Gill 2010), and innovative measures to mitigate these impacts. Forthwith we use the term "wind turbine–wildlife interactions", to capture all interactions wildlife species may have with wind turbines, associated infrastructure and human activity within the wind farm area throughout its entire life cycle.

Since the early 2000s, the number of peer-reviewed publications on wind energy generation and wildlife impacts has increased more than tenfold (Fig. 1). Especially after the first Conference on Wind energy and Wildlife impacts in 2011 there was a threefold increase in the number of publications. For example, while studies on behavioural responses of wildlife to wind farms increased from around 2005, mitigation studies started to become more common from 2010 onwards. A similar trend was also seen in the programs of the consecutive Conferences on Wind energy and Wildlife impacts. Reviews synthesizing the current knowledge have furthered our understanding of the behavioural and ecological mechanisms guiding

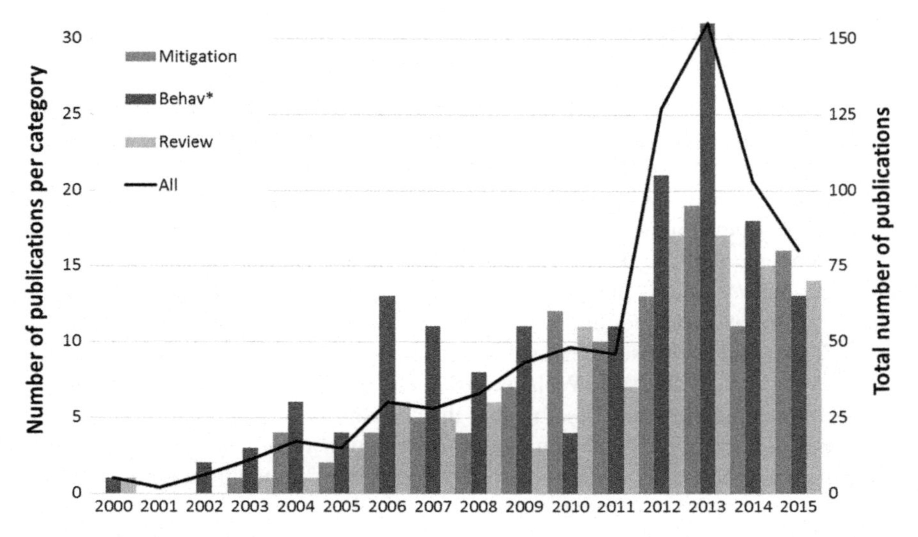

Fig. 1 Number of peer-reviewed publications recorded in the wind-wildlife impacts literature database (https://wild.nrel.gov/) maintained by National Renewable Energy Laboratory (NREL) within the period 2000–2015 (*black line*; *right* y-axis). *Stacked columns* indicate the number of publications within specific categories (*left* y-axis): publications focusing on mitigation (*orange*) or behaviour (*blue*), and review publications (*green*) (Color figure online)

species' responses and potential vulnerability to wind farms (Cryan and Barclay 2009; Inger et al. 2009; Marques et al. 2014; May 2015; Schuster et al. 2015). The enormous increase in publications on wind turbine–wildlife interactions during the last decade necessitates an evaluation of the current knowledge on these topics, especially with regard to underexposed topics. This chapter presents an overview of key knowledge gaps, consequent future research directions (see also Gill 2005; Kunz et al. 2007; Wang et al. 2015) and their significance for management and planning for wind energy generation.

Setting the Stage: Planning for Wind Energy Development

Improved understanding of the implications of wind energy for the environment and adequate implementation of mitigation efforts, require transdisciplinary approaches that are embedded in complex decision-making processes. The discussion of social acceptance in wind energy development as well as the trade-offs between ecosystems services (e.g. climate change mitigation vs. biodiversity benefits) have become paramount research challenges. The boundaries between science and policy in wind energy and wildlife research need to be considered more systematically, as science-policy transitions might prove further decisive to future research.

Green Versus Green Ethics

The rapid rate of wind energy development throughout the world has simultaneously led to stronger opinions from both opponents and proponents of wind energy development (IPCC 2011; Wolsink 2012). This requires considering arguments of 'green versus green' environmentalism, where proponents promote the benefits of wind energy development in reducing CO_2 emissions to mitigate climate change, and opponents point to the costs involved for biodiversity and ecosystem services through land-/seascape changes (Warren et al. 2005). The problem here not only lies in a simplistic misjudging of the biodiversity issue as NIMBYism (Not-In-My-Back-Yard), but may also be related to various forms of resistance and institutional settings (Cowell et al. 2011; Wolsink 2012; Huesca-Pérez et al. 2016). Differences between beneficiaries and those bearing the costs may thereby cause challenges for planning. This is due to uneven spatial distribution of equity, inter-generational dimensions (decisions made today affecting future generations) and scientific uncertainty of impacts (Gardiner 2011).

In addition, when decision-making is valued within an ethical framework of economic rationality (Cowell et al. 2011), short-term economic benefits and local societal costs often take precedence over environmental considerations (Kopnina 2013). Environmental impacts may then be perceived as a technological problem or economic cost that can be avoided, minimized, mitigated or compensated to reach no net loss (Wolsink 2012). On the one hand, given the benefits of wind energy development for climate change mitigation in the longer term, a case might be made for a certain level of acceptable environmental impact. On the other hand, given the benefits of efficiently implementing the mitigation hierarchy during wind farm development and the resulting long-term benefits for biodiversity, a case might be made for stronger restrictions in the planning of future wind energy projects. To address these spatio-temporal ethical challenges further research is required to support political decision-making processes and planning strategies.

Transdisciplinary Decision-Making

Top-down siting and consenting processes, coupled with mistrust among stakeholders and institutional settings, impede the ability to appropriately site wind farms in locations with acceptable impacts (Warren et al. 2005). The clue in reconciling arguments for and against wind energy development lies therefore in scaling decision-making processes and strategic planning to intermediate spatial and temporal scales. By assessing potential impacts (environmental or societal) of a specific project at a regional spatial scale transcending its life cycle, i.e. above the local and immediate opponent scale but below the national/global and diffuse proponent scale. Both the proponent and opponents must then upscale or down-scale their arguments to fit the planning level (i.e. 'think global, act local…but plan regional', cf. Warren et al. 2005). This requires that strategic environmental assessments are prioritized to assess all alternative strategies, are taken up in the decision-making process, and are tied to Environmental Impact Assessment (EIA) practice (Jay 2010; Geißler 2013; Geißler et al. 2013; Phylip-Jones and Fischer 2014).

Several authors have also stressed the need for collaborative and transdisciplinary strategic planning processes and transparent decision-making (Warren et al. 2005; Cowell et al. 2011; Wolsink 2012; Petrova 2016). The issues related to wind energy development and wildlife impacts have been and will continue to be addressed by policy-makers, regulatory agencies, industry, non-governmental organizations, and the scientific community. Perspectives of these groups on such issues vary, as do motivations, power, consistency and levels of engagement. The role of scientists in the science–policy–practice interface is to provide evidence-based and policy-relevant information upon which transparent decisions can be based. To improve this transdisciplinary interface, further research is required to evaluate approaches that may enhance participation and transparency in decision-making processes.

Planning and Management Approaches and Regulations

Planning and management of wind energy projects vary considerably among countries, and have evolved over the years. While early wind energy projects were constructed without clear strategic planning requirements, the monitoring and mitigation guidelines in permitting processes have recently become more commonplace. Implications for relevant regulations in relation to wind and wildlife issues also advance as more knowledge and experience is gained. These implications include, for example, problems associated with take permits, application of the precautionary principle, and transboundary effects (Voigt et al. 2012; Köppel et al. 2014). Comparative research will be needed to assess how various planning and management strategies are able to address human–wildlife conflicts relating to wind energy development. Such an evaluation should identify key planning components that moderate conflict levels, reduce uncertainty, and avoid delays in consenting processes. Investigations could be directed toward the consequences of regulations and guidelines that were prepared with more or less adaptive approaches (Köppel et al. 2014). Yet, the practicability and effectiveness of such adaptive approaches remain to be tested.

A relatively new opportunity for addressing offshore wind energy projects and their impacts on wildlife in particular is available through marine spatial planning (MSP). An adaptive approach can also be adopted for marine planning where it has the potential to reduce the loss of ecosystem services, help address or avoid conflict, and create economies of scale and efficiencies in enforcement and management (Portman 2015). MSP is a process that aims to rationalize the use of marine space and identify the compatibility between activities. This facilitates the identification of conflicts and synergies between uses. Such an approach incorporated in an MSP process could contribute to the integration of offshore wind energy projects and other co-uses while minimizing impacts to wildlife (Portman 2011).

The Mitigation Hierarchy

Reconciling wind energy development with conservation of the environment necessitates that mitigation measures are implemented to attempt to eliminate negative impacts caused throughout the life cycle of a wind farm. Implementation of effective and practical measures to mitigate impacts is paramount in order to achieve climate-change mitigation goals whilst protecting biodiversity (Madsen et al. 2006; Marques et al. 2014; May et al. 2015; Peste et al. 2015; Arnett and May 2016). Mitigation simultaneously decreases the general level of conflicts with wildlife and enables development at sites previously considered to pose too great a risk. The (proposed) mitigation of environmental impacts is a key stage within EIA process, where developers are to mitigate impacts following the so-called 'mitigation hierarchy'.

The prioritized steps of the mitigation hierarchy are tiered to the consecutive decision gates required for wind farm development. This tiered approach ensures that mitigation decisions are taken prior to the appropriate development phase when they are to be implemented. The hierarchy is as follows. (1) Impacts should foremost be avoided when planning prior to siting. (2) Unavoidable impacts should minimized during the design phase prior to construction. (3) Measures to further reduce impacts should be implemented during construction prior to operation. (4) Any residual impacts should thereafter be compensated during operation. (5) At the end-of-life of a wind farm, the area should be restored as part of decommissioning (May 2016).

Avoid: Consensus-Based Siting Approaches for Improved EIAs

Spatial planning that informs all involved parties on the wildlife species likely to be affected by wind energy projects is essential for avoiding these impacts through careful siting (e.g. Garthe and Hüppop 2004). Pre-construction sensitivity maps for locational guidance (e.g. Bright et al. 2008; Bradbury et al. 2014) can provide important input to multiple-criteria assessments for siting of wind farms (e.g. Tsoutsos et al. 2015). Geographic Information Systems (GIS) are valuable tools for this purpose, enabling different data layers to be overlaid with any proposed wind farm boundary, as part of the risk assessment. This requires up-to-date knowledge on the distribution and ecology of potentially affected species at appropriate geographic scales (Hammond et al. 2013). Pre-construction surveys, designed to collect data to record occurrence of sensitive species and predict impacts, are an essential part of the EIA procedure. Such surveys should be made available to enable access to cost-effective, incremental and updated information on the distribution of sensitive species and their habitats across wind energy projects. For migratory and highly-mobile species that shift their distributions to adapt to changes in prey availability or season (e.g. Baerwald and Barclay 2009; Hammond et al. 2013), there exists a need to increase our understanding on what drives their distribution to predict these shifts. For species that are mostly affected during the construction phase, such as Harbour porpoises (Madsen et al. 2006) or red grouse (Pearce-Higgins et al. 2012), prediction of high-density habitats and high-sensitivity time periods, such as breeding and nursing periods, will be crucial.

Another important issue with regard to the avoidance-effectiveness of spatial planning is data availability across spatial scales. Most surveys are done at the individual project level, but compilation of such fine-scale information on the occurrence and spatio-temporal distribution of relevant species is limited due to a lack of standardization and spatial coverage. Regional spatial planning can therefore only utilize available regional or countrywide occurrence data resulting in a higher degree of uncertainty (e.g. no-data vs. absence of occurrence). Future research should develop a systematic approach for the respective data requirements at the different spatial scales to enhance the effectiveness of sensitivity mapping (cf. van Kuik et al. 2016). This will help to avoid the licensing of wind energy projects in important areas for species of conservation concern (e.g. Bright et al. 2008).

Minimize: Project-Level Siting and Micro-Siting Tools

During the pre-construction design phase for licensed wind energy projects, but also when repowering, potential impacts can be minimized by adjusting the ecological footprint of wind farms or single wind turbines. Measures to minimise impacts through adjustments in turbine configuration, wind farm design and micro-siting are aimed at decreasing the potential hazard or exposure of the turbines to wildlife (May et al. 2015). Vertical axis wind turbines have recently been promoted as being environmentally-friendly (Islam et al. 2013), however the scientific evidence for this technology is still lacking (Santangeli and Katzner 2015). Utilizing fewer and larger turbines that are placed farther apart may contribute to reducing collision risk (Smallwood et al. 2009; Dahl et al. 2015), yet empirical data also suggests disproportionally higher fatalities rates for bats at larger wind turbines (Barclay et al. 2007).

Micro-siting of turbines within the landscape during the pre-construction design phase can optimize wind capture whilst simultaneously taking into account areas of high bird concentration or sites with increased collision risk (e.g. Bohrer et al. 2013). Although predictive tools have been developed to provide insights into possible impacts of different wind farm designs (e.g. Masden et al. 2012), it remains unclear to what extent micro-siting practice has actually resulted in adjusted design of wind farms. This requires that both wind engineers learn how siting decisions relate to collision impacts, and environmental scientists understand the impacts of siting on wind energy generation. There is great opportunity for siting to simultaneously optimize wind energy generation and mitigate wildlife impacts. To validate predicted risk zones in operational wind farms, comparative research is required to investigate how turbine configurations and wind farm designs are affecting wildlife. By employing e.g. meta-analyses, actual risk caused by disturbance potential, barrier effects and collision risk can be evaluated against siting options. This would require access to data from commissioned monitoring and technical data on energy yield of wind turbines at consented sites. Finally, development of guidelines for environment-friendly construction as well as limiting construction and maintenance activity in sensitive periods may contribute to reduced disturbance potential.

Reduce: In Situ-Innovative Techniques

The efficacy of post-construction measures to reduce impacts varies across taxa and geographic region (Madsen et al. 2006; Marques et al. 2014; May et al. 2015; Arnett and May 2016). Many impact reduction measures have been proposed, but only few have been tested and found to be effective (e.g. cut-in speeds for bats: Baerwald et al. 2009; Arnett et al. 2011). To test the effectiveness of onsite impact-reduction measures satisfactorily, tests should be consistent with experimental design principles. The five most important principles include clear articulation of the hypothesis associated with the impact reduction strategy, use of controls, replication and interspersion of treatments, and implementation at appropriate temporal and spatial scales. In addition, sample sizes of experimental units need to be decided in order to obtain a suitable effect size. Acoustic deterrence of marine mammals during pile-driving have so far shown varying results (Madsen et al. 2006). Further research needs to focus on ways to reduce the emitted noise from construction activities (e.g. bubble curtains, mode of operation).

Acoustic and visual deterrence, with or without detection systems, has also been proposed for birds and bats, although their efficacy, and level of habituation, has yet to be tested in situ. Similarly, more in situ testing is needed on wind-turbine design modifications (e.g. blade painting or safety illumination schemes). Curtailment has been shown to be effective in reducing bat fatalities (e.g. Arnett et al. 2011), but limited evidence exists in support of curtailment as a strategy to reduce bird fatalities (de Lucas et al. 2012). However, temporary shutdown of turbines has potential as long as effective algorithms can be developed to restrict shutdown to specific events of near-collisions (May et al. 2015). One research track that has received less attention so far is ecological management strategies to reduce wildlife impacts. On-site and/or off-site habitat management may dissuade wildlife to be attracted to the wind turbines and/or lure them away towards improved habitat refuges outside the wind farm site.

Compensate: Offsetting Methodology, Options and Tools

Compensation is still in its infancy with regard to offsetting residual impacts during a wind farm's life cycle. It requires good knowledge of the magnitude of population-level impacts and of the extent that the previous steps of the mitigation hierarchy have lessened the impact. There is also a pressing need to understand which compensation strategies may benefit wildlife species most, if at all, and which scaling methodology may be most appropriate to use (May 2016). However, also more fundamental questions need to be addressed, related to like-for-like compensation and scaling issues. Dependent on the nature of the impact (e.g. fatalities versus displacement), off-site habitat enhancement or out-of-kind compensation options may or may not be able to offset impacts. This will depend on the direct and indirect benefits the measure has on the species' demography.

Compensation may be further hampered by scaling issues when sufficient action can only be obtained over large geographical areas. Creation of artificial reefs, de facto refuges from fishing, and re-establishment of the undisturbed seabed in offshore wind farms (Wilson and Elliott 2009) could be considered compensation for potential impacts, however, only of a temporal nature. The spatial origin of the impact will also be a prerequisite for effective compensation as affected animals may originate from large geographic areas (Hüppop et al. 2006; Voigt et al. 2012; Hammond et al. 2013). As a last resort, research programmes should consider the potentials and drawbacks of compensation measures in areas other than the wind farm site, taking into account the like-for-like habitat paradigm.

Restore: Best-Practice from Planning to Decommissioning

At the end of the life of a wind farm decommissioning should ensure that the original status of a wind farm area is restored. Although best-practice guidance exists (Welstead et al. 2013), the long-term ecological restoration of the wind farm area is not regarded as an issue during planning. Life cycle assessment and environmental legislation do require consideration of decommissioning so this will require further research. More operational knowledge will be required on the long-term efficacy of restoration with regard to wind farms, including but not limited to removal of infrastructure, hydrology, vegetation re-establishment and ecosystem recovery (May 2016). What will also be an important facet is how shifting baselines affect restoration options as the surrounding landscape may have undergone more or less permanent changes during the 25-year operational phase of a wind farm. In the offshore environment, decommissioning of e.g. oil and gas platforms pre-suppose that the natural processes of the dynamic environment enable recovery (Schroeder and Love 2004), however how comparable these findings are is unknown. The consequences of artificial reef creation and recovery of the 'natural state' of the seabed due to reduced fishing within offshore wind farms for long-term ecological functioning of such artificially altered ecosystems provide a future research area (Inger et al. 2009; Wilson and Elliott 2009). Restoration will likely become more topical in the coming decade as more and more wind farms will reach their end-of-life and require decommissioning.

Species-Specific Responses to Wind Farms

Wildlife responses to wind turbines are highly variable and often very species-specific (Drewitt and Langston 2006). This variability arises from a range of behavioural, ecological and environmental risk factors (Box 1). There still remain significant gaps in our understanding on the relative importance of these risk factors and the underlying mechanisms that trigger responses (May 2015). Priorities for EIA, monitoring and research tend to focus on endangered species and species of conservation concern; however, these may not always be the most tractable species to study. The main effects arising from the construction and operation of wind farms on wildlife are additional mortality due to (sub)lethal injury, disturbance and displacement, as well as loss of habitat (Schuster et al. 2015). The empirical basis for understanding these effects, their species-specific responses, and the likelihood of impacts, is variable in both quality and quantity.

Box 1 *Cross-taxa ecological and environmental risk factors*

- life-history traits
- manoeuvrability, physiology
- behavioural patterns, movement behaviour
- seasonality and utilisation of space
- habitat preferences and connectivity
- topographic terrain/substrate
- weather conditions
- wind farm operation characteristics
- background anthropogenic effects

Understanding Movement Behaviour, Habitat Preferences and Connectivity

An improved understanding of the movement ecology of species in conflict with wind turbines are of eminent need, for mobile and migratory species but also for stationary species that may not be able to relocate. Important aspects of movement ecology include movement behaviour, habitat preferences and utilization, and connectivity with respect to potential interactions with wind farms. Although there have been in-depth studies of some species of concern, there remain key gaps in understanding. Inter-annual, seasonal and diurnal cycles notably require long-term studies to reveal temporal variability in ecology and behaviour. Furthermore, the effect that availability of alternative habitat and refuges may have on movement behavioural responses at wind farms are important to explore. Multiple-scale studies, connecting fine-scale movement behaviour to habitat associations, landscape connectivity and ultimately its consequences for populations are essential.

Research is also required on the proximate and ultimate causes of movement behaviour, focusing on how and why specific areas are utilized. From this perspective, obtaining increased insight into the morphological (e.g. wing morphology), physiological and cognitive mechanisms underlying movement decisions will be crucial (Nathan et al. 2008). How wind-turbine turbulence affects birds and bats with different aerodynamic and cognitive capabilities also requires further research. Greater understanding of wildlife behaviour and perception of wind farms (Martin 2012; Tougaard et al. 2015) will improve objective assessments of risk and assist with developing mitigation measures. To advance our knowledge base, a combination of in-depth and long-term studies is required on model species, theoretical reviews as well as meta-analyses. Together this will improve the prospects of discerning what aspects of a species' ecology and behaviour increase interactions with wind farms, leading to potential impacts, and may inform effective mitigation measures (May et al. 2015).

Understanding Avoidance/Attraction Mechanisms

Wildlife may respond to wind turbine-induced effects through fleeing, activity shifts or changed habitat utilization (either increased or decreased); usually termed avoidance/attraction. An increasing number of empirical studies have improved our understanding of avoidance, although significant knowledge gaps remain. Formalizing the different forms of avoidance facilitates the design of avoidance studies and ensures that all associated predictions are considered à priori. This in turn helps to minimize modelling bias in predictive risk models and enhances the potential for comparison across sites (May 2015). The effects of human-made aerial structures on birds and bats are not yet well understood (Drewitt and Langston 2008; Cryan et al. 2014; Walters et al. 2014) and subsea changes that occur are only just starting to be understood (Lindeboom et al. 2015). Studies teasing apart the relative impacts derived from wind turbine structures versus other features, such as vehicle/vessel movements associated with maintenance activities and powerlines or subsea cables, are therefore required. Disturbance of wildlife can occur at any stage during the lifetime of a wind energy project, indicating the need for preconstruction, during construction and post-construction studies (Pearce-Higgins et al. 2012). Displacement causes functional habitat loss, which may be total or partial, temporary or long-term. Whether or not displacement has population consequences will depend on a combination of the availability of alternative habitat, duration and magnitude of displacement, and the consequences for survival and productivity, all of which justify further study (Gill et al. 2001; May 2015). These aspects have so far not been addressed in studies related to the extent of displacement (May 2015).

Conversely, species may habituate (e.g. Madsen and Boertmann 2008) or even be attracted to wind turbines. Some species will actively associate with the wind-turbine structure and foundations and include them as alternative habitats to move between (for examples see: Schuster et al. 2015). Why and under which circumstances habituation or attraction may occur should be the focus of longitudinal studies. Whether wind farms may create novel communities (e.g. artificial reefs, foraging habitat) or encourage redistribution is important for both the species forming these assemblages and for predators that aggregate there. Information about the causes and consequences of barrier effects is sparse, but can be related to avoidance of structures, noise or electromagnetic fields from subsea electrical cables leading to increased energy expenditure and loss of connectivity (Masden et al. 2010b; Gill et al. 2012).

Effects of Noise

The construction of offshore wind farms, and in particular the noise generated during pile-driving, has been identified as adversely affecting the behaviour of marine mammals and fish, including displacement (Gill et al. 2012; Tougaard et al. 2015). Other hypothesized effects of sound introduced into the water require further research, including masking of communications, increased stress levels leading to reduced fitness and the occurrence of temporary or permanent threshold shifts in hearing. Operational noise levels are very unlikely to lead to injury to cetaceans or seals and there is no indication that they will lead to avoidance behaviour. Impact studies have demonstrated that the effects of sound on marine mammals range from negative impact, no change to an increase in abundance according to location (Scheidat et al. 2011), although the sound source and characteristic of sound may be similar. From this perspective, it will be important to get a better understanding on low-frequency noise-propagation conditions (e.g. water depth, seabed substrate, currents) and other anthropogenic noise-source properties (Madsen et al. 2006). Also, there is a great lack of understanding how these effects on individuals translate to a potential population impact. In planning future wind farms, this is a vital point to investigate, as it stresses the fact that results from one wind farm are not necessarily transferable to other wind farms located in different physical environments. The noise impact of construction and/or operation activities for other species such as birds (stress, displacement) or fish (barotrauma, behavioural changes) seems less evident (Kight and Swaddle 2011; Francis and Barber 2013).

Monitoring Risk for Planning and Management

Regulatory permitting processes typically involve, as part of an EIA, an assessment of the risks the development pose to wildlife. Although such risk assessments are often based on the limited information available pre-construction, they form the basis for consenting decisions comprising the entire life cycle of a wind farm. Recognizing and properly addressing the variability and uncertainty in risk assessments will contribute to evidence-based decision-making processes. This requires novel research designs and techniques, advancements on the strengths and weaknesses of predictive modelling, and tackling the spatio-temporal challenges in risk assessments.

Design and Techniques for Targeted Impact Monitoring

As can be deduced from the manifold of research themes, scientific investigation is needed using the appropriate metrics and methodology to answer clearly formulated hypotheses. Relevant metrics include fatality rates, area utilization, behavioural patterns, increased energy budgets, breeding success and survival (including density dependent effects). An often-overlooked candidate metric is social interactions in the vicinity to wind turbines. In particular, there remains a need for multiple-site longitudinal studies tailored to targeted individual species, which should apply well-designed Before-After-Control-Impact (BACI) or gradient methods. This allows for differentiation of short-term and long-term effects and to investigate the extent to which habituation may occur (Stewart et al. 2007). Assessing population-level consequences and effects on migratory species especially require long-term studies at larger geographic scales (e.g. countrywide, migratory flyways). Appropriate measuring of impacts caused by wind turbines is critical for comparing impacts among projects, wind-turbine attributes and sites, and mitigation strategies. This requires standardization of field and analytical methods. Future research must be directed towards reducing uncertainty in the type, magnitude and duration of effects, and their population consequences. In addition, research is needed on the strengths and weaknesses of various techniques such as observational, telemetry, radar, acoustics, video and thermal imaging for obtaining useful data with regard to accuracy, precision, consistency, completeness as well as species-specificity (Exo et al. 2003).

Animal-borne tracking devices, such as GPS tags, will become increasingly important for understanding behavioural responses to wind turbines and consequent collision risk/avoidance, including providing measurements of flight height and flight speed (e.g. Cleasby et al. 2015; Thaxter et al. 2015). Most evidence for collisions with wind turbines comes from onshore studies of birds and bats. Information is notably limited at offshore wind farms. There are pronounced technical challenges to overcome in order to obtain empirical data on collisions at offshore wind turbines, requiring innovative fatality detection technology. These technologies include the use of high-resolution animal tracking, drones with pre-defined flight patterns and high-resolution video cameras, and accelerometers in rotor blades to detect impacts. Research is also needed to evaluate and standardize monitoring procedures, including effects of study duration on patterns in occurrence, fatality rates, search interval, search radius, carcass removal and detection trials (Bernardino et al. 2013). Standardized fatality rate estimates are most valuable for comparative purposes (Loss et al. 2013; Smallwood 2013) to understand site or wind turbine attributes that contribute most to fatality rates and to direct mitigation measures.

Predictive Modelling and Forecasting Risk

Although it will be crucial to increase our knowledge base on the ecology of wind turbine–wildlife interactions, consenting decisions are based on current knowledge and pre-construction data coupled with predicted effects on species of concern. To enable consenting authorities to make evidence-based decisions, pre-construction monitoring needs to be well-designed in order to limit variability across projects as much as possible, as well as to attain the required power-of-analyses. Statistical testing should preferably give insight into both the magnitude and the likelihood of the estimated effects. Predictive modelling should explicitly be clear on assumptions of the chosen model as well as on the uncertainty relating to both data quality and model outcomes (Masden and Cook 2016).

Predictive models come in various forms and functions, such as collision risk models, habitat-based movement modelling, population modelling and individual-based models. Dependent on the model of choice, specific data is required to understand the impact metric. As yet there is only limited insight into how the forecasting of risk should be performed, including spatial and temporal scales, and especially how science can inform policy makers to set risk thresholds. Development of individual-based models simulating species-specific movement behaviour and responses to wind turbines are appealing due to their universal applicability. However, data scarcity for parameterization and being computationally intensive are still challenges. Probability-based techniques have been under-used in the past, with a reliance on deterministic approaches to measure the effect and describe the impact. Whilst statistically robust, these deterministic approaches do not inform an objective assessment of the inherent uncertainties and associated risks (Schaub and Kéry 2012).

Tackling Spatio-Temporal Uncertainty in Risk Assessments

Fundamental to risk assessment is to understand the likelihood of a certain event to occur. Hence, for assessing risks to wildlife it is essential to take into account that the majority of species move around either on a daily or seasonal basis.

Furthermore, their behaviour coupled with species-specific traits may bring them into encounter with wind turbines. Current knowledge on where and when the animals are is variable depending on the taxa of interest. Hence, this variable knowledge brings uncertainty to any risk assessment. Failing to include the dynamics of species-specific responses to wind turbines leads to assumptions that may be unrealistic (May 2015). Individual responses lead to individual vulnerabilities and unless a large proportion of animals responds in a similar way to the hazard, it is difficult to extrapolate responses of individual animals to biologically meaningful impacts (Boehlert and Gill 2010). Hence, the variability and uncertainty in risk assessment becomes more apparent.

Each hazard to the species of interest should be properly characterized and this must include the spatial and temporal characteristics and the probability of exposure, to ensure that the risks are properly evaluated and that assessment of uncertainties is undertaken. Risk assessments traditionally are based on single point estimates of the risk (i.e. deterministic), judged against some threshold (that is often difficult to define), as if the likelihood of occurrence of the risk is precisely known. However, the uncertainty in animals' spatial and temporal occurrence means that, although science-based, policy thresholds may become inappropriate (Johnson 2013). Whilst this adds to the complexity of the analysis, a simple chain of cause and effect to an 'average animal' is not sufficient. Obviously, uncertainty in the risk assessments may be reduced through appropriate study designs ensuring ample sample size, replicates and study duration. Probabilistic approaches attempt to address this by setting parameters in predicted risk assessments based on a range of risk probabilities and thus take into account variability and uncertainty.

Although science can empirically predict the likelihood and magnitude of change, societal acceptance determines the final threshold value(s). However, thresholds need not be binary, but could also include levels of acceptance using e.g. traffic light approaches incorporating inherent variability (e.g. confidence quantiles). Logically, perception of risk and the associated uncertainty rises with complexity that then leads to us being risk-averse according to the pre-cautionary principle. However, renewable energy is a useful contributory tool for a low-carbon future with the aim of reducing adverse environmental effects of climate change. Hence, we should seek to improve our understanding of spatio-temporal variability of wind turbine–wildlife interactions and treat the associated uncertainty as an optimization challenge to balance wind energy development with wildlife conservation.

Upscaling Impacts

With a piecemeal development where each wind farm may in itself present little conflict, multiple wind farms may in sum, however, seriously impact individual species or ecosystems over larger geographical areas. The increasing deployment of wind energy onshore and offshore poses ever more challenges to spatial and environmental planning systems, necessitating appropriately addressing cumulative and transboundary impacts. Here, environmental impacts will also need to be considered throughout a wind farm's life cycle to help attain no net loss.

Assessing Demographic and Cumulative Impacts

From a conservation point of view, population-level impacts of anthropogenic activity, such as wind energy development, are most relevant for long-term species persistence. However, this is not reflected in current legislative frameworks. One of the main challenges of future research will be to determine more precisely how individual animals react to wind farms, and whether or not this will have any population level effects or impacts for different species (e.g. Diffendorfer et al. 2015). The distinction between effects and impacts is an important one. Effects on individuals or groups of individuals are observed, or predicted, based on knowledge about a species' ecology and behaviour. Effects may or may not, lead to impacts on populations, acting on survival rate and/or breeding productivity (Boehlert and Gill 2010). The magnitude of population-level impacts may depend on a species' life-history strategies (e.g. longevity, recruitment rate, age structure) coupled with their ecology. Assessing population-level impacts may however prove to be challenging, particularly in cryptic and wide-ranging or migratory species. Additionally, population studies are difficult to perform in long-lived species because the longevity of focal species may require long-term commitment of researchers and funding agencies.

Further, relatively little is known about density-dependent effects on life-history traits that might counterbalance increased additional mortality by wind turbines. In practice, it may be complicated to tease apart impacts of a specific wind farm from other anthropogenic activity within the region or other regions in migrating species, or from regime shifts e.g. due to climate change. Such cumulative effects are debated, but implementation is hampered by lack of a clear definition as well as methods suitable to assess these. Integrating key pressures on populations and the contribution likely to be attributable to wind energy generation in population models may enable assessment of the cumulative impacts of multiple wind farms at different spatial scales (Heinis et al. 2015). Individual- and agent-based simulation models may prove to be best suited to addressing such cross-sector and transboundary effects on populations (Masden et al. 2010a; Schaub 2012; Nabe-Nielsen et al. 2014).

Ecosystem and Life-Cycle Impacts of Wind Farms

Research has so far primarily focused on effects on the behaviour, distribution and abundance of single species associated with single wind farms. Nevertheless, although the basic problems still remain seasonal and site-specific, some species' populations may be negatively impacted through disturbance or mortality (Schuster et al. 2015) while other populations may be relatively unaffected, or even benefit through the provision of novel habitat or refuges (Inger et al. 2009; Lindeboom et al. 2011). From an ecosystem perspective, however, species interact within communities and across trophic levels; impacts on one species may therefore in turn indirectly affect other species, ultimately affecting ecosystem function. While biodiversity loss and its consequences to ecosystem function are important in their own right, renewable energy systems may also degrade ecosystem services (Hastik et al. 2015; Papathanasopoulou et al. 2015). Given the complexity of ecosystems with interacting species within their environment, it is evident that there is a significant conflict potential if site-selection processes are not carried out carefully and holistically (Gill 2005; Stewart et al. 2007). Assessments of the total ecosystem load across species, and implementing this into an ecosystem service-logic, presents new research challenges. Ecosystem modelling exercises could enhance our understanding of long-term consequences of renewable energy for ecosystem processes on different trophic levels and at larger spatial and temporal scales (e.g. Burkhard et al. 2011).

Impacts, either for single species or for the ecosystem as a whole, will need to be addressed and mitigated throughout the life-cycle of a wind farm to contribute to the no net loss goal of lowest possible environmental costs per kWh from wind energy projects (cf. Gardner et al. 2013; May 2016). Determining what no net loss entails will be critical in assessing how wind energy generation can be reconciled with nature conservation laws, directives and policy (Cole 2011). This requires research assessing environmental impacts through all life cycle stages (construction, operation and decommissioning, possibly repowering) of a wind farm, considering impacts on human wellbeing, ecosystem quality and natural resources. However, life cycle impact assessments (LCIA) still lack the inclusion of impacts on biodiversity or land impacts (Arvesen and Hertwich 2012; Michelsen and Lindner 2015). Improved LCIA can highlight the main environmental impacts and identify trade-offs between different wind energy development options with regard to siting and modes of operation to attain no net loss.

Towards Consolidation

Despite the wealth of scientific information available, this paper has emphasized future research directions in relation to wind turbine–wildlife interactions. The defined sensitivity of animal species often relates to their conservation status and expert judgement of the species at risk (Furness et al. 2013). Inevitably, this leaves behind other sensitive species that may have traits that make them susceptible in the future over longer time scales or greater spatial extent. This will require formal meta-analyses to obtain a better understanding of which traits enhance a species' susceptibility to wind energy development. Understanding why species-specific responses occur, require also approaches inherent to species' ecology. Longitudinal and multiple-site studies, which apply common methods and incorporating Before-After-Control-Impact (BACI) approaches, will offer greatest insights. Such studies will render important information for decisions on whether measures can be taken to reduce or mitigate any predicted population-level impacts.

In view of the fast rate of wind energy development and political goals for further development, cumulative impacts will become more urgent. The challenge to science is not only to identify and measure the extent of cumulative impacts on vulnerable species and ecosystems but also to provide solutions to handle and mitigate these impacts. With increasing cumulative impacts, the pressure for implementing effective mitigation measures is growing in importance. Consequently, a key research priority should centre on the development, monitoring and continuous improvement of mitigation measures to counteract impacts of cumulative wind energy development on wildlife. Finally, all involved stakeholders in wind energy development must shift their stance in effectively sharing previous, current, or upcoming data and results, as knowledge sharing within transdisciplinary co-learning will be pertinent to avoid persistence of science-policy-practice gaps. This challenge to reconcile wind turbine–wildlife interactions has the opportunity to be substantially addressed in future research directions.

References

Arnett EB, Huso MMP, Schirmacher MR, Hayes JP (2011) Altering turbine speed reduces bat mortality at wind-energy facilities. Front Ecol Environ 9:209–214

Arnett EB, May RF (2016) Mitigating wind energy impacts on wildlife: approaches for multiple taxa. Hum Wildl Interact 10:28–41

Arvesen A, Hertwich E (2012) Assessing the life cycle environmental impacts off wind power: a review of present knowledge and research needs. Renew Sustain Energy Rev 16:5994–6006

Baerwald EF, Barclay RMR (2009) Geographic variation in activity and fatality of migratory bats at wind energy facilities. J Mammal 90:1341–1349

Baerwald EF, Edworthy J, Holder M, Barclay RMR (2009) A large-scale mitigation experiment to reduce bat fatalities at wind energy facilities. J Wildl Manage 73:1077–1081

Barclay RMR, Baerwald EF, Gruver JC (2007) Variation in bat and bird fatalities at wind energy facilities: assessing the effects of rotor size and tower height. Can J Zool 85:381–387

Bernardino J, Bispo R, Costa H, Mascarenhas M (2013) Estimating bird and bat fatality at wind farms: a practical overview of estimators, their assumptions and limitations. NZ J Zool 40: 63–74

Boehlert GW, Gill AB (2010) Environmental and ecological effects of ocean renewable energy development. A current synthesis. Oceanography 23:68–81

Bohrer G, Zhu K, Jones RL, Curtis PS (2013) Optimizing wind power generation while minimizing wildlife impacts in an urban area. PLoS ONE 8:e56036

Bradbury G, Trinder M, Furness B, Banks AN, Caldow RW, Hume D (2014) Mapping seabird sensitivity to offshore wind farms. PLoS ONE 9:e106366

Bright J, Langston R, Bullman E, Evans R, Gardner S, Pearce-Higgins J (2008) Map of bird sensitivities to wind farms in Scotland: a tool to aid planning and conservation. Biol Conserv 141:2342–2356

Burkhard B, Opitz S, Lenhart H, Ahrendt K, Garthe S, Mendel B, Windhorst W (2011) Ecosystem based modeling and indication of ecological integrity in the German North Sea—case study offshore wind parks. Ecol Ind 11:168–174

Cleasby IR, Wakefield ED, Bearhop S, Bodey TW, Votier SC, Hamer KC, Österblom H (2015) Three-dimensional tracking of a wide-ranging marine predator: flight heights and vulnerability to offshore wind farms. J Appl Ecol 52:1474–1482

Cole SG (2011) Wind power compensation is not for the birds: an opinion from an environmental economist. Restor Ecol 19:147–153

Cowell R, Bristow G, Munday M (2011) Acceptance, acceptability and environmental justice: the role of community benefits in wind energy development. J Environ Planning Manage 54: 539–557

Cryan PM, Barclay RM (2009) Causes of bat fatalities at wind turbines: Hypotheses and predictions. J Mammal 90:1330–1340

Cryan PM, Gorresen PM, Hein CD, Schirmacher MR, Diehl RH, Huso MM, Hayman DTS, Fricker PD, Bonaccors FJ, Johnson DH, Heist K, Dalton DC (2014) Behavior of bats at wind turbines. Proc Natl Acad Sci 111:15126–15131

Dahl EL, May R, Nygård T, Åstrøm, J, Diserud OH (2015) Repowering Smøla wind power plant. An assessment of avian conflicts. NINA Report 1135. Norwegian Institute for Nature Research, Trondheim, Norway

de Lucas M, Ferrer M, Bechard MJ, Muñoz AR (2012) Griffon vulture mortality at wind farms in southern Spain: distribution of fatalities and active mitigation measures. Biol Conserv 147:184–189

Diffendorfer JE, Beston JA, Merrill MD, Stanton JC, Corum MD, Loss SR, Thogmartin WE, Johnson DH, Erickson RA, Heist KW (2015) Preliminary methodology to assess the national and regional impact of U.S. wind energy development on birds and bats. Scientific investigations report 2015-5066. U.S.G. Survey, Reston, Virginia (USA)

Drewitt AL, Langston RH (2008) Collision effects of wind-power generators and other obstacles on birds. Ann N Y Acad Sci 1134:233–266

Drewitt AL, Langston RHW (2006) Assessing the impacts of wind farms on birds. Ibis 148, Suppl. 1:29–42

Exo KM, Hüppop O, Garthe S (2003) Birds and offshore wind farms: a hot topic in marine ecology. Wader Study Group Bull 100:50–53

Francis CD, Barber JR (2013) A framework for understanding noise impacts on wildlife: an urgent conservation priority. Front Ecol Environ 11:305–313

Furness RW, Wade HM, Masden EA (2013) Assessing vulnerability of marine bird populations to offshore wind farms. J Environ Manage 119:56–66

Gardiner SM (2011) A perfect moral storm. The ethical tragedy of climate change. Oxford University Press, New York

Gardner TA, Von Hase A, Brownlie S, Ekstrom JMM, Pilgrim JD, Savy CE, Stephens RTT, Treweek JO, Ussher GT, Ward G, Ten Kate K (2013) Biodiversity offsets and the challenge of achieving no net loss. Conserv Biol 27:1254–1264

Garthe S, Hüppop O (2004) Scaling possible adverse effects of marine wind farms on seabirds: developing and applying a vulnerability index. J Appl Ecol 41:724–734

Geißler G (2013) Strategic environmental assessments for renewable energy development—comparing the United States and Germany. J Environ Assess Policy Manage 15:1340003

Geißler G, Köppel J, Gunther P (2013) Wind energy and environmental assessments—a hard look at two forerunners' approaches: Germany and the United States. Renew Energy 51:71–78

Gill AB (2005) Offshore renewable energy: ecological implications of generating electricity in the coastal zone. J Appl Ecol 42:605–615

Gill AB, Bartlett M, Thomsen F (2012) Potential interactions between diadromous fishes of U.K. conservation importance and the electromagnetic fields and subsea noise from marine renewable energy developments. J Fish Biol 81:664–695

Gill JA, Norris K, Sutherland WJ (2001) Why behavioural responses may not reflect the population consequences of human disturbance. Biol Conserv 97:265–268

Hammond PS, Macleod K, Berggren P, Borchers DL, Burt L, Cañadas A, Desportes G, Donovan GP, Gilles A, Gillespie D, Gordon J, Hiby L, Kuklik I, Leaper R, Lehnert K, Leopold M, Lovell P, Øien N, Paxton CGM, Ridoux V, Rogan E, Samarra F, Scheidat M, Sequeira M, Siebert U, Skov H, Swift R, Tasker ML, Teilmann J, Van Canneyt O, Vázquez JA (2013) Cetacean abundance and distribution in European Atlantic shelf waters to inform conservation and management. Biol Conserv 164:107–122

Hastik R, Basso S, Geitner C, Haida C, Poljanec A, Portaccio A, Vrščaj B, Walzer C (2015) Renewable energies and ecosystem service impacts. Renew Sustain Energy Rev 48:608–623

Heinis F, de Jong CAF, Rijkswaterstaat Underwater Sound Working Group (2015) Framework for assessing ecological and cumulative effects of offshore wind farms. Cumulative effects of impulsive underwater sound on marine mammals. TNO 2015 R10335-A. M.o.E.A.M.o.I.a.t. Environment, Den Haag, the Netherlands

Huesca-Pérez ME, Sheinbaum-Pardo C, Köppel J (2016) Social implications of siting wind energy in a disadvantaged region—the case of the Isthmus of Tehuantepec, Mexico. Renew Sustain Energy Rev 58:952–965

Hüppop O, Dierschke J, Exo KM, Fredrich E, Hill R (2006) Bird migration studies and potential collision risk with offshore wind turbines. Ibis 148:90–109

Inger R, Attrill MJ, Bearhop S, Broderick AC, James Grecian W, Hodgson DJ, Mills C, Sheehan E, Votier SC, Witt MJ, Godley BJ (2009) Marine renewable energy: potential benefits to biodiversity? An urgent call for research. J Appl Ecol 46:1145–1153

IPCC (2011) IPCC special report on renewable energy sources and climate change mitigation. Cambridge University Press, Cambridge

Islam MR, Mekhilef S, Saidur R (2013) Progress and recent trends of wind energy technology. Renew Sustain Energy Rev 21:456–468

Jay S (2010) Strategic environmental assessment for energy production. Energy Policy 38: 3489–3497

Johnson CJ (2013) Identifying ecological thresholds for regulating human activity: effective conservation or wishful thinking? Biol Conserv 168:57–65

Kight CR, Swaddle JP (2011) How and why environmental noise impacts animals: an integrative, mechanistic review. Ecol Lett 14:1052–1061

Kopnina H (2013) Forsaking nature? Contesting 'biodiversity' through competing discourses of sustainability. J Educ Sustain Dev 7:51–63

Kunz TH, Arnett EB, Erickson WP, Hoar AR, Johnson GD, Larkin RP, Strickland MD, Thresher RW, Tuttle MD (2007) Ecological impacts of wind energy development on bats: questions, research needs, and hypotheses. Front Ecol Environ 5:315–324

Köppel J, Dahmen M, Helfrich J, Schuster E, Bulling L (2014) Cautious but committed: moving toward adaptive planning and operation strategies for renewable energy's wildlife implications. Environ Manage 54:744–755

Lindeboom H, Degraer S, Dannheim J, Gill AB, Wilhelmsson D (2015) Offshore wind park monitoring programmes, lessons learned and recommendations for the future. Hydrobiologia 756:169–180

Lindeboom HJ, Kouwenhoven HJ, Bergman MJN, Bouma S, Brasseu S, Daan R, Fijn RC, de Haan D, Dirksen S, van Hal R, Hille Ris Lambers R, ter Hofstede R, Krijgsveld KL, Leopold M, Scheidat M (2011) Short-term ecological effects of an offshore wind farm in the Dutch coastal zone; a compilation. Environ Res Lett 6:035101

Loss SR, Will T, Marra PP (2013) Estimates of bird collision mortality at wind facilities in the contiguous United States. Biol Conserv 168:201–209

Madsen J, Boertmann D (2008) Animal behavioral adaptation to changing landscapes: spring-staging geese habituate to wind farms. Landscape Ecol 23:1007–1011

Madsen PT, Wahlberg M, Tougaard J, Lucke K, Tyack P (2006) Wind turbine underwater noise and marine mammals: implications of current knowledge and data needs. Mar Ecol Prog Ser 309:279–295

Marques AT, Batalha H, Rodrigues S, Costa H, Pereira MJR, Fonseca C, Mascarenhas M, Bernardino J (2014) Understanding bird collisions at wind farms: an updated review on the causes and possible mitigation strategies. Biol Conserv 179:40–52
Martin GR (2012) Through birds' eyes: insights into avian sensory ecology. J Ornithol 153:23–48
Masden EA, Cook ASCP (2016) Avian collision risk models for wind energy impact assessments. Environ Impact Assess Rev 56:43–49
Masden EA, Fox AD, Furness RW, Bullman R, Haydon DT (2010a) Cumulative impact assessments and bird/wind farm interactions: developing a conceptual framework. Environ Impact Assess Rev 30:1–7
Masden EA, Haydon DT, Fox AD, Furness RW (2010b) Barriers to movement: modelling energetic costs of avoiding marine wind farms amongst breeding seabirds. Mar Pollut Bull 60:1085–1091
Masden EA, Reeve R, Desholm M, Fox AD, Furness RW, Haydon DT (2012) Assessing the impact of marine wind farms on birds through movement modelling. J R Soc Interface 9: 2120–2130
May RF (2015) A unifying framework for the underlying mechanisms of avian avoidance of wind turbines. Biol Conserv 190:179–187
May R (2016) Mitigation for birds. In: Perrow M (ed) Wildlife and windfarms: conflicts and solutions—volume 2. Onshore: Monitoring and mitigation. Pelagic Publishing, Exeter, United Kingdom
May R, Reitan O, Bevanger K, Lorentsen SH, Nygård T (2015) Mitigating wind-turbine induced avian mortality: sensory, aerodynamic and cognitive constraints and options. Renew Sustain Energy Rev 42:170–181
Michelsen O, Lindner J (2015) Why include impacts on biodiversity from land use in LCIA and how to select useful indicators? Sustainability 7:6278–6302
Nabe-Nielsen J, Sibly RM, Tougaard J, Teilmann J, Sveegaard S (2014) Effects of noise and bycatch on a Danish harbour porpoise population. Ecol Model 272:242–251
Nathan R, Getz WM, Revilla E, Holyoak M, Kadmon R, Saltz D, Smouse PE (2008) A movement ecology paradigm for unifying organismal movement research. Proc Natl Acad Sci U S A 105:19052–19059
Papathanasopoulou E, Beaumont N, Hooper T, Nunes J, Queirós AM (2015) Energy systems and their impacts on marine ecosystem services. Renew Sustain Energy Rev 52:917–926
Pearce-Higgins JW, Stephen L, Douse A, Langston RHW (2012) Greater impacts of wind farms on bird populations during construction than subsequent operation: results of a multi-site and multi-species analysis. J Appl Ecol 49:386–394
Peste F, Paula A, da Silva LP, Bernardino J, Pereira P, Mascarenhas M, Costa H, Vieira J, Bastos C, Fonseca C, Pereira MJR (2015) How to mitigate impacts of wind farms on bats? A review of potential conservation measures in the European context. Environ Impact Assess Rev 51:10–22
Petrova MA (2016) From NIMBY to acceptance: toward a novel framework—VESPA—for organizing and interpreting community concerns. Renew Energy 86:1280–1294
Phylip-Jones J, Fischer TB (2014) Strategic environmental assessment (SEA) for wind energy planning: lessons from the United Kingdom and Germany. Environ Impact Assess Rev 50:203–212
Portman M (2011) Marine spatial planning: achieving and evaluating integration. ICES J Mar Sci 68:2191–2200
Portman M (2015) Marine spatial planning in the Middle East: crossing the policy-planning divide. Marine Policy 61:8–15
Santangeli A, Katzner T (2015) A call for conservation scientists to evaluate opportunities and risks from operation of vertical axis wind turbines. Frontiers Ecol Evol 3:68
Schaub M (2012) Spatial distribution of wind turbines is crucial for the survival of red kite populations. Biol Conserv 155:111–118
Schaub M, Kéry M (2012) Combining information in hierarchical models improves inferences in population ecology and demographic population analyses. Anim Conserv 15:125–126

Scheidat M, Tougaard J, Brasseur S, Carstensen J, van Polanen PT, Teilmann J, Reijnders P (2011) Harbour porpoises (Phocoena phocoena) and wind farms: a case study in the Dutch North Sea. Environ Res Lett 6:025102

Schroeder DM, Love MS (2004) Ecological and political issues surrounding decommissioning of offshore oil facilities in the Southern California Bight. Ocean Coast Manag 47:21–48

Schuster E, Bulling L, Koppel J (2015) Consolidating the state of knowledge: a synoptical review of wind energy's wildlife effects. Environ Manage 56:300–331

Smallwood KS (2013) Comparing bird and bat fatality-rate estimates among North American wind-energy projects. Wildl Soc Bull 37:19–33

Smallwood KS, Neher L, Bell DA (2009) Map-based repowering and reorganization of a wind resource area to minimize burrowing owl and other bird fatalities. Energies 2:915–943

Stewart GB, Pullin AS, Coles CF (2007) Poor evidence-base for assessment of windfarm impacts on birds. Environ Conserv 34:1

Thaxter CB, Ross-Smith VH, Bouten W, Clark NA, Conway GJ, Rehfisch MM, Burton NH (2015) Seabird–wind farm interactions during the breeding season vary within and between years: a case study of lesser black-backed gull Larus fuscus in the UK. Biol Conserv 186: 347–358

Tougaard J, Wright AJ, Madsen PT (2015) Cetacean noise criteria revisited in the light of proposed exposure limits for harbour porpoises. Mar Pollut Bull 90:196–208

Tsoutsos T, Tsitoura I, Kokologos D, Kalaitzakis K (2015) Sustainable siting process in large wind farms case study in Crete. Renew Energy 75:474–480

van Kuik GAM, Peinke J, Nijssen R, Lekou D, Mann J, Sørensen JN, Ferreira C, van Wingerden JW, Schlipf D, Gebraad P, Polinder H, Abrahamsen A, van Bussel GJW, Sørensen JD, Tavner P, Bottasso CL, Muskulus M, Matha D, Lindeboom HJ, Degraer S, Kramer O, Lehnhoff S, Sonnenschein M, Sørensen PE, Künneke RW, Morthorst PE, Skytte K (2016) Long-term research challenges in wind energy—a research agenda by the European Academy of Wind Energy. Wind Energy Sci 1:1–39

Voigt CC, Popa-Lisseanu AG, Niermann I, Kramer-Schadt S (2012) The catchment area of wind farms for European bats: a plea for international regulations. Biol Conserv 153:80–86

Walters K, Kosciuch K, Jones J (2014) Can the effect of tall structures on birds be isolated from other aspects of development? Wildl Soc Bull 38:250–256

Wang S, Wang S, Smith P (2015) Ecological impacts of wind farms on birds: questions, hypotheses, and research needs. Renew Sustain Energy Rev 44:599–607

Warren CR, Lumsden C, O'Dowd S, Birnie RV (2005) 'Green on green': public perceptions of wind power in Scotland and Ireland. J Environ Planning Manage 48:853–875

Welstead J, Hirst R, Keogh D, Robb G, Bainsfair R (2013) Research and guidance on restoration and decommissioning of onshore wind farms. Commissioned report no. 591. Scottish Natural Heritage, Inverness, UK

Wilson JC, Elliott M (2009) The habitat-creation potential of offshore wind farms. Wind Energy 12:203–212

Wolsink M (2012) Wind power: basic challenge concerning social acceptance. In: Meyers RA (ed) Encyclopedia of sustainability science and technology. Springer, New York

Sharing Information on Environmental Effects of Wind Energy Development: WREN Hub

Andrea Copping, Luke Hanna and Jonathan Whiting

Abstract As the need for clean low carbon renewable energy increases worldwide, wind energy is becoming established in many nations and is under consideration in many more. Technologies that make land-based and offshore wind feasible, and resource characterizations of available wind, have been developed to facilitate the advancement of the wind industry. However, there is a continuing need to also evaluate and better understand the legal and social acceptability associated with potential effects on the environment. Uncertainty about potential environmental effects continues to complicate and slow permitting (consenting) processes in many nations. Research and monitoring of wildlife interactions with wind turbines, towers, and transmission lines has been underway for decades. However, the results of those studies are not always readily available to all parties, complicating analyses of trends and inflection points in effects analyses. Sharing of available information on environmental effects of land-based and offshore wind energy development has the potential to inform siting and permitting/consenting processes. WREN Hub is an online knowledge management system that seeks to collect, curate, and disseminate scientific information on potential effects of wind energy development. WREN Hub acts as a platform to bring together the wind energy community, providing a collaborative space and an unbiased information source for researchers, regulators, developers, and key stakeholders to pursue accurate predictions of potential effects of wind energy on wildlife, habitats, and ecosystem processes. In doing so, WREN Hub ensures that key scientific uncertainties are identified, tagged for strategic research inquiry, and translated into effective collaborative projects.

Keywords Sharing information · Wind development · WREN hub · Online knowledge management system · Wind energy · Land-based and offshore wind · Environmental effects · Knowledge sharing

---

A. Copping (✉) · L. Hanna · J. Whiting
Pacific Northwest National Laboratory, 1100 Dexter Ave N. Suite 400, Seattle, WA 98109, USA
e-mail: andrea.copping@pnnl.gov

J. Köppel (ed.), *Wind Energy and Wildlife Interactions*,
DOI 10.1007/978-3-319-51272-3_16

Introduction

As the demand for clean, renewable energy increases around the world, all available sources that promise reasonable rates of energy production and social acceptability are being examined. Wind energy is at the forefront of renewable energy sources in many nations and is under consideration in many more (GWEC 2014). Land-based wind has been well established in certain nations such as the US, Germany, and Denmark for many years, and is under development across Asia and other European nations such as Switzerland (IEA 2013). Offshore wind has passed the experimental stage in the UK and other European countries (EWEA 2015), heading towards commercial viability, while other countries such as the US are just launching their first projects (Smith et al. 2015). Resource characterization studies of wind energy provide information that atmospheric scientists and engineers use to design and optimize wind turbine blades and to design the other components that make up the balance of station for land based and offshore installations, calculated to maximize the power generated while minimizing costs and maintenance requirements. Although clearly an important part of establishing wind farms, the legal and social acceptability associated with potential effects on the environment continues to contribute considerable uncertainty to establishing wind energy as a preferred renewable energy source (Musial and Ram 2010; States et al. 2012).

Research and monitoring of wildlife interactions with wind turbines, towers and transmission lines has been underway for decades (Strickland et al. 2011). Interactions of marine wildlife, seabirds, and marine habitats with wind infrastructure have become a focus of scientific inquiry and monitoring programs during the past two decades (Williams et al. 2015; Bailey et al. 2014; Bergstrom et al. 2014). However, the results of those studies are not always readily available to all parties, complicating analyses of trends and inflection points in effects analyses. Furthermore, it is complicating planning studies for nations early in the establishment of their industry.

Sharing of available information on environmental effects of land-based and offshore wind energy development has the potential to inform siting and permitting/consenting processes. Sharing information would allow turbine and foundation developers to modify equipment and operations that are most likely to threaten wildlife or to alter habitat, and would allow the research community to assist the industry in smoothly expanding wind farms on land and at sea, while responsibly protecting the delicate environmental and social balances on which we all depend. Developing an organized and readily accessible window into the knowledge that has been gained worldwide can assist with moving the wind community towards these goals.

WREN Hub is an online knowledge management system that seeks to collect, curate, and disseminate scientific information on potential effects of wind energy development. WREN Hub acts as a platform to bring together the wind energy community, providing a collaborative space and an unbiased information source for researchers, regulators, developers, and key stakeholders to pursue accurate predictions of potential effects of wind energy on wildlife, habitats, and ecosystem processes. In doing so, WREN Hub ensures that key scientific uncertainties are identified, tagged for strategic research inquiry, and translated into effective collaborative projects.

Description of WREN Hub

WREN Hub has been built on a Drupal platform, hosted by the knowledge management system *Tethys* (https://tethys.pnnl.gov). *Tethys* has powerful database management capabilities, and allows for community tagging of information and discussion, while maintaining the integrity of the underlying information. All tagging and community involvement in *Tethys* is monitored and suggestions to add functionality and features that promote additional interaction opportunities are considered for additional development. The front page of *Tethys* links to WREN Hub and other aspects of the online knowledge management system (Fig. 1).

Tethys contains several thousand journal articles, technical reports, presentations, and other media on the environmental effects of wind and marine renewable energy. WREN Hub hosts a page that connects users to the activities of international collaborations. This includes surveys of wind energy stakeholders (developers, researchers, regulators and others) on their impressions and beliefs of the state of wind energy and wildlife, an interactive calendar of events such as conferences and workshops of interest, and also documents the activities of the WREN initiative and provides links to other wind energy and wildlife resources. In addition, WREN Hub acts as a collaborative space for development of white papers and other products specifically planned for the international WREN collaborative.

Origin

The concept for WREN Hub grew out of the WREN initiative (Working together to Resolve Wind and Wildlife Issues) under the International Energy Agency Wind Committee. WREN (also known as Wind Task 34) brings together nations with an interest in resolving effects on wildlife and their support systems, and the establishment of wind energy as an important renewable energy source (http://www.ieawind.org/). The US proposed the WREN initiative and leads the effort out of the US Department of Energy Wind and Waterpower Technologies Office and their national laboratories (National Renewable Energy Laboratory—NREL; and Pacific Northwest National Laboratory—PNNL). As of the end of 2015, ten nations have joined the initiative (France, Germany, Ireland, Netherlands, Norway, Spain, Sweden, Switzerland, United Kingdom and United States), to collaborate on

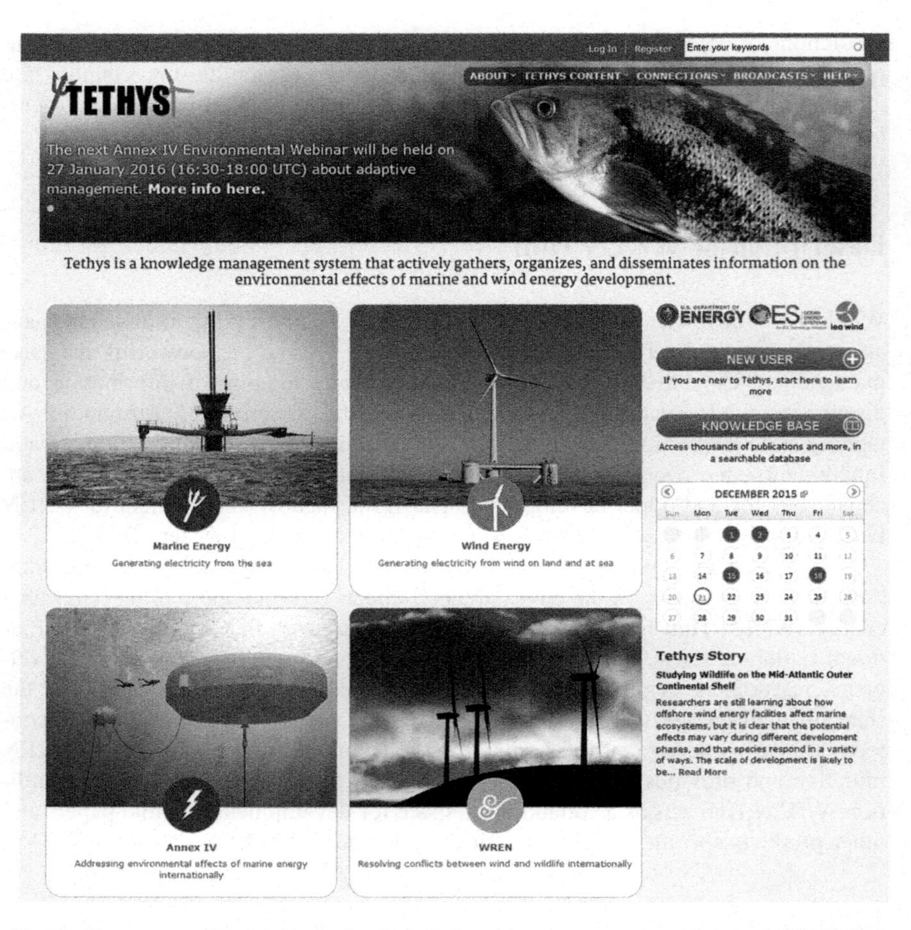

Fig. 1 Home page of *Tethys*, the online knowledge management system that houses WREN Hub. WREN Hub can be accessed from the lower right hand colored tile. All papers and other media can be accessed from the knowledge base button above the calendar

environmental effects of land-based and offshore wind. Some nations have a particular interest in land-based wind alone, while others are focused on land-based and offshore wind.

Content

The heart of *Tethys* is the Knowledge Base, a searchable, annotated database of scientific papers, as well as monitoring and mitigation reports. It also includes other media that are directly related to wildlife and wind energy, as well as content related

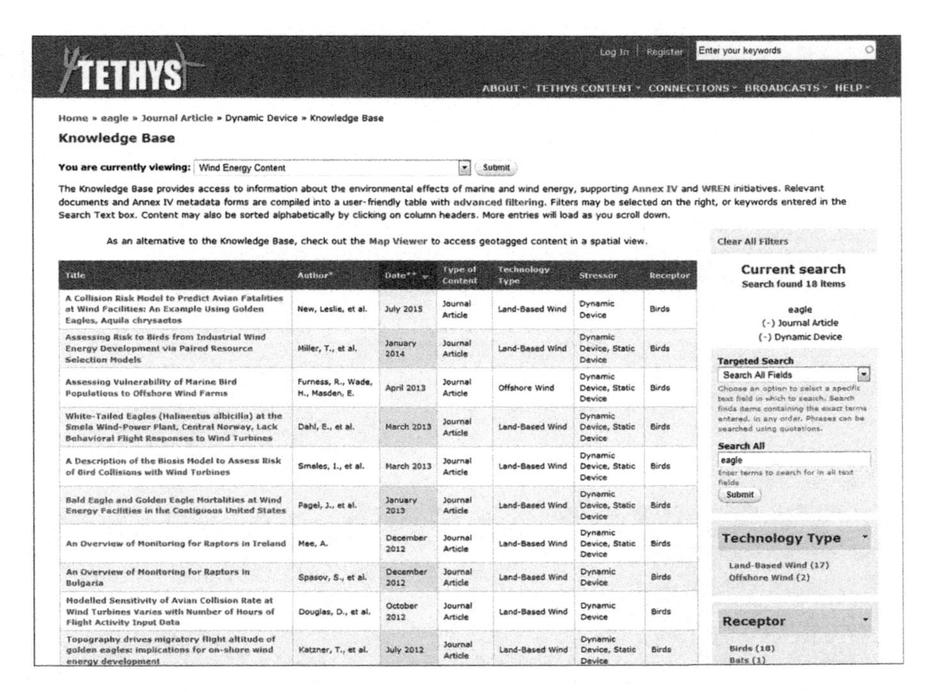

Title	Author*	Date**	Type of Content	Technology Type	Stressor	Receptor
A Collision Risk Model to Predict Avian Fatalities at Wind Facilities: An Example Using Golden Eagles, Aquila chrysaetos	New, Leslie, et al.	July 2015	Journal Article	Land-Based Wind	Dynamic Device	Birds
Assessing Risk to Birds from Industrial Wind Energy Development via Paired Resource Selection Models	Miller, T., et al.	January 2014	Journal Article	Land-Based Wind	Dynamic Device, Static Device	Birds
Assessing Vulnerability of Marine Bird Populations to Offshore Wind Farms	Furness, R., Wade, H., Masden, E.	April 2013	Journal Article	Offshore Wind	Dynamic Device, Static Device	Birds
White-Tailed Eagles (Haliaeetus albicilla) at the Smøla Wind-Power Plant, Central Norway, Lack Behavioral Flight Responses to Wind Turbines	Dahl, E., et al.	March 2013	Journal Article	Land-Based Wind	Dynamic Device, Static Device	Birds
A Description of the Biosis Model to Assess Risk of Bird Collisions with Wind Turbines	Smales, I., et al.	March 2013	Journal Article	Land-Based Wind	Dynamic Device, Static Device	Birds
Bald Eagle and Golden Eagle Mortalities at Wind Energy Facilities in the Contiguous United States	Pagel, J., et al.	January 2013	Journal Article	Land-Based Wind	Dynamic Device, Static Device	Birds
An Overview of Monitoring for Raptors in Ireland	Mee, A.	December 2012	Journal Article	Land-Based Wind	Dynamic Device, Static Device	Birds
An Overview of Monitoring for Raptors in Bulgaria	Spasov, S., et al.	December 2012	Journal Article	Land-Based Wind	Dynamic Device	Birds
Modelled Sensitivity of Avian Collision Rate at Wind Turbines Varies with Number of Hours of Flight Activity Input Data	Douglas, D., et al.	October 2012	Journal Article	Land-Based Wind	Dynamic Device	Birds
Topography drives migratory flight altitude of golden eagles: implications for on-shore wind energy development	Katzner, T., et al.	July 2012	Journal Article	Land-Based Wind	Dynamic Device, Static Device	Birds

Fig. 2 An example of the WREN Hub knowledge base, search results for: land-based and offshore wind (technology types), and birds (receptor)

to marine renewable energy (wave and tidal development). The Knowledge Base is viewable as a spreadsheet view (Fig. 2), from which further filtering, sorting, and choosing of papers can be accomplished using filters that represent:

- Technology types (land-based wind turbines, offshore bottom-mounted turbines, offshore floating turbines, etc.);
- Receptors of interest (aspects of wildlife and habitats that might be harmed by wind energy development, like bats, seabirds, raptors, passerines, etc.); and
- Stressors (those specific aspects of a wind turbine or balance of station that might cause harm or stress, like movement of the turbine blades, noise from the devices, electro-magnetic fields from undersea cables, etc.); and
- Content type (journal articles, conference proceedings papers, reports, book chapters presentations, etc.).

In many cases, the paper, report or presentation cited in the Knowledge Base can be retrieved through *Tethys*, subject only to copyright laws. When a copyrighted entry is cited, a link is added to the entry that will allow the user to access or purchase the entry from the appropriate external source. A typical page that a user might find when choosing a paper in *Tethys* can be seen in Fig. 3.

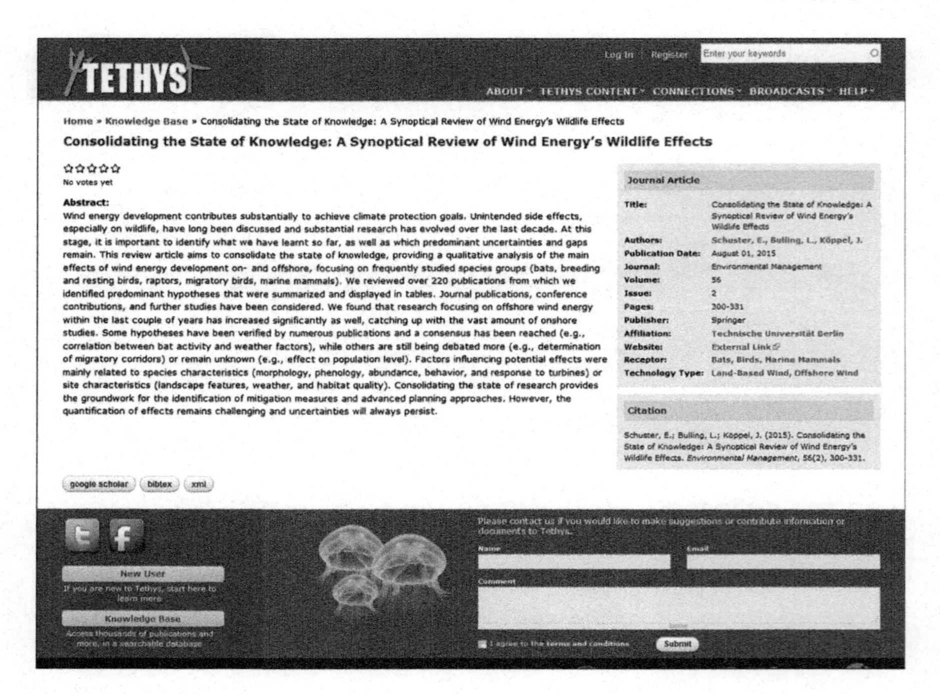

Fig. 3 A typical media page that describes a journal article selected from the search

Table 1 Summary of content of the *Tethys* knowledge base that is applicable to wind energy development and wildlife, as of January 1, 2016

Content type	Number of *Tethys* entries		
	Land-based wind	Offshore wind	Both LBW and OSW
Journal articles	350	340	60
Conference proceedings	44	48	14
Monitoring reports	325	312	45
Theses/dissertations	24	21	2
Other media	26	63	8
Total	756	797	129

As shown in the figure above, the title, authors, and abstract are generated for the user when the search terms are entered. The tile on the right summarizes the attributes of the paper, including the technologies, stressors, and receptors chosen, as well as links to the authors, their organizations, and the location of the paper for download. As a journal article, this particular paper is not directly available on *Tethys* due to copyright considerations. The program automatically also generates the article citation in ALA format.

As of April 1, 2016, the *Tethys* Knowledge Base contains over 3100 documents, a large proportion of which are applicable to wind energy and wildlife. Documents that are specific to wind and wildlife are shown in Table 1. Many additional documents in *Tethys* are applicable to important components of wind energy development (for example, addressing challenges and progress in marine spatial planning for energy). But these may not be reflected in the wind-specific content documented here. New content is constantly being added to the Knowledge Base.

The content above is further divided into resources that are aimed at land-based wind (LBW), offshore wind (OSW), and/or which are applicable to both. The content of the *Tethys* Knowledge Base can also be displayed geographically through the Map Viewer. Based on Google Maps, all papers and other media that are found in the Knowledge Base, for which a georeference tag can be gleaned, are displayed on the world map. In many cases, papers or monitoring reports are specifically tied to a location; for other entries, the center of a region or country to which the work applies is tagged. However, some Knowledge Base entries do not lend themselves to geotagging, so the number of entries in the Knowledge Base will always be greater than that in the Map Viewer.

Entries in the Map Viewer can be sorted and filtered as they are in the Knowledge Base (Fig. 4). The figure below displays wind projects and/or geographically focused research studies. The large blue icons indicate that there are multiple resources available on that site, while the smaller white bubbles indicate a single resource. The dialogue box appears when the user clicks on a resource and

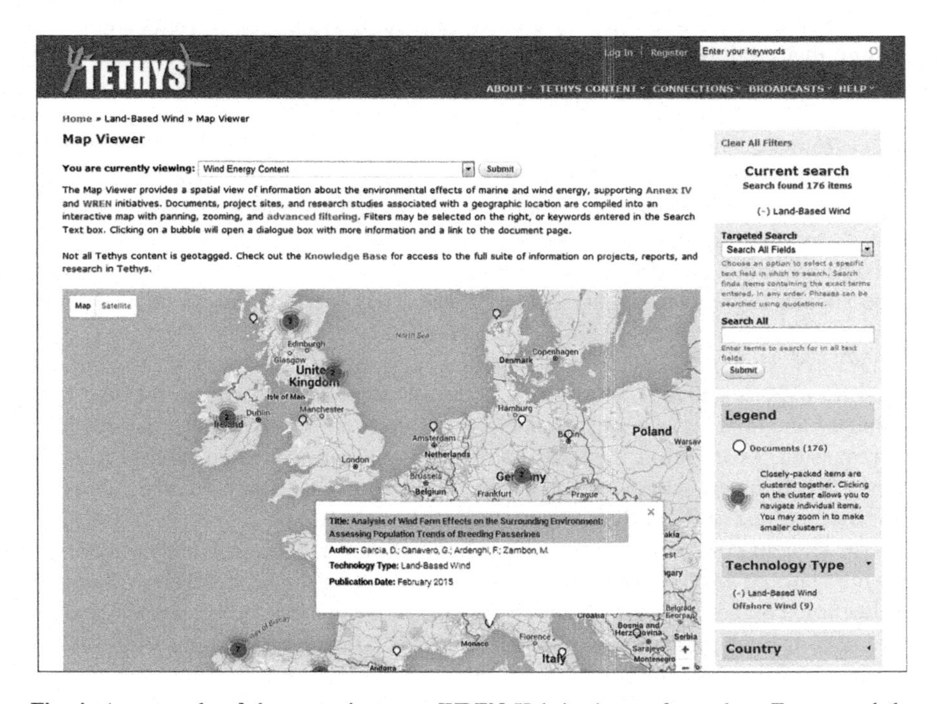

Fig. 4 An example of the map viewer on WREN Hub is shown, focused on Europe and the location of resources

indicates the type of material available, and in many cases, the ability to directly download the report or paper.

Collection of WREN Hub Content and Quality Control

Documents for WREN Hub are collected using several methods, including using the Web of Science and other online tools for peer reviewed papers. WREN Hub relies on assistance from wind project developers, researchers, consultants, and member nations to identify monitoring reports and other "grey" literature that is not easily picked up by online databases. In addition to annual systematic searches of database and other sources, WREN Hub receives up-to-date notifications of papers as they are published. WREN Hub is also very grateful for the partners at the National Renewable Energy Laboratory who collect US-based papers and reports on wind energy and wildlife interactions in their WILD database. Input from WILD is collected on an ongoing basis. As WREN Hub has gained sufficient profile, users who are aware of the WREN Hub effort have been sending suggestions for content.

To ensure that the content of WREN Hub is as accurate and complete as possible, a stringent quality assurance/quality control procedure is conducted before papers or reports are added, and periodic checks are made to ensure that the tags and filters are accurate. The key criteria for adding content to WREN Hub are to ensure that the topic area is appropriate, and that the paper or report is reasonably publically available. Wherever possible, papers or reports are directly downloadable or available for download following appropriate permission (sometimes including fees) from the copyright holder. Each paper or resource is tagged for content, to allow for easy and accurate searching. Over time, the tagging effort has become more complex, dividing receptors into specific groups of birds or bats that may be at risk from wind turbines, and specifying habitats and interactions between turbines or offshore wind foundations with marine or avian animals. The tagging effort is also subject to frequent quality control checks, with periodic larger tagging efforts to ensure continuity throughout WREN Hub.

WREN Hub engages with researchers and other knowledgeable wind and wildlife experts annually in a formal peer review process to ensure that the functionality and content is of high quality. Through a set of questions formal feedback is requested, which forms the basis for changes and corrections on WREN Hub. We have also initiated a broad feedback mechanism using Survey Monkey that asks more casual users about their experiences and suggestions.

All content collection, quality control efforts, and tagging accuracy is overseen by our staff; however WREN Hub users are key to ensuring that the large document collection, contacts and organizational details, and other resources, remain accurate and useful. Each page of WREN Hub has a feedback box and contact information for our staff is readily available; these tools help to encourage users to tell us about any errors they find, improvements they would like to see, or other types of feedback.

Support for International Collaboration on WREN Hub

The availability of papers, reports and other media on *Tethys* supports the broad wind and wildlife community of researchers, developers, regulators, and other stakeholders. It ensures that the state of information is widely shared and that duplication of research effort is minimized to answer important questions. In addition, WREN Hub supports the activities of the international WREN collaborative, intended to facilitate international collaboration and advance global understanding of potential environmental effects of wind energy. While most WREN information and content is publicly accessible on WREN Hub, some space is reserved for WREN member nations for collaborative work, under password-protection. These materials support the work programme set out for WREN, including the collaborative development of white papers that seek to document, better understand, and further advancements in common challenges associated with wind energy development and wildlife.

The first papers under development concern are: (1) the role that adaptive management can and does play in decreasing scientific uncertainty that slows LBW and OSW development; and (2) extrapolation of results of individual animal collisions (birds and bats primarily) with wind turbines, to effects on the underlying populations. These and other work products are developed by teams of WREN's member country representatives and their colleagues, using the platform of WREN Hub to enable direct collaboration. Once the white papers have been prepared in draft form, they will be reviewed and made available through WREN Hub. Final white papers and a series of derivative outreach products, such as illustrated summaries, one-page information sheets, and downloadable slide presentations, will also be made public on WREN Hub.

Outreach Activities Built on WREN Hub

The primary value of WREN Hub to the greater wind and wildlife community may be the collection, annotation, and access to a comprehensive set of papers and other media. However, WREN Hub also serves as a forum or gathering place for practitioners to come together to learn more about new and emerging research topics, to communicate with one another, and to further joint research agendas. The *Tethys* FaceBook page and Twitter feed allow information to be disseminated widely, including seeking feedback on the site.

As an outreach and engagement tool, WREN Hub hosts webinars on a quarterly basis, which are archived on WREN Hub (presentations on YouTube, including the audio of presentations as well as question and answer periods). Each webinar has two to three international speakers chosen to speak on topics relating to interactions between wind energy and wildlife. Typical topics to date are shown in Table 2. Details on the

Table 2 WREN Hub webinars presented to date

Webinar title	Date presented	Description of content
Monitoring bat activity offshore	March 01, 2016	This webinar details techniques for monitoring bat activities offshore, in order to determine the impact that offshore wind may have on bat populations
Wildlife monitoring and wind energy	December 02, 2015	This webinar details monitoring tools and large-scale monitoring efforts related to offshore wind energy
Mid-atlantic baseline study	November 17, 2015	The goal of the mid-atlantic baseline studies (MABS) project was to provide comprehensive baseline ecological data and associated predictive models and maps to regulators, developers, and other stakeholders for offshore wind energy
Avian sensitivity mapping and wind energy	August 27, 2015	The speakers will present on tools they have developed to assess the potential vulnerability of birds to wind energy development
Understanding avian collision rate modeling/population context at land-based and offshore wind farms	April 02, 2015	Wind farms can impact birds through collisions, barrier effects and displacement and habitat loss. Of these, collisions have attracted the most attention as the effects of direct mortality are considered to have the greatest potential to impact bird populations
Attraction and interaction of marine mammals and seabirds to offshore wind farms	December 09, 2014	The presenters provide examples of research methodologies for the detection of animal populations in the vicinity of offshore wind farms, describe the technologies used to detect the animals, and, where available, discuss any existing mitigation measures or post installation/operation monitoring
Bats and wind energy	September 03, 2014	This webinar addressed the issue of reducing bat fatalities from land-based wind in the US and central Europe
Wildlife monitoring and baseline studies for offshore wind development	April 24, 2013	This webinar focused on presenting initial results of the first year of data collection for two large-scale studies in the US that establish baseline data to aid in the siting and permitting of future projects

presentations, authors, and follow up can be found at https://tethys.pnnl.gov/environmental-webinars?content=wind and the content can be downloaded for viewing.

Other Renewables Content on *Tethys*

In addition to supporting information gathering, curation, and dissemination on interactions between wind energy and wildlife, *Tethys* also supports activities aimed at potential environmental effects of marine renewable energy, with an emphasis on wave and tidal energy development (https://tethys.pnnl.gov/marine-energy). Information is also collected on potential effects of hydrokinetic energy harvest in large rivers, as well as that of ocean current energy harvest, ocean thermal energy conversion (OTEC), and salinity gradients. Approximately 1500 documents are specific to marine renewable energy and environmental interactions. *Tethys* supports the international initiative Annex IV, under the International Energy Agency Network Ocean Energy Systems collaborative (http://www.ocean-energy-systems.org/). The work of Annex IV has many similarities to that of WREN, but includes activities that are tailored to the much younger marine renewable energy industry.

Value of Sharing Information

Renewable energy sources are gaining traction worldwide as alternatives to fossil fuels, although many of the carbon-based sources are likely to play some role in energy production for decades to come. As countries transition towards renewable energy sources, the ability to predict potential consequences can help to accelerate the process. While the effects of fossil fuel exploration, extraction and combustion are reasonably well known, the many uncertainties surrounding new renewable sources have translated into regulatory and stakeholder concerns in relation to increased risk to environment and social systems. Sharing information broadly on potential effects of wind energy generation will serve to put the risks in context with broader risks from known sources. For example, concerns for wildlife at potential risk from land-based or offshore turbines is generally focused on populations that are already severely depleted and threatened due to a range of impacts associated with climate change. These impacts include rising temperatures, changes in habitat distribution and quality, shifts in populations of predators, and alterations of prey availability.

Climate change impacts are likely to affect threatened populations as much as losses due to collisions with or avoidance of wind turbines (Crick 2004; Masden et al. 2010; Sovacool 2012). However, our national laws and regulations seldom take this shifting baseline into account, or consider the beneficial effects of reducing carbon emissions into the oceans and atmosphere. Through sharing of research results and linking researchers around the world, better estimates of proportional risks associated with wind energy development on wildlife can be developed. Armed with these scientifically based estimates of risk, regulators and stakeholders can more readily grasp the proportionality of risk associated with wind energy development, in relation to other threats to the species of concern. Particularly in nations where wind energy makes up a relatively new source of energy, collective scientific thinking and analysis will inform the development of public policy and governing regulations. Statistics gathered on outreach activities, webinar attendance and downloads, and the dissemination of white papers through WREN Hub shows that this platform has been engaging scientists worldwide. It sets the stage for new collaborations and research directions that will help usher energy mixes for many nations that lean heavily on wind generation.

Sharing of information provides a key resource for the next generation of scientists, engineers, policy makers, and influential stakeholders. The use of WREN Hub documents, connections to active researchers, and downloads of archived presentations by students and teachers, indicate that the material on WREN Hub is reaching a broad audience. This will have a great effect on the next generation of employees, research leaders, and voters. Future WREN Hub and other *Tethys* outreach programs explicitly target graduate degree programs in a range of scientific, policy and engineering fields to provide tools for exploring and engaging more fully in the wind and other renewable energy fields.

As a relatively new energy source, wind has benefitted from lessons learned from other renewable and non-renewable energy sources. WREN Hub documents the most relevant lessons, including estimates of effects on marine species around moored and floating offshore oil and gas platforms, and provides insights into collision of bird and bat species with turbines. Lessons about offshore wind development effects has also been informed by bird and bat interactions studied for land-based wind, where studies are less costly and risky than at sea studies. These lessons and analyses are documented and explored on WREN Hub.

Conclusion

The purpose of WREN Hub is to enhance understanding about the interactions of wind energy development with wildlife, and to act as a tool to help decrease concerns and resolve conflicts about the development of this low-carbon energy source. The US has committed to continuing support for the *Tethys* database of documents and archived resources as long as Federal policies and programs continue to support wind energy development. Through this, WREN will be a stable, reliable resource that provides a measure of assurance that, in this digital age, collections of curated materials will not be lost.

References

Bailey H, Brookes KL, Thompson PM (2014) Assessing environmental impacts of offshore wind farms: lessons learned and recommendations for the future. Aquat Biosyst 10:1–13. doi:10.1186/2046-9063-10-8

Bergström L, Kautsky L, Malm T, Rosenberg R, Wahlberg M, Åstrand Capetillo N, Wilhelmsson D (2014) Effects of offshore wind farms on marine wildlife—a generalized impact assessment. Environ Res Lett 9:034012. doi:10.1088/1748-9326/9/3/034012

Crick HQP (2004) The impact of climate change on birds. Ibis 146:48–56. doi:10.1111/j.1474-919X.2004.00327.x

EWEA (European Wind Energy Association) (2015) The European offshore wind industry—key trends and statistics 1st half 2015. Available via http://www.ewea.org/fileadmin/files/library/publications/statistics/EWEA-European-Offshore-Statistics-H1-2015.pdf. Accessed 22 Dec 2015

GWEC (Global Wind Energy Council) (2014) Global wind report annual market update (2014O). Available via http://www.gwec.net/wp-content/uploads/2015/03/GWEC_Global_Wind_2014_Report_LR.pdf. Accessed 22 Dec 2015

IEA (International Energy Agency) (2013) Technology road map. Available via http://www.iea.org/publications/freepublications/publication/Wind_2013_Roadmap.pdf. Accessed 22 Dec 2015

Masden E, Fox A, Furness R, Bullman R, Haydon D (2010) Cumulative impact assessments and bird/wind farm interactions: developing a conceptual framework. Environ Impact Assess Rev 30(1):1–7

Musial W, Ram B (2010) Large-scale offshore wind power in the United States: assessment of opportunities and barriers. NREL/TP-500-40745. NREL, Golden

Smith A, Stehly T, Musial W (2015) 2014–2015 Offshore wind technologies market report. NREL/TP-5000-64283. NREL, Golden

Sovacool B (2012) The avian and wildlife costs of fossil fuels and nuclear power. J Integr Environ Sci 9(4):255–278

Strickland MD, Arnett EB, Erickson WP, Johnson DH, Johnson GD, Morrison ML, Shaffer JA, Warren-Hicks A (2011) Comprehensive guide to studying wind energy/wildlife interactions. Prepared for the National Wind Coordinating Collaborative, Washington

States J, Brandt C, Copping A, Hanna L, Shaw W, Branch K, Geerlofs S, Blake K, Hoffman M, Kannberg L, Anderson R (2012) West coast offshore wind—barriers and pathways. PNNL 21748. PNNL, Richland

Williams K, Connelly E, Johnson S, Stenhouse J (2015) Wildlife densities and habitat use across temporal and spatial scales on the mid-atlantic outer continental shelf (2012–2014). Report by Biodiversity Research Institute, p 184